Hot Equations

HOT EQUATIONS

Science, Fantasy, and the Radical Imagination on a Troubled Planet

Jesse S. Cohn

University Press of Mississippi / Jackson

The University Press of Mississippi is the scholarly publishing agency
of the Mississippi Institutions of Higher Learning: Alcorn State University,
Delta State University, Jackson State University, Mississippi State University,
Mississippi University for Women, Mississippi Valley State University,
University of Mississippi, and University of Southern Mississippi.

www.upress.state.ms.us

The University Press of Mississippi is a member
of the Association of University Presses.

"The Fantastic from Counterpublic to Public Imaginary: The Darkest Timeline?"
appeared in *Science Fiction Studies* vol. 47, no. 3 (2020), pp. 448–63.

"The Pursuit of Rhetorical Sovereignty in Indigenous Futurisms"
appeared in *SFRA Review*, vol. 51, no. 4 (2021).

Copyright © 2024 by University Press of Mississippi
All rights reserved

∞

Library of Congress Cataloging Number: 2024931685
Hardback ISBN 978-1-4968-5015-7
Paperback ISBN 978-1-4968-5016-4
Epub single ISBN 978-1-4968-5017-1
Epub institutional ISBN 978-1-4968-5018-8
PDF single ISBN 978-1-4968-5019-5
PDF institutional ISBN 978-1-4968-5020-1

British Library Cataloging-in-Publication Data available

To Rosa and her friends

"'Him?' Ivorene, why won't you—Good Boy's not real! Admit it!" . . .

"Define real," Ivorene said, then sagged in her seat. She was too tired to argue. "No, never mind. Don't. Whether Good Boy or Aunt Lona or any of them are 'real' doesn't matter in the end. Just act like they are and everything will work out fine."
—Nisi Shawl, "Good Boy"

The real, the possible, and the political are all joined at the hip. It is precisely because other possibles have been turned into "impossibles" that we find it so difficult to imagine other realities.
—Arturo Escobar, *Pluriversal Politics*

Contents

Acknowledgments ix

Part I: Escape from the Great Sorting Machine 1

Introduction 3

1. Predictive Analytics for Monstrous Times 9

2. Sublime Machines: Valves of the Heart 31

3. Methodological Metafetishism 61

4. Unbarring the Other 83

Part II: The Fantastic within and beyond Modernity 99

5. Notes toward a Rhetoric of the Fantastic 101

6. Rhetorics of the Impossible 121

7. Rhetorical Sovereignty 145

Part III: Toward a Multiplicative Realism 157

8. The Rhetoric of Fictive Science;
   Or, The Academy of Outrageous Books 159

9. How the Gordin Brothers Escaped Western Gravity 179

Part IV: Fantastic Politics in an Imperiled Pluriverse 219

10. Unicorn Rhetoric. 221

11. The Fantastic from Counterpublic to Public Imaginary 233

12. Acting Supernaturally (with Notes on the Monster of Anarchy) 249

Appendix: A Provisional Manifesto for Nonmodern Anarchisms 261

Notes 265

Bibliography 311

Index 343

Acknowledgments

I would be remiss not to give proper recognition to the many people of color, particularly women and Two-Spirit people of color, whose scholarship has made mine possible. I hope my citational practices bear witness to these in full, but some especially important scholarly companions on this journey have included adrienne maree brown, Seo-Young Chu, Grace L. Dillon, Madhu Dubey, Trey Ellis, Walidah Imarisha, Zakiyyah Iman Jackson, Daniel Heath Justice, Arriana Planey, Nisi Shawl, Ebony Elizabeth Thomas, Gerald Vizenor, Alexander G. Weheliye, Sylvia Wynter, and Kevin Young. Much gratitude to all of them.

I also owe a serious debt of thanks to Eugene Kuchinov for opening up the world of Russian Cosmism to me; to James Gifford for wise counsel and being my nonelectric Electric Monk; to Chantelle Gray, Bettina Escauriza, Andrew Culp, and Brent Ryan Bellamy for important conversations and inspiration; to Sherryl Vint and Gerry Canavan for their great vote of confidence in "The Fantastic From Counterpublic to Public Imaginary"; to Ania Aizman, Daniel Runnels, Anna Elena Torres, Toru Oda, and all the participants in the ACLA Seminar on Speculative Fiction and Decolonial Thought for true scholarly mutual aid; to Taylor Collins, Alex Vrbanoff, and Clare Marcotte for crucial feedback; to the warm support of my colleagues at the Institute for Anarchist Studies and the *Perspectives* collective; and most profoundly, to Darlene for listening to and challenging my ideas, and for reminding me that I know how to do this.

I would like to remember with love the lives of Josh Lukin and Sol Neely.

PART 1
Escape from the Great Sorting Machine

Introduction

> The matter of which materialists speak, matter spontaneously and eternally mobile, active, productive, matter chemically or organically determined and manifested by the properties or forces, mechanical, physical, animal, and intelligent, which necessarily belong to it . . . this matter has nothing in common with the *vile matter* of the idealists.
> —MIKHAIL BAKUNIN, *God and the State*

A carnival of political violence that some insist on calling an *insurrection* (January 6, 2021) has intervened between the first draft of this book (October 2020) and the present draft (May 2023). Who knows what will happen between now and its eventual publication? One problem of living and writing in a period of dramatic decline: the ground keeps slipping out from under your feet.

This book began with a series of disordered observations, observations disordered by the great disorder of the plague year 2020. I do not mean that this was a series in the chronological sense; actually, it is impossible for me to say which came first, then second, and so on. The sense of dislocation in time that the lockdown created—those days without contact, the entirety of social life reduced to screens and furtive encounters at stores with bare shelves, dire and directionless days—has left its mark on my writing, I'm afraid. The only chronology that really pertains here is the terrible chronology of our times: the overshadowing apocalyptic atmosphere of burgeoning fascism, climate change, and pandemic throwing us out of any normative relationship to time.

I wondered if this sense of being unmoored in time, which seemed to be shared by many, might indicate something larger about the kind of times we were in. It seemed to me that the two great global institutions that modernity had birthed—the capitalist system and the system of nation-states—were failing miserably even by their own standards, failing to uphold human rights or even to secure the indefinite future progress they promised, creating increasingly unlivable conditions and then tightening their borders against the refugees fleeing them. Maybe modernity itself had run its course. And maybe that wasn't a bad thing in itself, if you looked at it from an anarchist perspective (the form of radical imagination that best expresses my hopes); maybe, instead, we should seek solutions outside of modernity as such, in the kinds of society that had learned to live happily without capitalism or nation-states. But living in those systems for so long had made anything outside of them seem unthinkable, impossible. We were stuck in an in-between period. No wonder time seemed out of joint.

It was under these shadows that new science fiction and fantasy by writers like N. K. Jemisin, Nalo Hopkinson, Nnedi Okorafor, Rivers Solomon, and Nisi Shawl took on new meaning for me. First of all, I marveled at the rapidity with which these Black women's voices had appeared in a field from which they had been rigorously excluded for so long. Secondly, it struck me that much of their work argued pretty explicitly against the notion that fantasy can be rigorously excluded or even distinguished from science fiction, that they no longer paid heed to the distinction between (modern) science and (ostensibly premodern) magic. And it struck me as significant that the misogynist and white supremacist backlash incurred by the influx of Black women's voices in science fiction, fantasy, and horror, and the way that it transformed the institutions of fandom into arenas of political struggle in 2009–2015, had predicted in such a direct and terrifying way the shape of things to come in 2016–2020.

I remarked that, in the meantime, popular culture had become utterly saturated with the imagery of the fantastic in ways that profoundly altered the mass/minority dynamics that had seemingly always structured this field, leading me to ask: What does it mean that the very moment that minority voices had risen to prominence in science fiction, fantasy, and horror (SFFH)—and encountered such vicious backlash—was the moment when SFFH had seemingly became a majoritarian taste, a new "cultural dominant"? What might it mean that, in this in-between time, when the familiar was vanishing, but we still feared what might come next, the fantastic had seemingly fused with the real world?

As I puzzled over these phenomena, my historical research on radicalism was leading me to some surprising discoveries about the reception of scientific knowledge in anarchist journals around the first half of the twentieth century. This led to further questions. Why, for instance, in the pages of the journal *l'anarchie* in 1906, could one find a six-part series about "The Question of the Creation of an Artificial Living Being"? This series, drawn from a lecture given to a working-class audience at the Université Populaire by biologist Raphaël Dubois, led me to scores of related articles in French, Spanish, Mexican, Cuban, and Argentine journals written by a sprawling network of scientists and militant intellectuals, centered around biologist Alfonso L. Herrera's notion of "plasmogeny." Plasmogeny, an ostensible "new science," since all but forgotten, aspired to bridge the worlds of chemistry and biology to explain the origins of life from nonliving matter. This forgotten chapter in the history of science seemed to resonate, a century in advance, with the preoccupations of the New Materialists and Actor-Network Theorists, who seek precisely to bridge the gaps between the realms of inert matter and conscious mind, passive object and active subject, the nonhuman and the human.

It appeared to me that this confusion of realms was very much at the heart of science fiction and fantasy. Both are modern genres that are repeatedly drawn to imagine objects taking on the vitality and intentionality of human subjects—the Golem, the Android, Frankenstein's Creature—as well as to imagine human subjects taking on the rigidity, automatism, and inertia of objects. Viewed from a certain angle, this tendency started to look like a deep hostility toward the distinction between the human and nonhuman realms that modernity was supposed to have firmly established.

And so I started to think of certain writers of the fantastic in juxtaposition with the plasmogenists, both modernity's discontents, both attracted to animist heresies—radical imaginings, utopian fantasies, and scientific theories that blended elements of futurism with recollections (or rather, in Native author Gerald Vizenor's term, *survivances*) of premodern worlds. This, in turn, had some interesting implications for the relation between modernity and the worlds it has attempted to destroy, and that it is still in the process of destroying—and, in particular, the spectrum of Indigenous worlds wherein society and nature, the human and nonhuman, have never been separated, where animism and fetishism are not obsolete follies but guiding assumptions.

Here, I feel that the fantastic and its history can come to the rescue of these worlds, which is to say, to the rescue of the common world within which all these worlds fit. In *Capitalist Realism* (2009), Mark Fisher pointed to the shrunken state of our social imagination: after decades of neoliberalism, Margaret Thatcher's slogan, "TINA" (There Is No Alternative to the continued extraction of profits), had become second nature to us, and even though economic growth appeared poised to destroy the environment and us with it, we could no longer imagine any other way to live. I'll argue in this book that we might see Fisher's "capitalist realism"—and its corollary, what Chantelle Gray calls "statist realism," i.e., the viewpoint that "posits the State as the horizon of human possibility in such a manner that a coherent and viable replacement for it is virtually inconceivable"[1]—as instances of the "subtractive realism" powering so much of the literary fantastic. "Subtractive" realities, according to Kathryn Hume, are premised on the "destruction of normal order through removal of some element": the forward flow of time (H. G. Wells' *The Time Traveller*), gender differences (Joanna Russ' *The Female Man*), social constraints (William S. Burroughs' *Wild Boys*), logic (Lewis Carroll's *Alice Through the Looking Glass*), freedom (George Orwell's *Nineteen Eighty-Four*), etc.[2] Capitalist realism and statist realism likewise rely on the suppression of certain realities, using some of the same rhetorical tricks that science fiction, fantasy, and horror use to propel us into other worlds (and, significantly, back again).

Once we recognize the fantastic and rhetorical dimensions of politics, we are better poised to understand certain challenges to the capitalist and statist "realism" of modernity. For instance, the forgotten utopian writings of Russian revolutionaries Abba and Wolf Gordin, directing the powers of fantasy against the authority of science, show us how to resist the kind of rhetoric that compares the merest programs of social democracy to "free unicorn rides." We can also call into question the dominance of a narrow form of scientific realism (scientism) over fantastic discourse, showing how deeply science and the fantastic are entangled, which may point to ways out of the sterility of the "Science Wars" that have never ceased to threaten the humanities with irrelevance and inconsequence.

It is my hope that by weaving connections among these observations, we might obtain a different perspective on the present. Science fiction, fantasy, and horror have become central to the public imaginary, the rhetorical realm wherein futures are contested, where references to the fantastic serve both to define what is impossible and to delineate the new normal. To that end, I want to accumulate, snowball-style, a new vocabulary—*postnormal times, New Black Fantastic, sociotechnical disaster, Great Sorting Machine, biotechnical sublime, metafetishism, sociomagic*, etc.—that encodes this different understanding of things. This different understanding does not yet really have a name, but it bears strong affinities to the anarchism which, having emerged from its late-twentieth-century eclipse, is now widely shared by those who stand in the way of the new fascism. It emerges from the interstices of that complex political tradition, from certain moments of potentially fruitful ambivalence and aberration that never received their full development in the course of history. Now that history has run off its normal tracks, now that every day seems more science fictional, more terrifyingly fantastic than ever, perhaps it is time to see what a *nonmodern anarchism* would look like.

How can I speak of any sort of radical vision in a world that is gaudily collapsing around our ears? This is the question that the ostensibly cold calculations of political realism ask me. But the central argument of this book is that the increasingly hot reality in which we find ourselves warrants a thoroughgoing reconsideration of modern "realism" in politics, aesthetics, science, and theory.

1

Predictive Analytics for Monstrous Times

> An apocalypse is a relative thing, isn't it?
> —N. K. Jemisin, *The Stone Sky*

Apocalypse When?

What kind of times are we living in?

Google is ready to frame this question for me: if I type *are we living in* . . . it will suggest:

are we living in a simulation
are we living in a computer simulation
are we living in the matrix
are we living in a black hole
are we living in a dystopia
are we living in idiocracy
are we living in 1984
are we living in a dream
are we living in the 21st century
are we living in an ice age

But also:

are we living in a hologram
are we living in black mirror
are we living in bad times
are we living in base reality

are we living in brave new world
are we living in the cenozoic era
are we living in children of men
are we living in contagion
are we living in cyberpunk
are we living in end times
are we living in fahrenheit 451
are we living in the future
are we living in gilead
are we living in the handmaid's tale
are we living in the hunger games
are we living in jumanji
are we living in kali yuga
are we living like in those days
are we living in modernity
are we living in an orwellian society
are we living in a police state
are we living in postmodernism
are we living in present
are we living in quarantine still
are we living in a recession
are we living in the twilight zone
are we living in unprecedented times
are we living in v for vendetta
are we living in a video game
are we living in westworld
are we living in which century
are we living in which era
are we living in which yuga
are we living in which dimension
are we living in which age
are we living in which galaxy
are we living in the worst of times

Although the "predictive analytics" informing algorithmic search suggestions do just as much to shape and direct searches as to reflect what users are looking for,[1] presumably, some of these are among the most common questions typed into a screen late at night, perhaps by Americans living under the cloud of constant low-grade anxiety that became our epochal baseline affect after 9/11 and the conditions of precarity deepened by the "perpetual economic emergency" following the subprime crash of 2008, then deepened again by the advent of COVID-19.[2] Apparently, then, the question that Brian McHale (1992) thought signaled "postmodernism's shift of focus to ontological issues and themes"—Dick Higgins' "Which world is this?"[3]—has taken on new dimensions, particularly in the last of these nearly three decades since. Works of the fantastic imagination no longer merely *reflect* this condition of ontological uncertainty or anxiety; they serve to *define* it—and to define how we ask the second of McHale/Higgins' questions about "the world": "What is to be done in it?"[4]

Though set in other worlds—either realms entirely distinct from ours, like Middle Earth, or our own world but for a quantum of otherness (Darko Suvin's *novum*), or something in between, like the notoriously ambiguous Panem of Suzanne Collins's *The Hunger Games*—works of the fantastic, as Cory Doctorow says, are now part of a common "toolkit" for thinking about the worlds in which we actually find ourselves, metaphors we live by.[5] This seems to be the case in spite of, or even *because of*, the fact that they are, by nature, resistant to the ordinary process of metaphor, which should flow smoothly from vehicle to tenor, from familiar source to unfamiliar target; they operate in a "*counterfigurative*" direction, as Seo-Young Chu writes, "displacing the ordinary attributes of figurative language—its weightlessness, virtuality, as-if-ness, dependence on cognitive labor—with the vivacity, solidity, persistence, and givenness that characterize the perceptible world of literal facts."[6] Works of the fantastic do not only defamiliarize the familiar world, making it new; they bear witness to its growing *unfamiliarity*.

What we might once have taken to be a world composed of well-defined "objects," like the specimens in museum collections, is instead populated by a proliferation of much stranger stuff that resists representation. Theorists have struggled to cobble together new vocabularies for these "cognitively estranging referents," objects that aren't quite objects, such as Timothy Morton's "hyperobjects" (objects that defy any attempts to localize them, like global warming), Bruno Latour's "quasi-objects" (objects that are simultaneously natural and cultural), Peter Sloterdijk's "nobjects" (intangible and unownable objects, like the very air we breathe), Calvin O. Warren's "~~objects~~" (human beings reduced to the utterly abject status of not-even-an-object), or Timothy Ingold's non-objectal "things" (conceived as being constantly in the process of formation).[7] These strange forms of being are said to typify the "pluriverse" we inhabit, a world containing many overlapping but different worlds. Accordingly, the fantastic increasingly *is* how we ask ourselves: which world are we in, and what can be done in it?

This loss of the object and confusion of worlds perhaps already marks our time as the time of a global *constitutional* crisis, a crisis of what Bruno Latour dubbed "the Modern Constitution."[8] Modernity is founded upon the distinction between subjects and objects, a human realm and a natural realm, each with its own distinctive "powers." It is as if modernity had a single founding document (this is, of course, a kind of counterfactual, science-fictional proposition): like the US Constitution, it would decree a "separation of powers" between those two realms, imagined as ontologically separate but equal. What some have called the Anthropocene is precisely an epoch when that foundational distinction has become visibly untenable.[9] Theorists have increasingly complained that this invisible document isn't worth the paper it isn't written on, for the symmetry of subject/object, human/nonhuman, culture/nature has always functioned as a "metaphysical hierarchy" in which meaning and personhood has been reserved for the upper house, inertia and insignificance accorded to the lower house.[10] Moreover, it has always underwritten the most brutal physical hierarchies, as some human beings, their humanity being unrecognized, have been relegated to the category of objects, of natural resources (to be used) or obstacles to progress (to be eliminated), subjected to "bestialization" or "thingification."[11]

Here, it must be said that people subjected to bestialization and thingification have a distinctive relation to the question, *which world are we in*. People allowed to be "human" have had a privileged relation to normative realism—the kind of representations of the world that assure us that we have a place in it. Drawing on horror scholar Eugene Thacker's distinction between the terrifying "world-in-itself," the horrifying "world-without-us," and the comforting, familiar "world-for-us," Travis Linneman more precisely calls this last world the "world for *some* of us."[12] Others have struggled to find themselves represented there. In this sense, as Greg Tate remarks, "Black people live the estrangement that science fiction writers imagine," and thus may not share the assumptions of the normatively "real" world that white people bring with them to a science fiction story, for instance; they are subject to a "double estrangement," in Joy Sanchez-Taylor's words.[13] It is for this reason that we need alternatives to the customary theory of science fiction, according to which it estranges the familiar, representing it to us in the guise of the other. Seo-Young Chu suggests instead that we begin with the strangeness of the world itself, which science fiction writers struggle to pull into representation, making it comprehensible. Hers is a theory well suited to the doubly estranged—a point that even such a perceptive critic as Gerry Canavan misses in his review of Chu's book.[14] This is one reason why the creative explosion of the last two decades that I call "the New Black Fantastic" cannot be reduced to a case of new artists taking up old tools. These writers walk the same Earth but do not necessarily share the same world (the world of the Same). For Black SF authors, as Mark Bould puts it, "the ships landed long ago"; for Indigenous SF authors, the apocalypse has already taken place (and is still ongoing); for trans SF authors, gender dystopia is already here.[15]

Yet the question, *which world are we in, and what can be done in it*, is not just a matter of description and perspective; it is a constitutional kind of question. It is not just a matter of correctly identifying objective conditions but of deciding how the world order that is coming apart should be reconstituted. It requires an expanded political imagination. And so I would like to think a bit about our world with the help of an exemplary exercise of contemporary political imagination: N. K. Jemisin's magisterial Broken Earth Trilogy (*The Fifth Season*, 2015; *The Obelisk Gate*, 2016; *The Stone Sky*, 2017).

Rifting

Let us begin, as Jemisin suggests, with the *end* of the world. In *The Stone Sky*, Alabaster observes that for some people, the end of the world is always here and now—in a grim echo of J. B. Lenoir's "Down In Mississippi" ("The season was always open on me"): "Every *season is the Season for us*."[16] That is to say: for social pariahs, both in The Stillness, the fictional world of Jemisin's *Broken Earth* trilogy, and in our own reality, all times ("season[s]") might as well be the end times ("*the* Season"), threatening life with extermination, threatening all meaning with meaninglessness. In this way, Jemisin answers the *which world?* question: ours is the world that shot Trayvon Martin and Michael Brown, that left them unidentified in the morgue for a day and in the street for four hours, that rewarded their killers with wealth and social approbation, "a period in which it felt as if a black person was getting extrajudicially killed by police every day . . . and so many racists were coming out of the woodwork to basically say that they liked living in a world so filled with injustice and wrongness, and that they didn't want it to change."[17] Unlike more conventional literary representations of contemporary racial injustice and police brutality, however, Jemisin's fantastic narrative, opening as it does with an apocalyptic "Rifting," simultaneously addresses the "metabolic rift" that characterizes our world's relationship to the Earth.[18] A "Fifth Season," in Jemisin's grim echo of our ongoing "Sixth Extinction" event, is "an extended winter—lasting at least six months per Imperial designation—triggered by seismic activity or other largescale environmental alteration."[19] As such, it signals a disruption not only of the natural order but of the temporal order. The uniformity of time assumed by the phrase "*the* end of *the* world" is radically called into question: *for whom* is the world ending? And this, in turn, unsettles the unity of place: *which* world is ending?

The political geography of the Broken Earth is made strange for us, partly by the lack of a map as detailed as that created by Christopher Tolkien for his father's creations; what Tim Paul's map of The Stillness gains in geological realism (unlike Middle Earth, it has no land of evil surrounded by a conveniently rectangular fence of mountains), it loses in the demarcation of human settlement and governance (few cities are identified, and no political boundaries are marked). As a planet of just one supercontinent, it bears no physical resemblance to the physical geography of our world, and unlike Collins's Panem (a postapocalyptic echo, perhaps, of "pan-American" as well as of the Roman *panem et circenses*, "bread and circuses"), none of its place-names resonate with those of our world.[20] We learn that Sanze, a civilization that has largely disintegrated but left its "imperial governance system" behind in something like the manner of a defunct Rome, linked individual "comms" together in "quartents" within "regions,"[21] but there seem to be no hard borders, as in a medieval Europe marked by little fiefs with uncertain boundaries between them. The imperatives of survival make the condition of the "commless" uniquely precarious.[22] By contrast, the world that we habitually call "*the* world," as Richard J. F. Day and Adam Lewis point out, is constituted by nation-states and their borders, almost entirely without remainder, even if borders, notoriously debatable, are also a principle of war and disorder, creators of refugees and other "commless" people.[23] Even if the United Nations High Commissioner for Refugees has estimated in recent years that more than ten million men, women, and children worldwide are stateless, *in theory*, everyone is supposed to be sorted, as if by Hogwarts' magic hat, into States that grant them recognition either in virtue of their ethnicity or under the supposed secularity of a generic "citizenship," granting them rights, making them persons of some sort under a set of published laws, even if nearly a quarter of the world's population lives in States too "fragile" to guarantee this.[24] A proper place for everyone. The failure of this world order is reflected less clearly in the discourse of political science than in science fiction (and the fantastic, more broadly speaking), which—along with postcolonial literature, significantly—has repeatedly imagined the ends of this world: the times when it falls apart for some of us, the places where it fails to cohere at all. More recently, as the fantastic has dramatically expanded in its authorship and audiences, it has begun to imagine the actual situation of pariahs within the world's *non*-"fragile" spaces, the industrialized nations of the Global North: Black lives, Indigenous lives, queer and transgender lives, the lives of migrants and immigrants. There, too, the world is perpetually about to end, or it has ended, or it might end. The time and space of the modern world, as a certain pariah in Nazi-occupied Poland once observed, is full of gaps and rifts, intervals and in-betweens.[25] This world is itself suspended in an interval, a dreadful in-between: its order is indeed ending, but whatever will replace it is as of yet unable to fully emerge, so that we live in "the time of monsters."[26]

The metabolically and politically rifted world of the Broken Earth books, the continent called The Stillness, is populated not only by "stills" (the ordinary, traumatized remnants of the human race, their lives organized by caste and ritual adherence to the "stonelore" of survival) but by "broken monsters" of many kinds: some of them (the Stone Eaters) were once human, some (the orogenes) have never been treated as human, and some (the Guardians) have had their humanity drained away in the service of a project of control.[27] If, as Jemisin reflects, SF has traditionally "centered on white male protagonists," monstrosity in Jemisin's world is generally outward, an externally conferred condition: "We are the monsters they created," reflects Hoa.[28] "They kill us," Alabaster remarks of the stills, "because they've got stonelore telling them at every turn that we're born evil—some kind of agents of Father Earth, monsters that barely qualify as human."[29] "Why would he do such a thing?" muses Schaffa, remembering "the monster that he was" in the service of the Guardians.[30] If anything, internalizing one's own monstrosity is possibly a form of self-repair: "she has been called *monster* so many times that she finally embraces the label."[31] "'Perhaps,' Schaffa tells her as she sobs these words . . . 'But you are *my* monster.'"[32] "Look at you, little one. if you are the monster they imagined you to be . . . you are also glorious."[33] It is the stills, of course, who are most monstrous and unnatural of all, repaying the orogenes' service with fear and hatred: the "end of the world" for Essun arrives when her still husband, Jija, on discovering that their child is an orogene like his mother, kills him in cold blood. At the same time, as Alan Moore might put it, empires need their monsters: orogenic slave labor, controlled by the anti-orogenic Guardians, is all that "stills" The Stillness.[34] The coming of a time of monsters is not a natural event, or rather, it is a prolonged moment in which, as Bruce Sterling observes, unnatural, "monstrous" things "have gone past sin and become necessity."[35]

What *is* a "time of monsters"? Monsters, according to Noel Carroll, are "figures that cannot be (cannot exist) according to the culture's scheme of things," "impure and unclean," "unnatural composite[s]."[36] Edward Ingebretsen adds that "Monsters have, or seem to have, freedoms we lack. They transgress, cross over, do not stay put where—for the convenience of our categories of sex, race, class or creed—we would like them to stay."[37] They stand outside the tidy scheme we imagine "nature" to be, a "scheme of things" in which everything has its proper place. In a time of monsters, nothing is simply "natural" in that sense (which may also amount to saying that *everything*, no matter how monstrous, is "natural"). There are no (or only) "natural" disasters in The Stillness: a Fifth Season, too, is an "unnatural composite" of human and natural agencies. Every one of the Seasons is a salvo in an ongoing war between a wounded, enraged Earth and the human beings who once "tried to put a leash on the rusting planet."[38] Not only are the quakes and eruptions that have reduced human life to a constant struggle for survival direct expressions of the Earth's hatred, but they are subject to the countering influence of orogenes, human beings with the ability to intervene in the Earth's processes, to "reach for the fire within the earth, or suck the strength from everything around [them]" and thereby either cause or quell the tremors.[39]

Here, indeed, Nature appears as the Big Other that the Lacanian psychoanalytic theorist Slavoj Žižek insists "does not exist": to hear the voice of the Earth saying "*hello, little enemy*" would be "a psychotic projection of meaning into the real itself."[40] For the psychotic, according to Jacques Lacan, this is precisely what happens: "Everything has become a sign for him. Not only is he spied upon, observed, watched over, not only do people speak to, point, look, and wink at him, but all this . . . invades the field of real, inanimate, nonhuman objects."[41] The stills' terror and resentment toward "Father Earth," the orogenes' ceaseless activity to combat it, can only appear under the psychoanalytic categories of "psychotic projection" and the "frenzied activity" of the obsessional neurotic who "works feverishly all the time . . . to avoid some uncommon catastrophe that would take place if his activity were to stop."[42] But this would be to simply dismiss not only the "vivacity, solidity, persistence, and givenness" of Jemisin's fictive world (and that of other works of the fantastic) but also the real cosmologies of any number of traditional societies. For such societies, the world itself and any number of nonhuman actants actually *do* return our gaze and speak to us, and the notion of an ethical, intersubjective relationship with the Earth is no metaphor. Whereas for Lacan, the natural (the real) and the social (the symbolic) are utterly incommensurable, modern works of the fantastic and nonmodern cultures both, significantly, often regard these domains as existing within a single continuum, as inextricably entangled.[43]

Such works are themselves the product of entanglement, of processes of hybridization. At the same time, they are also caught up in the "purifying" activity that constitutes modernity—not least a struggle over genre self-definition. Thinking about questions of the demarcation of fantasy from science fiction, or even the more obviously political questions of who should be allowed to write and read them, as "struggles" in the political sense might have seemed like a stretch prior to 2016 when it became glaringly apparent that the past decade of contestation for ownership of gamer culture and comics fandom (in attacks on women in the video game and comics industries, culminating in the "Gamergate" and "Comicsgate" campaigns), and of the SF and fantasy fields (the white reaction to "RaceFail '09," leading to the "Sad Puppies" and "Rabid Puppies" campaigns to sabotage the Hugo and Nebula Awards), and so on—had been a harbinger of the rise of the alt-right.

All of this reveals a deep homology between the processes of genre "purification" and projects of racial and sexual "purification." The question of who owns the social imaginary, the power to conceive worlds, has always been a question of *who owns the future*.

The Great Sorting Machine

Academia, where I live and work, is a world that perpetually fears its own end, although it has begun belatedly to suspect that this threatens it not from without, in the form of barbarous hordes of supposedly unworthy students, but from within the walls of the city it once guarded, as the Empire loses interest in its continued existence (and after all we've done for it!). It bids us dress in mock-medieval gowns once or twice a year as a reminder of our origins, but of its old traditions of self-management, it retains only the outward semblance, real control having been ceded to a caste of administrators. Its ability to confer status is seriously eroded; its "rich, proud tradition of serving as an engine of social mobility" in the US context lasted only as long as the money did, spanning just about two generations, from an expansion funded by the G.I. Bill to its Reagan-era retrenchment.[44] Under neoliberalism, it is increasingly reduced to its gatekeeping function, as even in the purportedly meritocratic US system, defunded public universities seek the tuition money of wealthier students; the income-enhancing effects of higher education get preferentially distributed to students from higher-income families, reproducing and further entrenching inequality.[45] "Higher education," Clark Kerr once smugly wrote, "acts as a great sorting machine. It rejects as well as selects and grades."[46]

As such, the university forms just one part of a much larger Great Sorting Machine—a process that continually selects out the "worthless" from the "worthy,"[47] "unentitled" noncitizens from "entitled" citizens, and the "illegal" immigrant from the "lawfully present,"[48] the "law-abiding" from the "criminal," the "criminal" (subject to civilian law) from the "prisoner of war" (subject to the Geneva Conventions) and the "prisoner of war" from the "enemy combatant" (subject only to the determinations of the US State Department),[49] the "abnormal" from the "normal,"[50] "nonpersons" from "persons,"[51] "ungrievable life" from "grievable life,"[52] "nonlife" from "life" as such,[53] "being" itself from abject "nonbeing."[54] While it may be true that life itself entails a membranous sorting of self from other,[55] it is also possible to regard permeability, the transmission of material across the membrane—respiration, nutrition, excretion, the exchange of genes, the very activity of metabolism—as equally intrinsic to life; "association," assembly, the creation of collectivities, are as fundamental to being as disassociation, disassembly, "selection."[56] The goal of the Great Sorting Machine is *always* "purification," as Bruno Latour says, even if it is also perpetually creating "hybrids" that belie any imagined purity.[57] It sometimes claims to serve life, but what it calls "life" is a very narrow construction indeed, a constriction of life's possibilities: a *Selektion* in the word's most terrible sense.

If, in coming to grips with the strangeness of our times, we are turning more and more to literatures of intense estrangement—if we, like the protagonist of Charles Yu's novel of the same name, are seeking to learn "how to live safely in a science fictional universe"—we must also recognize that many of the narratives crisscrossing this universe are really the self-justifications of the Great Sorting Machine. "I'm *in conversation with* most of traditional science fiction, postapocalyptic fiction," remarks Jemisin:

> Most science fiction—post-apocalyptics, traditionally—throughout the genre's history, centered on white male protagonists. They were often focused on those men using the opportunity of the apocalypse to either build a new world or to unleash their own inner monster, or warrior, or whatever.[58]

Alfred Bester once tried to send up all such masculine fantasies of Being Chosen or Being Special, to reveal them as "baby dreams"—"To be the last man on earth and own the earth . . . To be the last fertile man on earth and own the women . . ."[59] Jemisin references the baby dream in Robert Heinlein's *Farnham's Freehold* (1964): "this is a book about a middle-aged white guy who basically somehow knew the apocalypse was coming, built a shelter, survived the apocalypse, and happened to have a small group of people with him, which he just thought was the perfect small group of people to dominate and force to rebuild society in his image, in the image that he wanted it to be."[60] It is thus also a utopia, a utopia of one man, not unlike the one described by a certain "real estate developer and president of the Trump Organization" in an what's-your-personal-utopia interview in *Omni* (1988): "When I picture my Utopia, I envision a city . . . It would be a big city, and I would be in charge."[61] We now (2020) live in Trump's personal utopia, certainly a "no place" (but not a good place) for immigrants, for transgender and Black people, for radical dreamers of any kind. That is to say: too many of the stories we tell about possible worlds in this mode of estrangement turn out just to present a rationale for the too-familiar world, the world in which some (of a particular gender and race and class and sexuality and ability . . .) are chosen, made special, and others treated as dispensable.

How to Make an "Unwanted Factor" Disappear

In questioning how science fiction can function as a "toolkit for thinking" (or a "blueprint for disaster"), Cory Doctorow nominates, along with *Farnham's Freehold*, an even more canonical work of American SF: the thirty-eight-times anthologized short story, "The Cold Equations," by Tom Godwin, first published in John W. Campbell, Jr.'s *Astounding Science Fiction* (1954). Perhaps it can be considered a distant precursor of the particularly suspect Theater of Cruelty that has become a new cultural dominant in the age of Peak Television,[62] Grimdark fantasy,[63] Revisionary or Dark Superhero comics,[64] Torture Porn cinema, and Survival Horror videogames,[65] wherein the more willing an auteur is subject protagonists to torture, maiming, rape, or murder (and/or to make the viewer/reader/player complicit in such), the more they accumulate the cultural capital of "authenticity" and "gritty realism." Unlike these spectacles of cruelty, however, "The Cold Equations" does not assume the reader's consent but muddles the issue with pathos and equivocation. Beneath the muddle, Godwin's plot—a young woman stowing away on a landing craft bringing medical supplies to a frontier planet is forced to jump overboard, for otherwise, it will crash from the extra weight, and both she and the pilot and the colonists depending on the delivery will die—amounted, as Campbell privately crowed to Isaac Asimov, to a carefully contrived means for "forc[ing]" readers to "agree that there is a place for human sacrifice."[66]

It is seen as a classic of the "Campbell Era" in US science fiction history, i.e., an exemplary text for the SF produced under the editorship of the man Jeannette Ng correctly named, on receiving the Campbell Award for her *Under the Pendulum Sun*, as "a fucking fascist," a vocal believer in masculine white supremacy.[67] Although critiques of "The Cold Equations" are at least as canonical as the story itself—its blatant sexism alone earned it a poetic sendup in *Asimov's Science Fiction* as recently as 2008[68]—no one, to my knowledge, has identified "The Cold Equations" as a fascist narrative, although its ending was apparently dictated by Campbell's demands.[69] On the contrary, the story keeps proclaiming its humane, universalist ideals: "Wherever you go, human nature and human hearts are the same," the pilot, Barton, assures Marilyn, and to show that he is no exception, he quietly cries a little ("Something warm and wet splashed on his wrist").[70] Indeed, this insistent performance of "human" values helps to *mask* the story's cruelty.

If the fascism of our present moment seems to revel in undisguised, spectacular cruelty ("the cruelty is the point," as the saying goes), it does not yet do so without certain performances of humaneness, dispensed in the manner of an Emperor Nero who happens to be in a generous mood at the gladiatorial arena today. Moreover, it does not fail to supply quasi-deontological and quasi-utilitarian rationales for inflicting pain, suffering, and death on those to whom the Emperor has given the thumbs-down sign: *there'd be no problem if they just came here the legal way; if you haven't done anything wrong, you have nothing to worry about; there aren't enough jobs to go around as it is* . . . So, too, "The Cold Equations" spends most of its energy in efforts to persuade us that Marilyn is simply framing things incorrectly when she objects, "You're going to make me die and I didn't do anything to die for."[71] The correct framing, the story insists, is that Marilyn is a victim of her own innocence (or ignorance), which fatally exposes her to "forces that killed with neither hatred nor malice":[72] the meteorological forces that destroyed the colonists' medical supplies, the biological force of the alien disease threatening the colonists' lives, the physical forces governing falling bodies, perhaps even the forces of "human nature" propelling Marilyn to take an unknown risk to visit her brother on the frontier.[73] There is no one but Marilyn (who "didn't know what she was really doing") to be blamed for the situation; ultimately, it is just "the universe" that forces a choice between the sacrifice of just one life and the loss of that life with many more.[74] Only, as many have pointed out, this moral dilemma is entirely contrived: that the door to the landing craft is barred by nothing more substantial than a "KEEP OUT" sign is a matter of criminal negligence; that no one thinks to throw one hundred ten pounds of *anything else* overboard is a matter of unthinkable stupidity; that the landing craft is not engineered to carry one hundred ten extra pounds is a matter of outrageously bad design, unscrupulous cost-cutting, or both; that human beings are, in the first place, colonizing a world already populated by

Indigenous people is a political matter—all of these are very much within the realm of human responsibility.[75] *Most of the realm of human responsibility is made invisible within the frame of the story.* Or, as Bruno Latour would say, it has been "blackboxed."[76]

If we are intent on locating guilty parties, Latour might seem like the last person we should turn to for help. Indeed, his insistence that agency (and hence responsibility) may as likely be found among nonhuman objects, as well as human subjects, has drawn withering criticism from scholars such as Andreas Malm. Isn't the problem in this story, for instance, really just the moral *abdication* of the subject? What does it mean to say that "Objects Too Have Agency" if it is not some kind of irrational return to animism or panpsychism—a "fetishistic" relationship to the object writ large?[77] It is tempting to say, instead, that "The Cold Equations" expresses ideology in something like the manner of what Roland Barthes called "myth": that is, it disguises "History," the social realm, characterized by signification and change (and for Malm if not for Barthes, by human intentionality and agency), as "Nature" or mere "closed, silent existence," regarded as basically "unchangeable"[78]— precisely Godwin's "laws of nature, irrevocable and immutable."[79] In this way, responsibility is elided. Another possible critical path would follow Slavoj Žižek's insistence that "Nature does not exist":[80] Since the Real cannot be represented, any representation of "the laws of nature" as a kind of quasi-Other who punishes transgressors is an ideological fantasy. In this case, too, Latour's notion of object agency appears fetishistic.

Yet the nonhuman agency that proliferates across the space of "The Cold Equations"—in spite of all protestations to the contrary, Godwin *does* personify nature so that storms "strike" helpless human encampments with "fury"[81]—in no way precludes the many forms of human agency that the story seeks to erase, nor does it disavow the responsibility of human subjects for their participation in "alliance" with objects like the landing craft and systems like gravitation, their role in networks of agency.[82] The more insistently Godwin tries to separate out the components of the network, to "purify" his narrative by parceling out the social from the physical, to relegate human subjectivity to a helpless witness of (albeit a precious repository of value within) an inhuman object world, the more the lines are blurred: human actions and reactions appear as "natural" as the natural processes that increasingly appear to take up the space evacuated by the shrinking vanishing human. The title phrase is yielded up in a strange mixture of ideological confirmation and self-subverting confession: "To himself and her brother and parents she was a sweet-faced girl in her teens; to the laws of nature she was *x*, the unwanted factor in a cold equation."[83] The narrative voice here undoes its own intentions by imputing a subjective standpoint to nature itself: "*to the laws of nature she was . . .* " And how cold can the equation be when it is seemingly invested (the narrator equivocates about the locus of agency here) with the power to desire, to *want* or *not want*?

"The Cold Equations" demonstrates how to blackbox, how to invisibly remove "unwanted factors" by engineering a murder without a murderer—or, in Cory Doctorow's phrase, how to beat the rap by carefully constructing "a crisis of your own making . . . [that] never ends," since "a crisis . . . isn't the time to lay blame." And so it is that "The Cold Equations" might serve as the aptest illustration of two Latourian theses: insofar as modernity itself is a project of purification, the construction of a disenchanted or non-anthropomorphized world, "we have never been modern," and it is precisely because the social and the natural are inextricably intertwined with one another that "there are no natural disasters, only sociotechnical ones".[84]

Two Sociotechnical Disasters: (1) "Hurricane Katrina"

Let the cracks in the system of levees, drainage canals, and floodwalls protecting New Orleans on August 29th, 2005, stand as a physical analog and exemplar of *the monstrous condition*, an augury of the dissolution of the system that kept people and places in their places: the end of the world. At the same time, the death and destruction wrought by the post-hurricane flooding offered a vivid example of that same system in operation: in the way of *that* world, the deaths of poor Black people, their being warehoused in leaky stadiums or left stranded for days in the sodden wreckage of their homes, was an expected and acceptable loss, to be followed by some judicious relocation of the displaced, the reclamation of devastated neighborhoods as fresh real estate for the speculators, the redesign of pulverized school systems as spiffy new opportunities for private operators. Capitalism was not in the least worried by ruins; it had a new "world" ready to download and impose upon the Earth.

In short, Katrina discourse, the dominant set of interpretations circulated after the event, stands as exhibit A for Hilgartner's proposition, now become a commonplace, almost a cliché, of social geographers and sociologists of risk, so much so that *There Is No Such Thing as a Natural Disaster* provides the *title* of a book about Katrina (Hartman and Squires). As one group of contributors to that collection wrote,

When the levees were topped and New Orleans began to flood, the mainstream media continued to fixate on the destructive force of Katrina's winds to the exclusion of everything else, including race and racism. The media presented an endless array of equations for calculating and assessing potential storm damage, but none of these formulas included race as an input variable.[85]

This insistent *naturalization* of nearly two thousand deaths functioned to absolve human beings and their institutions of responsibility, belying the complex entanglement of human and more-than-human agencies at work.[86] The hurricane was a "natural" event—but the heat energy that powered it may well have been increased by human carbon emissions; the collapse of flood protection systems, both artificial (the levees) and natural (the coastal wetlands), was due to bad engineering, special dealing to real estate developers, and decades of governmental neglect of infrastructure; the vulnerability of so many residents, particularly poor Black New Orleanians, was due to their lack of access to transportation (the City asked them to evacuate themselves, ignoring the fact that tens of thousands had no cars) and, more fundamentally, to inequalities of power, to centuries of white supremacy and slavery; the chaotic shambles of relief efforts was due to decades of hollowing out of the federal, state, and local government, increasingly reduced to their repressive apparatuses, shorn of the ability to coordinate and deliver the most basic guarantees of survival to their own citizens and constituents.

Almost none of these causes of the suffering and deaths of thousands were really addressed in the aftermath. Instead, we saw concerted efforts at blaming the victims for their own victimization: they had been too stubborn to leave, they didn't listen to the warnings, they should have gone to the official shelters, they were asking for it by living on a flood plain, they were lazy and stupid, they were responsible for their own poverty . . . In short, Katrina discourse is an exercise in *blackboxing the causation*, making the "unwanted factor" of power disappear.

Two Sociotechnical Disasters: (2) The "Border Crisis"

Just before Independence Day, 2019, a report from the Department of Homeland Security's Inspector General found a scene of squalor in the US Border Patrol's jails on the southern border: "standing-room-only cells, where some detainees have been held for a week or more, detainees not having access to showers, kids not having access to hot meals."[87] Confronted with this report by an NPR interviewer, Brandon Judd, head of the Border Patrol's labor union, said that he "agree[d] with it 100%":

> JUDD: Our facilities, 100%, are absolutely overcrowded. When we're dealing with the number of people that are crossing the border right now, when we take them into custody, *we have to* take them back. *We have to* process them. *We have to* put them through a certain process to verify who they are, make sure that the children that they have with them are, in fact, their children. And then *we have to* wait for ICE [Immigration and Customs Enforcement] or *we have to* wait for HHS [the Department of Health and Human Services] to take these individuals off of our hands. If HHS and ICE doesn't take them off of our hands, then *we have to* hold them.
>
> I mean, think about unaccompanied children that we take into our custody. If HHS . . . do[es] not come and take these children off of our hands, we can't just release them to the street. *We have to* hold them until the proper authorities come and take them off of our hands.
>
> NOEL KING [NPR]: You're saying there is, essentially, nothing that you can do about these conditions that people are being held in. Is that right?
>
> JUDD: There's absolutely nothing we can do.[88]

This masterclass in delinquency frames CBP as helpless witnesses to lamentable "conditions" imposed on them by other agencies, by Congress, perhaps even by those sadly improvident immigrants who, in this representation, are mysteriously appearing in CBP's jails without anyone hunting them down, handcuffing them, placing them in the backs of jeeps and vans, and incarcerating them—certainly not CBP. The magical incantation that makes this breathtaking disappearing act possible is a phrase uttered in the voice of a child to whom the Law is given: *We have to . . .*

I am, of course, being slightly unfair. As another kind of "unwanted factor," the immigrant women, men, and children do not really appear as *agents* in Judd's account at all. They are, rather, another helpless group, almost as helpless as CBP itself. If the Katrina discourse is characterized by victim-blaming, the Border Crisis discourse is rather a matter of disingenuously assuming the position of victimhood from a position of authority. It is the voice of the abuser saying *Look what you made me do*. It is the voice of the narrator in "The Cold Equations" assuring us: "there could be no alternative."[89] What is locked away inside the black box is the exercise of power over others' lives.

Policing Boundaries

One waits in vain for an individual story so heart-wrenching, an individual victim so clearly innocent and blameless, that these strategies of black-boxing no longer work. The victims of school shootings and racist police murders disappear into the black box as if it were an oubliette, and the public grows numb. If this mechanism for the redistribution downward-and-elsewhere of blame and responsibility is to fail, it will have to be by deliberate sabotage.

Here, perhaps we can once again learn from the case of "The Cold Equations." Much as if the fictive death of Marilyn Cross had been a real-life sociotechnical disaster, it has been litigated repeatedly after the "fact." The fact that all the facts are fictive makes this all the more difficult. As some have pointed out, objecting that the story gives no adequate explanation for why there is no alternative to her fatal step out of the airlock may seem question-begging: even if such clearcut Trolley Problems, *pace* Philippa Foot, do not often arise naturally, surely there have been and may yet be real-life situations in which a similar crisply bounded moral dilemma with similarly fixed and foreknown outcomes arises (one life or many lives, without even a third possibility). However, as John Huntington points out, while the story could have been revised in such a way as to remove all the other occluded possibilities represented, for instance, by the other objects that could have been thrown overboard—"the door of the closet, the blaster, the people's clothes, the pilot's chair, the closet itself, its contents, the sensor that registers body heat, the bench she sits on"—"in the process of such a rewrite . . . as we 'purify' the story we would deprive it of its imaginative texture. Without the blaster, the clipboard, the heat sensor, and so forth, what do we have?"[90] In other words, the story's counterfictive properties, "the vivacity, solidity, persistence, and givenness that characterize the perceptible world of literal facts," would be lost.[91] But imagining such a purification process should not distract us from the purification that the story *as it is* already enacts.

We have already seen how the story purifies motivations. As Brian Attebery points out, while the pilot must ultimately pull the lever that opens the airlock door, it seems "nice that he feels bad about the girl," even if "she brought her own fate upon herself."[92] The exaggerated sentimentality of the story, shared by its narrator and protagonists, masks a desire to punish, a desire briefly revealed before we find out that "It was a girl":

The stowaway was not a man—she was a girl in her teens, standing before him in little white gypsy sandals with the top of her brown, curly head hardly higher than his shoulder, with a faint, sweet scent of perfume coming from her and her smiling face tilted up so her eyes could look unknowing and unafraid into his as she waited for his answer.

Now what? Had it been asked in the deep, defiant voice of a man he would have answered it with action, quick and efficient. He would have taken the stowaway's identification disk and ordered him into the air lock. Had the stowaway refused to obey, he would have used the blaster. It would not have taken long; within a minute the body would have been ejected into space—had the stowaway been a man....

Why couldn't she have been a man with some ulterior motive? A fugitive from justice, hoping to lose himself on a raw new world; an opportunist, seeking transportation to the new colonies where he might find golden fleece for the taking; a crackpot, with a mission—

Perhaps once in his lifetime an EDS pilot would find such a stowaway on his ship; warped men, mean and selfish men, brutal and dangerous men—but never, before, a smiling, blue-eyed girl who was willing to pay her fine and work for her keep that she might see her brother.[93]

The "girl" is feminized with a shower of adjectives (*little, curly, faint, sweet, unknowing*) in sharp contrast with the masculinity (*deep, defiant, quick, efficient, brutal, dangerous*) of the stowaway that the pilot has fantasized about. It is her girlness that provokes narrative desire, avoiding the libidinal short-circuit that would have ended the story prematurely. Yet the story quietly disavows the narrative desire that the girl is meant to arouse and, above all, its proximity to a desire for punishment, a violent desire that must be sublimated into sentimentality.[94]

Another purification is accomplished at the level of genre. The story's dominant reception through at least the 1970s (when it drew new scrutiny from critics like Leslie Fiedler and Thomas Disch) positions it as an exemplar of what came to be known, after 1957, as "hardcore science fiction" or "hard SF."[95] That is to say, the story's insistence on segregating the emotional from the scientific, values from facts, "safe-and-secure Earth" from "the hard, bleak frontier," "manmade law" from laws "not of men's making or desire," helps to call hard SF into being.[96] That is to say, it helps to undo the hybridization process that created science fiction (or "scientifiction") *in the first place*—the speech-act by which Hugo Gernsback, in 1926, defined the new genre as "the Jules Verne, H. G. Wells and Edgar Allan Poe type of story—a charming romance intermingled with scientific fact and prophetic vision."⁹⁷ The original hybridity of SF, a compound of American Gothic, British "scientific romance," and French extrapolation, among other (and older) elements, is only erased⁹⁸ after decades of programmatic purification, the erection of "hierarchies of taste":

> the hierarchies of taste within a specific category, which would for example privilege *Astounding*'s hard-SF over *Planet Stories*'s supposedly lesser science fantasy, created a similar illusion that some texts within a genre are more pure than others. In this context, it made commercial sense to create a category called SF and, when it proved successful, for SF magazines to differentiate their product not only from other fiction categories but also from other SF magazines. Just as Barton ejected Marilyn from the airlock, so this new category ejected horror, fantasy, aviation stories, detective stories, and so on. Likewise, *Astounding* ejected science fantasy and other kinds of SF deemed insufficiently "rational," such as the more colorful and exotic planetary romances and space operas; and such expulsions continue to be made by many readers of SF.⁹⁹

Nonetheless, by the dawn of the Reagan era, Godwin's story, equipped with Campbell's ending, had assumed the status of a cultural "touchstone": to be a *real* SF fan for those not seduced by the mass spectacle of *Star Wars*, *Close Encounters of the Third Kind*, or *Alien*, was to be a hard SF fan, and that, in turn, entailed an affective stance of detachment; "getting" "The Cold Equations" was not a matter of understanding any actual math or physics but of a wish to imagine oneself as "part of a technologically-minded elite, someone who can contemplate the real workings of the universe without fuzzy thinking or sentiment."¹⁰⁰

For all of its would-be chilliness, "The Cold Equations" thus summons into being a community of readers as well as a genre: a community of would-be technocrats, aloof from the masses, moved more by the beauty of physics than by physical beauty—and here the imperatives of gender and sexuality become most important. If "The Cold Equations" tends, on a deeper level, to conflate physics with politics, midcentury US gender roles with the fundamental laws of the cosmos itself, this is because the story takes part in a historic "purification" of SF. Justine Larbalestier recounts how, from the mid-1950s through the mid-1970s, a debate raged in US science fiction fandom over "whether women, love, and sex have a place in science fiction":

> One set of arguments that is articulated within these debates clearly demarcates a division between public and private, where the public becomes the masculine

space of science fiction from which women are excluded. In this division, intelligence (the mind) is located within the field of science fiction and is thus associated with men, and sex (the body) is relegated to the private sphere of women outside science fiction. This imaginary masculine space of science fiction is conjured up in the pages of science fiction magazines . . . To keep this imaginary space masculine and populated only by "real" men the borders of the field have to be policed and women have to be excluded.[101]

This is precisely the scenario of "The Cold Equations," in which a mere "girl" cannot be expected to know or understand the harsh, tough realities of life in space: since her very presence entails danger, her body must be ejected from the masculine realm. Godwin—and, importantly, Campbell, as an editor with power and influence within the field—exorcise the demons of "women, love, and sex" from the space of SF. Yet even so, the story is haunted by a trace of the feminine, of emotion, of desire renounced:

> A cold equation had been balanced and he was alone on the ship. Something shapeless and ugly was hurrying ahead of him, going to Woden where its brother was waiting through the night, but the empty ship still lived for a little while with the presence of the girl who had not known about the forces that killed with neither hatred nor malice. It seemed, almost, that she still sat small and bewildered and frightened on the metal box beside him, her words echoing hauntingly clear in the void she had left behind her:
> *I didn't do anything to die for—I didn't do anything—*

Reassembling the Universe

N. K. Jemisin's fictions forcefully reinstate the hybridity that "The Cold Equations" (in sexually purifying SF space) and Border Patrols (in racially purifying US territory) deny. And yet, as Rebekah Sheldon suggests, this is not only a matter of "messing up binaries" or "breaching boundaries": Jemisin's "hopeful monsters" are hopeful precisely because they are constantly engaged in "practices of doing, making, caring, building, fostering, and speculating," composing new worlds, not only destroying the old.[102] In the Broken Earth trilogy, despite the preeminence of other developments in the long arc of the plot (the macroscale story of how Father Earth loses and regains his child), a central place is occupied by one such unlikely project: the construction of Castrima, a hybrid "comm" shared by orogenes, Stone Eaters, and ordinary humans.

In rewriting the script of postapocalyptic narrative and recomposing the world on monstrous terms, Jemisin's fictions, neither strictly fantasy nor strictly SF, not only reaffirm SF's origins as a hybrid but remind it of the origins of that hybridity. The broad literary imaginary from which SF, horror, and fantasy take their life, which John Clute calls "fantastika," is not only "a child of Romanticism in Europe," and so of revolutions scientific, industrial, and political, but of colonialism: the movement of bodies, armaments, and goods across borders, the movement of borders as such.[103] John Rieder finds the conditions for SF's emergence in the construction of territorial empires, which was also the construction of an infinity of confrontations between imperial subjects and "alien" others.[104] In this way, although hybridization and purification can operate in tandem, modernity's fundamental project is perhaps one of purification. A genuine curiosity about others and the universe of possibilities they represent is entangled, as Sylvia Wynter writes, with a deep and abiding racism, as "SF . . . ritually excludes or marginalizes the 'Lesser breeds without the law,' outside of technological rationality—what Ursula Le Guin has called the social, sexual, and racial aliens."[105] In other words, the fantastic often operates as *another* discourse of purification. Notwithstanding the irony, the "literature of alienation," a counterpublic space habitually referred to by its community of writers and readers as a "ghetto," also routinely abuses and deports its "social, sexual, and racial aliens."[106]

These aliens have refused to stay on the other side of the border, however. Walidah Imarisha and adrienne maree brown, with considerable justice, spoke of "Octavia's Brood," the literary progeny of the pathbreaking Octavia Butler; I would like to identify their work as part of an even broader phenomenon that I will call (with a nod to Trey Ellis' "New Black Aesthetic"[107]) the New Black Fantastic. The arrival of this wave of people of color, and particularly of Black women, is changing literary science fiction in ways that are already being felt in other media (notably comics, but incipiently film and TV[108]): Nnedi Okorafor, Nalo Hopkinson, Nisi Shawl, Rivers Solomon, Tananarive Due, and others have become insistent and transformative presences in the field. As habitual border-crossers, their very writing already a transgression, as A. O. Scott indirectly suggests, their writing sometimes seems to freely recombine the constituent elements of all three genres, the "elements of allegory, folk tale, Gothic and romance"—and satire, we should add—that Black experience seems to require for its literary self-representation.[109] And so they often undo the cold calculations of hard SF, dissolving (or redissolving) the distinctions between science fiction, horror, and fantasy. With these, they tend to dissolve (or redissolve) the distinctions between human and nonhuman, between the social and the natural, between world and Earth. Like Percy Shelley's unacknowledged legislators, they are dissolving—and perhaps rewriting—the Modern Constitution. And it is for this reason, perhaps, that they are uniquely positioned to tell us what times we are living in and to imagine the monsters, both terrifying and hopeful, who can live in times to come.

Sublime Machines
Valves of the Heart

> Why are we led to think of life in terms of mechanism? In what way actually are we, as men, parents of the machine?
> —Jacques Lacan, *Seminar II*

It is easy enough to say that one wants to think beyond the Kantian split between subjects and objects to dispel the illusion that these are distinct worlds. Of course, *we are hybrids of both. What does it mean, though, for us to really accept our being* both subject and object, social and natural, person and thing: human and other-than-human? This chapter will draw on science fiction to illuminate some moments of transition from one side of the great dichotomy to the other and to note something peculiar that happens in that process: turned one way, it looks like a process of *disenchantment*, and turned the other way, like a process of *reenchantment*.

Isaac Asimov's *The Caves of Steel* (1953), the first of his robot novels, first appeared in serial form in *Galaxy Science Fiction*, sandwiched between ads promising "The Unpublished Facts of Life," "SECRETS ENTRUSTED TO A FEW" (write to the Rosicrucians, AMORC, for more information) and other such heretofore hidden ways of access to a hidden universe—a Platonic notion interestingly echoed by the novel itself, with its underground dwellers who can no longer bear the light, the open air, of the Earth's surface.[1] There is a moment in *Caves of Steel*'s second installment in which R. Daneel Olivaw, the robot partner assigned to detective Lije Baley, casually opens up his arm to reveal "under a thin layer of fleshlike material . . . the dull blue-gray of stainless-steel rods, cords and joints":

"Would you care to examine Daneel's workings more closely, Mr. Baley?" asked Dr. Fastolfe politely.

Baley could scarcely hear the remark for the buzzing in his ears and the Commissioner's sudden hysterical laughter.[2]

The human detective's response—ours, too, if we are responding to him as narrative forces are prompting us to do—is to recoil:

> The minutes passed and the buzzing grew louder and drowned out the laughter. The Dome and everything it contained and Baley's time-sense all wavered.
>
> He found himself sitting in an unchanged position, but with a definite feeling of lost time. The Commissioner was gone; the trimensic receiver was milky and opaque; and R. Daneel sat at his side, pinching up the skin of Baley's bared upper arm. Baley could see, just beneath the skin, the thin dark outline of a hypo-sliver. It vanished as he watched, spreading away into the intercellular fluid, from that into the bloodstream and the neighboring cells, and then into all the cells of his body.
>
> "Do you feel better, partner Elijah?" asked R. Daneel.
>
> Baley did. He rolled down his sleeve and looked about. Dr. Fastolfe sat where he had been, a small smile softening the homeliness of his face.
>
> "Did I black out?" Baley asked.
>
> Dr. Fastolfe said, "You received a sizable shock, I'm afraid."[3]

If the characterization is a little stilted and the prose a little cliché-ridden (less redolent of a realistic response to surprise than recalling the obligatory blackouts of the hardboiled detective story on which *Caves of Steel* is partially modeled[4]), that is perhaps par for the course. *Par for the course* is itself a cliché that popped unbidden into my head as I was composing this paragraph, *of course*, having woken from an apparently dreamless sleep with the urge to write. It is not entirely appropriate to the circumstance, which is one of the marks of a slightly obsolescent robot in *Caves of Steel*; Olivaw's less advanced counterparts also devise inappropriate responses to present situations:

> Lije Baley had just reached his desk when he became aware of R. Sammy watching him expectantly.
>
> The dour lines of his long face hardened. "What do you want?"
>
> "The boss wants you, Lije. Right away. Soon as you come in."
>
> "All right."
>
> R. Sammy stood there with his unchanging blank grin.

Baley said, "All right, I told you! Go away!"

R. Sammy turned and left to go about his duties. Baley wondered irritably why those duties couldn't be done by a man.[5]

The cliché, as "a fixed bit of language," is another piece of "unchanging" verbal gadgetry, a *device*—figuratively speaking—that refuses to adapt to the new.[6] The very word "cliché" is a defunct literary device, a *dead metaphor* comparing the "predictable or unoriginal" in language or behavior to the operation of *un cliché*, that is, in French, the mechanism a printing press uses to stamp two, twenty, or twenty thousand copies of the same page on identical sheets of pulp: *le stéréotype*.[7] Classic science fiction writers write a little bit like those machines created to satisfy the very mass audiences they, in turn, help call into being; they let the mechanism show, even (or particularly) when they are trying to write "human" reactions such as shock.

Editor Herbert L. Gold is but too aware of this embarrassment, which he attempts to head off in an editorial affixed to the same issue of *Galaxy Science Fiction* as the second installment of Asimov's *Caves of Steel*: "With a little practice, anyone can be a cliché expert. There is a choice of diplomas: either a look of pained contempt or amused pity. Some sophisticates own both. The ability to recognize a cliché and scorn it is a necessary passport to literate circles."[8] It is by this acceptance of its "contempt" and "pity" that Gold and his readers, in 1953, reject the authority of "sophisticates" in "literate circles," i.e., the literary world that rejects them, for he cannily recognizes in the cliché a technology: it is a laborsaving device for thought, a feat of verbal engineering. "A successful cliché is (or was originally) the keenest view of a subject and the most concise way of expressing it. Psychology textbooks need chapters to state: *What can't be cured must be endured*."[9] What Kenneth Burke called "the bureaucratization of the imaginative" is just this engineer-like streamlining of imaginative language, here embraced—even if Gold admits that the "intolerance is justified," since "through incessant repetition, clichés become mere automatic responses"—as a positive virtue.[10] Perhaps Gold would have approved of Burke's suggestion that literary works are really only "equipment for living."[11]

A markedly similar scene of device-baring appears in Philip K. Dick's "The Electric Ant" (1969), in which unassuming businessman Garson Poole, recovering from an accident, wakes up in a hospital to find that his assumption that he was a human being is incorrect:

"You shouldn't be foning quite so soon," the doctor said as he studied his chart.

"Mr. Garson Poole, owner of Tri-Plan Electronics. Maker of random ident darts that track their prey for a circle-radius of a thousand miles, responding to unique enceph wave patterns. You're a successful man, Mr. Poole. But, Mr. Poole, you're not a man. You're an electric ant."

"Christ," Poole said, stunned.

"So we can't really treat you here, now that we've found out. We knew, of course, as soon as we examined your injured right hand; we saw the electronic components and than we made torso x-rays and of course they bore out our hypothesis."[12]

Poole, the electronics manufacturer, is himself a piece of electronics, a manufactured thing; together with the replicant business secretary Rachel in *Blade Runner*, Poole might bitterly remark: "I *am* the business." Perhaps, lacking a human brain, Poole would not be trackable by his own "ident darts," like a vampire who, lacking a soul, cannot appear in a mirror. But it is the actual scene of revelation that is stuck in my memory:

"Scan me visually," he instructed the computer. "And tell me where I will find the programming mechanism which controls my thoughts and behavior." He waited. On the fone's screen a great active eye, multi-lensed, peered at him; he displayed himself for it, there in his one-room apartment.

The computer said, "Remove your chest panel. Apply pressure at your breastbone and then ease outward."

He did so. A section of his chest came off; dizzily, he set it down on the floor.

"I can distinguish control modules," the computer said, "but I can't tell which—" It paused as its eye roved about on the fone screen. "I distinguish a roll of punched tape mounted above your heart mechanism. Do you see it?" Poole craned his neck, peered. He saw it, too.[13]

Since cliché has it that the heart is the seat of the emotions, of that which makes us vitally human, it is especially chilling to imagine finding a "programming mechanism" there in ourselves, insofar as we are led to imagine ourselves in Poole's place. Dick is not as insistent on that as Asimov is with his protagonist (a muttered "Christ," the adverb "dizzily" are about as much dramatics as Dick will offer us); rather, Poole's alienation, conveyed by a somewhat flat affect, bleeds over into our relationship with him as a character, so that, paradoxically, we are invited to feel *with* him all the more the less he feels; it is his feeling of estrangement that seems (strangely) to be the most accessibly "human" thing about him.

"Man," writes Max Stirner, laughing at all of us who think that is what we are, "you have wheels in your head!"[14] In other words, just when we think that we are thinking for ourselves, we are, in fact, thinking in a conformist way, mechanically reproducing received ideas, like poor Garson Poole. In Jacques Lacan's "The Instance of the Letter in the Unconscious, or Reason Since Freud" (1957), the domination of the "fixed idea," the "idea that has subjected the man to itself" over the creative imagination, the "the material medium [*support*] that concrete discourse borrows from language" over discourse and thought itself, is also at stake.[15] If Stirner's wheels-in-the-head image (1844) seems crude even when placed beside Freud's hydraulic model of mind, let alone Asimov's quaint positronic brains, Lacan seems well ahead of his time when he describes this dominance of the stuff of language over meaning, of "signifier over signified," as an "algorithm."[16] More than six decades later, we are all familiar with the capacity of algorithms to do our thinking for us as well as our shopping and media consumption; we have begun to think of ourselves as algorithmic creatures, as "abstract machines."[17] We might very well feel *more* for Poole than Dick's original readers did.

Let us call what Dick offers here the *biotechnical sublime*: the swinging open of the chest panel, the revelation that you, too, are a machine. It may be easier to think of the mechanical as marking the failed sublime of a Rick Deckard, in the director's cut of *Blade Runner*, finding that the unicorn in his dream is duplicated in the little silver origami figure left behind by the detective who will now be tasked with tracking *him* down as a replicant: what had seemed both intimate and fantastical is a corporate simulation, a "soulless image" of the kind that vexed William Wordsworth who, when he "first/Beheld the summit of Mont Blanc," having already gazed at too many engraved images in tourists' guides, "grieved/To have a soulless image on the eye/Which had usurped upon a living thought/that never more could be."[18] However, we could also think of the great humanizing moment of young Lydgate in George Eliot's *Middlemarch*, in which the boy, like Keats first opening Chapman's Homer, opens an encyclopedia at random and discovers the life of the mind:

> The page he opened on was under the head of Anatomy, and the first passage that drew his eyes was on the valves of the heart. He was not much acquainted with valves of any sort, but he knew that *valvoe* were folding doors, and through this crevice came a sudden light startling him with his first vivid notion of finely-adjusted mechanism in the human frame . . . the moment of vocation

had come, and before he got down from his chair, the world was made new to him by a presentiment of endless processes filling the vast spaces planked out of his sight by that wordy ignorance which he had supposed to be knowledge. From that hour Lydgate felt the growth of an intellectual passion.[19]

For Lydgate, it feels as if the signified has jumped over the bar and taken the place of the signifier for good: the "passion" of the mind in action has suddenly displaced the dead letter of "wordy ignorance." He is in the grip of powerful forces, forces that name his being, his calling ("the moment of vocation had come"); the abyss ("vast spaces") opens up before him, and he swoons. He stands in stark inverse to Eliot's Casaubon, the seeker after the "Key to All Mythologies" (reminiscent of what Lacan calls University Discourse, a paranoid and bureaucratic system of signs): "I feed too much on the inward sources," laments the obsessive scholar, "I live too much with the dead."[20] For Casaubon, the instance of the letter is too strong, demanding too much of both the mind and the world; for Lydgate, it offers a heady sense of visionary power, "the inward light which is the last refinement of Energy, capable of bathing even the ethereal atoms in its ideally illuminated space."[21] The moment of analogy, even as it willfully risks oversimplification, is powerful in its promise to bring together everything to summarize the entire universe.[22] If analogy is indeed a demon, as Mallarmé suggested, it whispers to us that it is Maxwell's Demon, a source of infinite energy. It tempts us to boldness. "Let us, therefore, conclude boldly," as La Mettrie writes, "that man is a machine, and that the entire universe contains only one single diversely modified substance."[23] Or rather, in the words of Gilles Deleuze and Félix Guattari, that "*Partout ce sont des machines*," it's machines everywhere.[24] Or: "one Nature for all bodies," "the infinite power of God, or Nature."[25]

How is it, in this short exploration of the biotechnical sublime, that the valves of our hearts can swing so easily either way:

- toward *disenchantment*—disappointment at the soulless image, the hollow chest, the replicant (it's all *machines* = there is *only* one kind of thing in the universe, *just* one plane of being . . .), *or*
- toward *reenchantment*—exhilaration, passion, delight (it's *all* machines = there is just *one* plane of being, this is *everything there is* . . .)?

How is it that a discourse designed to elicit one of these responses can sometimes so easily produce the other? I would like to look at two such paradoxical examples to see how they can take us beyond stock responses and clichés of thought.

First Paradox: The Chinese Room

"We like to believe," wrote Alan Turing in a famous paper of 1950, "that Man is in some subtle way superior to the rest of creation."[26] On the contrary, Turing suggested, given "about fifty years' time," computers of the sort he had pioneered might be able to think, or at least to duplicate the efforts of human thought well enough to fool a human being at least seventy percent of the time.[27] He then proposed putting this to the test that now bears his name, but which he dubbed "the 'imitation game'":

> It is played with three people, a man (A), a woman (B), and an interrogator (C) who may be of either sex. The interrogator stays in a room apart from the other two. The object of the game for the interrogator is to determine which of the other two is the man and which is the woman. He knows them by labels X and Y, and at the end of the game he says either 'X is A and Y is B' or 'X is B and Y is A.' The interrogator is allowed to put questions to A and B thus:
>
> C : Will X please tell me the length of his or her hair ?
>
> Now suppose X is actually A, then A must answer. It is A's object in the game to try and cause C to make the wrong identification. His answer might therefore be, 'My hair is shingled, and the longest strands are about nine inches long.'
>
> In order that tones of voice may not help the interrogator the answers should be written, or better still, typewritten. The ideal arrangement is to have a teleprinter communicating between the two rooms. Alternatively the question and answers can be repeated by an intermediary. The object of the game for the third player (B) is to help the interrogator. The best strategy for her is probably to give truthful answers. She can add such things as 'I am the woman, don't listen to him!' to her answers, but it will avail nothing as the man can make similar remarks.
>
> We now ask the question, 'What will happen when a machine takes the part of A in this game?' Will the interrogator decide wrongly as often when the game is played like this as he does when the game is played between a man and a woman?[28]

The famous Turing Test is thus, as Judith Halberstam notes, modeled on a "sexual guessing game ... show[ing] that imitation makes even the most stable of distinctions (i.e., gender) unstable":

> By using the sexual guessing game as simply a control model, however, Turing does not stress the obvious connection between gender and computer intelligence: both are in fact imitative systems, and the boundaries between female

and male, I argue, are as unclear and as unstable as the boundary between human and machine intelligence.[29]

Posing the question of whether humans are superior to or fundamentally different from machines apparently raises the question of men's supposed superiority to or fundamental difference from women. The latter is a question on which John R. Searle, author of an almost equally famous refutation of Turing's hypothesis, seems to be convinced that women are his inferiors, or at least that women under his authority may be called on to gratify him sexually, as attested by the court documents in *Ong v. Regents* (2017). It is, of course, possible to dispute Turing without being a rapist and sexual harasser, but Searle's wish to overlook what he calls the "accidents" of "your ethnic, racial, class, and gender background"[30] appears in a strange relation to his invidious reinscription of these hierarchies in the offices of the John Searle Center for Social Ontology. These facts and the questions they raise call for us to reread Searle's argument against Turing, the Chinese Room thought experiment, in a new and sobering light.

Writing just three decades after Turing, in the springtime of the computer revolution, Searle proposes to demonstrate not only that "humans have beliefs, while thermostats, telephones, and adding machines don't," but that nothing that lacks the "causal powers" of the human brain *can* have beliefs.[31] To reinscribe this ontological boundary between human and nonhuman, subject and object, Searle tells a strange little science fiction story, "The Chinese Room":

> Suppose that I'm locked in a room and given a large batch of Chinese writing. Suppose furthermore (as is indeed the case) that I know no Chinese, either written or spoken, and that I'm not even confident that I could recognize Chinese writing as Chinese writing distinct from, say, Japanese writing or meaningless squiggles. To me, Chinese writing is just so many meaningless squiggles. Now suppose further that after this first batch of Chinese writing I am given a second batch of Chinese script together with a set of rules for correlating the second batch with the first batch. The rules are in English, and I understand these rules as well as any other native speaker of English. They enable me to correlate one set of formal symbols with another set of formal symbols, and all that "formal" means here is that I can identify the symbols entirely by their shapes. Now suppose also that I am given a third batch of Chinese symbols together with some instructions, again in English, that enable me to correlate elements of this third batch with the first two batches, and these rules instruct me how to give back certain Chinese symbols with certain

sorts of shapes in response to certain sorts of shapes given me in the third batch. Unknown to me, the people who are giving me all of these symbols call the first batch "a script," they call the second batch a "story," and they call the third batch "questions." Furthermore, they call the symbols I give them back in response to the third batch "answers to the questions," and the set of rules in English that they gave me, they call "the program." Now ... after a while I get so good at following the instructions for manipulating the Chinese symbols and the programmers get so good at writing the programs that from the external point of view—that is, from the point of view of somebody outside the room in which I am locked—my answers to the questions are absolutely indistinguishable from those of native Chinese speakers. Nobody just looking at my answers can tell that I don't speak a word of Chinese.[32]

This rather odd scenario is proposed as a proof-by-analogy: since the person in the room doesn't understand Chinese (but is only following a set of rules to manipulate sets of symbols), a computer (which can only follow a set of rules to manipulate sets of symbols) is also incapable of understanding, i.e., of thinking, having intentions, being sentient, etc.

This story prompts a few reflections:

1) *First reflection.* This scene that Searle stages is a kind of bureaucracy—that is to say, following the great scholar of bureaucracies, Max Weber, it is a machine for the elimination of magic (*Entzauberung*), for "disenchantment." All those rules and symbol-manipulating procedures—surely this is as soulless, as mindless, as the transit of a loan application through the guts of a bank. If this is not Burke's "bureaucratization of the imaginative," it certainly draws on an imaginary of the bureaucratic. Perhaps it is even drawing on associations of China with bureaucracy—both the modern State, with its Party apparatus, and the ancient States that inspired Kafka's "Chinese stories." Certainly, as Margaret Rhee and Long Bui remind us, the white imaginary is replete with racialized images of Asians as "passive, stoic cogs who merely mimic, copy, and toil relentlessly."[33] *Those eternal Others—who can understand them?* Searle whispers into our ears.[34]

2) *Second reflection.* We might observe that in constructing two "point[s] of view" (inside the Room/outside the Room) so that they are rigorously "external" to one another, is how many philosophical thought experiments involve not only deprivations of various kinds—staging what Kathryn Hume calls "subtractive worlds," fantasies founded on the elimination of some taken-for-granted bit of familiar reality—but specifically deprivation

of liberty, imprisonment.[35] What could be more scientific than locking down all the variables? The carceral imagination of Western philosophy that runs from Plato's Allegory of the Cave to Searle's Chinese Room reveals its will to truth as a will to power, a desire for mastery.

3) *Third reflection.* If the Chinese Room system seems mindless, it is because it has been made artificially "*poor in world*."[36] It has been rendered insufficiently articulate, lacking "articulation," in Latour's terms: "entities are not by definition either mute or capable of speech. Rather, they are better or worse *articulated*"—experientially connected with other entities.[37] Like the Allegory of the Cave once again, this thought experiment is a fantasy of disembodiment: in both of these scenarios, the body's freedom to move through the world and interact with it is removed. Both, therefore, rely on and reproduce the notion that human experience consists in a series of representations.[38] In the case of the Cave, the representations arrive in the form of shadows on a wall. In the case of the Chinese Room, the representations arrive in the form of illegible "batches" of symbols that refer only (via the rules for their manipulation) to other symbols. If this closed, "homosemiotic" world of signs referring to other signs seems vacuous, bereft of meaning, it is because "language is fundamentally *heterosemiotic*," i.e., a matter of "transcoding," "translation."[39] Horst Ruthrof asks us to imagine if the Chinese Room were *embodied*: give it not only sensory input (the ability to feel, taste, smell, hear, and see the world for itself—and perhaps, Ruthrof suggests, other modalities "beyond human capability, such as infrared and Xray vision, electron microscope optics and so on") but also the ability to act in the world (to move, touch, grasp, feed, flee, fuck, fight, freeze . . .) as well as the ability to be affected by the world (to need, to suffer, to enjoy . . .), and then see if the homunculus locked inside it is really a *caput mortuum*, a "dead head" lacking any capacity to understand the meanings of a symbol like 愛.

4) *Fourth reflection.* The process of abstraction by which Searle has constructed the Chinese Room is strikingly similar to the process of *alienation* that Mikhail Bakunin describes as follows: thanks to our "faculty of generalization and abstraction,"

> human beings can project themselves by means of thought, examining and observing themselves as external, alien objects. By cognitively elevating themselves above themselves, and so above the surrounding world, they attain the perfected representation of abstraction, absolute nothingness. This last limit of the highest abstraction of thought, this absolute nothing, this is God.[40]

Just as the theological imagination of nineteenth-century Europe distills "God" from the negation of all particular, material, sensuous, living qualities, Searle's ignoramus is abstracted from "the surrounding world," locked inside a "representation" of purified cognition: the nothingness that Searle imagines a machine is.

5) *Fifth reflection.* But what is truly strange to observe is that the desires for purity and mastery that animate the Chinese Room scenario only achieve their goal by eliminating a possibility it encounters as monstrous: what if, even though the human "I" in the Room remains ignorant (as would be a single neuron in a brain), *the system as a whole* may understand, know, believe, intend, speak? As Katherine Hayles notes, theorists of "distributed cognition" reading Searle's piece can simply say that "the cognitive properties of the person in the room are not the same as the cognitive properties of the room as a whole," yet another demonstration of the fact that "every day we participate in systems whose total cognitive capacity exceeds our individual knowledge."[41] Oh, but this way lies madness indeed. Theories of distributed cognition are closely tied to materialist concepts with a lineage extending back to Marx, theorist of a "general intellect" arising from the process of labor, and Proudhon, theorist of a "collective reason" arising from the discursive entanglement of contradictory dialogue.[42] Searle does not wish us to pursue this line of thought at all. He will have to reconcentrate the cognitive processes that seem to be leaking out of the enclosure he has created for them.

Searle attempts to dispatch this "systems reply" to the Chinese Room Argument in two pages of argument, starting with a variation on his fantasy:

> let the individual internalize all of these elements of the system. He memorizes the rules in the ledger and the data banks of Chinese symbols, and he does all the calculations in his head. The individual then incorporates the entire system. There isn't anything at all to the system that he does not encompass. We can even get rid of the room and suppose he works outdoors. All the same, he understands nothing of the Chinese, and a fortiori neither does the system, because there isn't anything in the system that isn't in him. If he doesn't understand, then there is no way the system could understand because the system is just a part of him.[43]

This brings to mind Slavoj Žižek's remark about Defense Secretary Donald Rumsfeld's infamous defense of the Iraq invasion, the rationale for which had hinged on intelligence that the government "knew" to be false: the war was a debacle not because of any "unknown unknowns," hazards of war that simply could not have been anticipated, but because of its "unknown knowns," the things the government *did not "know" that it "knew"*—in other words, its collective military-bureaucratic "unconscious."[44] "Consciousness," as Gregory Bateson remarks, "for obvious mechanical reasons, must always be limited to a rather small fraction of mental process." He proposes this analogy: "Consider the impossibility of constructing a television set that would report upon its screen *all* the workings of its component parts, including especially those parts concerned in this reporting."[45] Perhaps, then, in Searle's second scenario, the memorized ledgers, symbols, and procedures would constitute something like the massive "cognitive unconscious" that consciousness itself presupposes, the unthought background to all thought. Perhaps this monumental act of memorization would create a split in subjectivity, effectively installing a second, Chinese-speaking self. But Searle does not engage in such speculations, preferring instead the rhetoric of *reductio ad absurdum*:

> Furthermore, the systems reply would appear to lead to consequences that are independently absurd. If we are to conclude that there must be cognition in me on the grounds that I have a certain sort of input and output and a program in between, then it looks like all sorts of noncognitive subsystems are going to turn out to be cognitive. For example, there is a level of description at which my stomach does information processing, and it instantiates any number of computer programs, but I take it we do not want to say that it has any understanding . . . But if we accept the systems reply, then it is hard to see how we avoid saying that stomach, heart, liver, and so on, are all understanding subsystems, since there is no principled way to distinguish the motivation for saying the Chinese subsystem understands from saying that the stomach understands.[46]

But the "absurd" result only operates as a deterrent if the imagination is suitably self-restraining to prevent us from taking one step farther. David Chalmers, among others, *has* taken this a step farther, asking, essentially, *why not*? If cognition is distributed across a system, why not imagine, for instance, that "the organization of our brain might be simulated by the people of China"?[47] This is very nearly the exact scenario explored, on a smaller scale, by Russian SF author Anatoliy Dneprov in "*Igra* [The Game]" (1961): fourteen hundred students are ushered into a stadium and assigned a set of rules for a "game" in which they convert strings of binary numbers into other strings of binary numbers. After a tedious afternoon spent shouting things like "Zero-zero-one-one!" the students are told that they have just simulated the functions of a computer in order to translate a Portuguese phrase into Russian.[48] With a much greater number, Ned Block argues—he suggests "the billion people in China (I chose China because it has a billion inhabitants)"—the functions of a human brain might be simulated.[49]

If the stadium of students could, hypothetically, do the work of Google Translate, could a billion people function as an infinitely bulkier, slower, more cumbersome version of me? For some, the "China Brain" argument is yet another *reductio ad absurdum* against Turing: "But surely, says the argument, this baroque system would not be *conscious*!" Chalmers observes. The "but surely" derives its power from intuition rather than empirical evidence or logical demonstration: "Many people have a strong feeling that a system like this is simply the wrong sort of thing to have a conscious experience. Such a 'group mind' would seem to be the stuff of a science-fiction tale, rather than the kind of thing that could really exist."[50] To say "science fiction" here is to say *preposterous, inconceivable*—although, as one of Dneprov's characters points out, this scenario is only as science-fictional as Hegel's notion of a World-Spirit.[51]

And there it is: the hinge between the disenchanted or contemptuous response (of *course*, the stomach doesn't think, of *course*, the heart doesn't either, in spite of figures of speech and folk beliefs, of *course*, a billion people can't have a single consciousness . . .) and the reenchanted, vertiginous, wonderstruck response (what *if* the stomach, heart, and liver think; what *if* cells and mitochondria think; what *if*, in the other direction, flocks of birds and cities and entire ecosystems think . . .). Disenchantment: a symbol-manipulating Universal Turing Machine would not think in the same way that we do, given the "connectionist" architecture of our brains, evolved not to calculate but to recognize patterns and enter social relationships. Reenchantment: perhaps such a symbolic mind would be an entirely new kind of intelligence, an *alien* form of life. Science fictionality, hitherto invoked as a specter to frighten the imagination back within safe limits, returns with a vengeance.

But science-fictional speculations like these are not what the Chinese Room was built for; used as intended, as Searle asserts, "the Chinese room *shows what we knew all along*."[52] Daniel Dennett famously dubbed the Chinese Room Experiment, along with the Allegory of the Cave, an "*intuition pump*," a kind of abstract machine that reproduces the known knowns, the things we *think* we know:

cunningly designed to focus the reader's attention on 'the important' features, and to deflect the reader from bogging down in hard-to-follow details ... Intuition pumps are the enduring melodies of philosophy, with the staying power that ensures that they will be remembered by our freshmen, quite vividly and accurately, years after they have forgotten the intricate contrapuntal surrounding argument and analysis.[53]

It may be so. Only like many other such machines, this one has a second, unauthorized use. The Chinese Room, like Doctor Who's TARDIS, is larger on the inside than it seems from without. It contains the possibility of the Extended Mind Hypothesis, of the Gaia Hypothesis, of a new panpsychism, of a renewed animism, of a biotechnically sublime world in which everything *really does think*.

Second Paradox: The Plasmogenic Sublime

The second of our paradoxical two-way valves appears in the same issue of *Galaxy* as the final installment of *Caves of Steel*, arriving at the end of the year (December 1953). In his column, science popularizer Willy Ley, best known for his tireless promotion of spaceflight to a pre-Sputnik American public, follows up on his recap of one of last year's top biology stories, the Urey-Miller experiment that had demonstrated the plausibility of the Oparin-Haldane hypothesis about the origin of life ("Is Artificial Life Possible?") by turning his attention to one of the last contenders against Alexander Oparin's and J. B. S. Haldane's account of abiogenesis, the emergence of living beings from nonliving matter. Under the title ("Homemade Pseudo-Cells"), Ley dispatches what was perhaps the last serious contender for a theory of abiogenesis, an answer to the question of how life emerged from nonliving matter. This rival theory is here represented by the work of Moritz Traube (1826–1894, erroneously referenced as "Johannes Traube"), "the French physician Stephan[e] Leduc" (1853–1939), "a Russian physician with the very un-Russian name of Martin Kuckuck" (birth and death dates unknown), and "Prof. Alfonso L. Herrera of Mexico" (1868–1942), all scientists who had engaged in their own experiments with artificial life using inorganic as well as organic compounds.[54] In 1905, Herrera had placed their work within a tradition culminating in his own theory, a new science of "the lifelike behavior of artefacts," that he named "plasmogeny."[55] The plasmogenists "aspire[d] to mimic living organic structures using reagents," but their "FUNDAMENTAL PROPOSITION" was that "all material phenomena of organisms, in the past and present, have been or are caused by known physicochemical forces."[56] Herrera's plasmogeny, as Jorge Quintana-Navarrete observes, flattened the distinctions "between the mineral kingdom and the vegetable kingdom, between inorganic matter and organized organic matter": "There is no such thing as living matter and dead matter, because everything in the Universe is alive."[57]

From the outset, Ley is clear that he regards the plasmogenists' science as pseudoscience, not because their experiments were unrepeatable (they can be repeated *ad libitum*), but precisely because of the element of mimicry, which means that they cannot mean what their authors took them to mean:

> Take a shallow glass dish and fill it with a solution of tannin. Then add one drop of old-fashioned carpenter's glue. Immediately a tough skin will form around the drop of glue. But the skin is not impervious to water; the glue drop absorbs water from the solution, "grows" and finally bursts the skin. Of course a new skin forms immediately, which then again grows too tight for the water absorbed and so forth.
>
> Even more astonishing in appearance is [Leduc's] "dividing cell" . . . All it involves is a few drops of salt water on a piece of glass. One drop, placed in the center, consists of a weak salt solution to which a dye has been added. The two other drops, placed to the right and left of the first, contain no dye, but are somewhat stronger salt solutions. Then you make the three drops touch and observe the result with good magnifying glass. The result looks precisely like pictures of a real cell in division with "chromosomes" splitting off in both directions. Or maybe you would like to see an am[o]eba eating a small alga. Use a sliver of glass, about a quarter of an inch in length, coat it with shellac, put it in very shallow water and add a drop of chloroform . . . The chloroform drop will exhibit the most lifelike properties, finally throwing out the "indigestible" glass.
>
> Finally, there are the "colpoids" of Prof. Alfonso L. Herrera of Mexico. The nutrient of the "colpoids" is a mixture of olive oil and gasoline, two parts of gasoline (by volume) for one part of olive oil. The "cells" are a solution of washing soda in water that might be dyed to show things more clearly. The principle is the same as the cell of Prof. Traube—namely, osmotic pressure—but Herrera's are more active by far. They will move rapidly across the oil-gasoline "lake," pursue each other, eat each other and behave, in general, like irritated and hungry animalcules.[58]

That these combinations of nonliving materials can exhibit such lifelike behavior, for Ley, should no more convince us than watching the growth of sugar crystals in a solution.[59] Two sentences suffice: "No, it doesn't prove anything directly. It merely shows that those physical forces that are utilized by an actually living cell can also put on a fine show by themselves."[60] It is the "fine show," finally, that is damning (while faintly praising).

An underlying theme in the discourse of the criticisms to which Herrera was subjected: "imitation," "seem[ing]," "appearance," "superficial resemblance."[61] In fact, Herrera's plasmogeny, with its emphasis on the genesis of *form* (*plásma*, Greek: "something formed"), seemed to emanate from another epistemological regime altogether: the pre-Enlightenment sciences of "similitude," the world not of modern biochemistry, with its concern for invisible deep structures (ultimately gratified by the discovery of RNA and DNA), but the early modern world of alchemy, with its concern for the resemblances between visible surface structures as exemplified by the old formula: *as above, so below*.[62] This world is, in fact, the premodern world of magic, ruled as it is by Frazer's "Law of Similarity,"[63] the world that Victor Frankenstein thought he had abandoned along with the "sad trash" of his youth (the forbidden books of Renaissance magicians Cornelius Agrippa, Paracelsus, and Albertus Magnus).[64] Herrera's treatment at the hands of the international scientific establishment is perhaps a reflex of horror at the return of a monster it had thought to have banished, a "form" that was a "filthy type" or version of its own, "more horrid from its very resemblance": the monster of similitude, once the key to the knowledge of the universe, but since the eighteenth century a figure of falsehood, an analogical demon.[65]

Herrera et al. labored before the Xray crystallography of Watson, Crick, and Franklin first laid bare the genetic code, a submicroscopic mechanism underlying the cell itself. The questions they raised, however, are not so easily dispelled: what were the actual mechanics of abiogenesis that allowed the amino acid chains to assemble and form proteins in the first place? In spite of Oparin's disdain for Herrera's work—"it is doubtful whether any contemporary biologist would admit that these structures are endowed with life"—his candidate for an appropriate proto-cellular architecture, the coacervate droplet, is no longer universally considered promising.[66] As Sidney Fox and Nicols Fox recount,

> Oparin's experiments in the production of artificial cells, however, were actually far behind Herrera's. Oparin and his colleagues used polymers from already evolved organisms [i.e., gum arabic and gelatin] to make a kind of cell known as the coacervate droplet, whereas Herrera used materials that might indeed have served as precursors in the progressive evolution to cells.[67]

Moreover, they note, "Decades after these experiments, new investigators found that the materials he used were organic substances abundant in interstellar matter, that formed in tremendous quantities around the stars ... In this, he was far ahead of his time."[68] In short, even if his theory could not solve the problem of abiogenesis on Earth, it might well have great relevance to abiogenesis elsewhere. Herrera had offered one of the first substantive clues to the puzzle of exobiology—the very stuff of *The Magazine of Science Fiction and Fantasy*.

The stars were, in fact, very much on Herrera's mind. For him, plasmogeny was, ultimately, the study of the emergence of form in nature *as such*, the formation of planets, stars, galaxies, and the universe itself fell within its scope. Nor was this merely by analogy with the birth, growth, and death of living systems: rather, since "what is chemical is life," "no star, force or being is made according to special laws or with special elements or particular purpose."[69] "From the molecule to 'the constellations of the Zodiac,'" DeFord writes, Herrera found that "'all the beings of the Universe are brothers' in the most literal sense, since all derive ultimately from a parent nebula, by way of its suns and their planets. The microcosm is, in its origin and nature, only an 'imitation' of the macrocosm whence in the last analysis it emanated, by 'a chain of union and causality.'"[70] Around 1917, Herrera seems to have begun folding plasmogenic chemistry into a speculative physics of "ether," "a physical medium that fills the universe and to which everything is reduced, for modern science," according to which "Everything is ether ... worms and nebulas, minds and storms, evil and good, love and desire, what is, what has been, what will be, everything, in the end, is, has been, will be, a property, a manifestation, a phenomenon of the ether."[71] In 1935, writing for the Spanish anarchist journal *Estudios* (fig. 1.2.1), Herrera acknowledges the limits to his expertise in astronomy and cosmology but nonetheless "dares to publish a very modest cosmogonic theory, which he submits in good faith to the opinion of astronomers and mathematicians, warning that he does not cultivate the sciences of the sky and of numbers and signs":

> If we deposit a very light lycopodium powder on an iron plate that is subjected to the attraction of two electromagnets from below, and which, because of its elasticity, tends to rise after being attracted, we see that the lycopodium grains group together and give the form of moving spheres. By increasing the intensity of the vibration with a wedge that lifts the plate, the dust receives a greater impulse and forms a persistent cloud, like a nebula. In other words, under the influence of an appropriate energy, very small particles imitate sidereal gases.

What is below and what is above mutually imitate one another, a correspondence of cosmic forms that suggests a single underlying process: a primordial energy or "proto-lightning [*proto-rayo*]" that organizes all substances, constituting the single plane of composition.[72]

Herrera was not shy about advertising "the gigantic aspiration of plasmogenesis"; on the contrary, he was given to grandiose statements that, despite his relative isolation from the international scientific community as a researcher from the Mexican periphery,[73] doubtless harmed his reputation in those circles:

> Should Plasmogeny succeed, Medicine will be able to cure or prevent all diseases, old age and death.
>
> Agriculture, livestock and, in general, animal husbandry, will be replaced by industrial artificial foods and raw materials, as is already the case with dyes, artificial silk, synthetic petrol, rubber, etc. Men will devote their energy to intellectual endeavor instead of debasing themselves, plowing fields and taming colts or tending to poultry.
>
> Philosophy will extend life and its problems to the Universe, and perhaps, Philosophy and Science will be perfected by man or by artificial brains...
>
> Physical and moral pains will be extinguished forever! Forever![74]

Herrera's heresy leaps forward in time, linking him, as an unacknowledged ancestor, to the transhumanists, at the same time that it also reaches backward to Romanticism, to Goethe's *Faust* and Percy Bysshe Shelley's *Prometheus*:

> I wish to be the dreamer aflame
> with enormous desire, powerful in flight,
> forehead encircled by a starry crown,
> and brandishing the lamp of heaven...
> And to fly in this way, to soar between worlds
> and over the uncharted, starless abyss
> and in the face of life's adverse fate
> on which his ship sails without a pilot
> to ease the suffering of the Universe
> with songs of love and hope.[75]

No wonder, then, that Herrera should be both praised and buried in *another* US science fiction publication: *The Magazine of Fantasy and Science Fiction* (August 1954). Whereas Ley, having "felt tempted to insert a few words"

mas o de las ondas últimas, llámense electrones, eterobios, eteriontos u ondas de probabilidad, como dicen actualmente.

Prodújose entonces una cantidad inconcebible de electricidad, surgió un rayo infinito, materialízóse en millones de millones de años y dió origen a la Nebulosa Primitiva de Jeans, de donde vinieron las galaxias o vías lácteas, que, a su vez, sufren condensaciones que son soles, y de ellos provienen los planetas.

Figura 3

Si el rayo terrestre que salta entre dos nubes tiene de veinte a treinta kilómetros, y dura un instante, el primer rayo o Protorrayo fué de las dimensiones del Universo y persiste todavía, con cien mil modalidades de energía, radiación, rayos cósmicos, calor, electricidad, luz, gravedad, etc.

No es tan descabellada la teoría, porque, según Phillips, la materia proviene de los electrones y protones

Figura 4

positivos y negativos que forman los átomos, unidos a otros elementos (*Scientific American*, enero de 1933, páginas 14-16).

Tales de Mileto decía que la electricidad es el alma del Universo y a ella lo reduce todo la ciencia moderna.

El positrón, parte del átomo, es la nueva partícula de materia o electrón positivo. (*Science News Letter*, febrero 25 de 1932, pág. 115.)

J. Ch. Chadwick, de Cambridge, admite que el neu-

trón es la materia embrionaria, el primer paso en la evolución de la materia a partir de la electricidad elemental o sustancia última del mundo.

En fin, por medio de descargas eléctricas muy intensas se transmutan hoy los elementos, se crean nuevos y se produce artificialmente materia y energía, así como radioactividad artificial. (*Science*, New York, Vol. 75, N. 1.947.)

Otra teoría cosmogónica muy en boga ha sido propuesta por el abate Lemaître, y supone que el Universo viene de la explosión de un átomo colosal, que reventó a la manera de los átomos radioactivos (figura 7).

Sin desarrollar más mi modesta teoría, que a otros corresponde destruir o confirmar, me pregunto:

¿Para qué sirve el infinito?

¿De dónde viene esa nada primordial?

Como Pitágoras y Jeans, lo reducimos todo a lo abstracto, al número, a las matemáticas, pero aun así subsisten las mismas preguntas.

Y, ya sea de un átomo que revienta o de un rayo desplegado como palacio de estrellas, nada hay que esperar para el hombre: ni misericordia, ni premio, ni castigo, ni caridad, ni revelación. No hay un Creador, sino una descarga. La

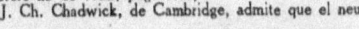

Figura 5   Figura 6   Figura 7

Figure 1.2.1. Illustrations from "Nueva teoría cosmogónica del Proto-Rayo: Una tempestad en la nada," *Estudios* vol. 13, no. 141, 1935, p. 28. Image courtesy of the International Institute of Social History.

about Herrera's work in his account of the Urey-Miller experiment (*Galaxy*, November 1953), had consciously elected to treat it *separately*, in what Bruno Latour would call a "purifying" gesture—"I decided to postpone this for a month so that nobody might be confused"[76]—Miriam Allen deFord, a science fiction writer herself as well as the widow of one of Herrera's correspondents and champions, the late science writer Maynard Shipley, felt less obligation to debunk plasmogeny. However, she begins by describing Herrera in terms borrowed from Mary Shelley:

> Victor Frankenstein, that pioneer protagonist of science fiction, formed from organic materials an adult man-monster, and by a sort of chemical sleight-of-hand vitalized the creature, to its and his undoing.
>
> Dr. Alfonso L. Herrera of Mexico, being a nonfictional character and a responsible scientist, naturally never had any such gaudy performance to present. All he did was to experiment, with startling results, in the borderland between the inorganic and the organic—and possibly between the inanimate and the animate.[77]

DeFord's language hesitates between the disenchanted and reenchanted responses:

Reenchantment	*Disenchantment*
"Their apparent *uncanny* imitations of life-processes . . ."	
	" . . . were the result of osmotic pressure."
	"Herrera never claimed they were more—"
"—though just once he was almost ready to do so."	
"Herrera's 'colpoids,' compounds of simple chemicals, imitated independent motion, respiration, nutrition, pursuit, flight, combat, the extrusion of pseudopods, and reproduction by fission . . ."	

" ... Yet their constituents were *only* such simple things as olive oil, gasoline, washing soda, tannin, glycerin, oleic acid, lime water, albumen, nitric acid, collodion, and linseed oil, in various combinations, and in essence they were *only* saponified particles."

"We are a far way by now from blobs of soda lye in olive oil and gasoline!"

"In a smaller volume published in Valencia in 1932, and in *The Universal Biology*, published in Madrid in 1933, he defined life as 'the physicochemical activity of protoplasm,' and added . . ."

"'There is no [dichotomy of] living matter and dead matter, because *everything* has life within the Universe . . .'"

"'everything that exists can be *reduced* to the mass or quantity of matter which a body contains, and its life, which is motion.'"

"(This seems to come perilously close to animism)."[78]

Oscillating between inflationist and deflationist interpretations, drawn to the pathos of a seemingly oxymoronic "scientific imagination," DeFord seems to collapse back into the purifying, debunking stance of Willy Ley: "From the fringe area of science, and the safe company of chemical experimenters, [Herrera] passed inevitably into that still more dubious area, the fringes of philosophy."[79]

It is hard not to hear some echoes of the novel deFord invoked at the outset: the young Victor Frankenstein is also tempted out of the "safe company" of scientists and into the "dubious area" of alchemy and magic, a subject of considerable concern to his parents and teachers.[80] Ironically, it is only after he has been sucked into the cloistered, stifling realm of University Discourse, removed from external social relations, that Frankenstein becomes and creates a monster. Herrera's pathos was to be forever on the margins of global University Discourse, to be ultimately expelled from it by Oparin, Haldane, Urey, and Miller, vanishing from the web of citations:[81]

> Herrera is persecuted by the Autonomous University of Mexico and is forced to ask for his retirement . . . every study, article, and attempt to investigate and discover the origin of protoplasm, the physical basis of life, is repudiated by the Paris Academy of Sciences, the Nobel Institute, the *Revue Scientifique* in Paris, *La Nature* of Paris, and *Nature* of London, *Scientific American* of New York, and, in general, by every scientific institution and journal.[82]

But rather than accepting his marginality, Herrera founded discursive networks of his own. In addition to linking himself to a tradition of other chemists working on the same problem ("plasmogeny" as a lineage or research program), and while active, via the Grand Lodge "Ignacio Ramirez" no. 20 of Tampico,[83] in freethinking Freemasonry (the Sixth National Masonic Congress declared "the need to promote studies of Plasmogeny"[84]), he created and placed himself at the center of:

- The Institut International de Plasmologie et de Biomécanique Universelles in Brussels, with its journal, Archives de Plasmologie Générale (1912);
- The *Dirección de Estudios Biológicos*, a scientific institution of the new revolutionary government (1915), attaching his own "gigantic aspirations" to the project of the nation, publishing its own *Boletín* and imparting direction to the management of the Museo Nacional de Historia Natural, with its zoo, botanical garden, and aquarium;
- The Société Internationale de Plasmogenie (1932–), which spawned two journals, the Gaceta de Plasmogenia and the Bulletin du Laboratoire de Plasmogénie (1932–);
- The Congreso internacional de Plasmogenia y Cultura General, Especialmente Mexicana, a kind of odd echo of the World's Fair of two years prior (1938);
- An "International Congress of Biophysics, Biocosmics and Biocracy" held on September 11–16, 1939, in New York.

But perhaps the largest and strangest network into which Herrera was inserted, though perhaps the most ephemeral as well, was not his own creation: the *Association Internationale Biocosmique*, founded by J. Estour and Félix Monier, with its own house organ, *La Vie Universelle* (1926–1939). There, he was joined by

- The aforementioned Martin Kuckuck in Switzerland;
- Luis Arístides Fiallo Cabral, Dominican astronomer and physician;
- Konstantin Eduardovich Tsiolkovskii, Russian rocketry pioneer;
- Wilhelm Reich, the unorthodox German Marxist, sexologist, and biologist;
- Russian astronomer Leonid Leonidovich Andrenko, author of speculative works on astrobiology;
- Italian philosopher Antioco Zucca;
- Gerard de Lacaze-Duthiers, French anarchist and self-styled "artistocrate";
- His fellow French individualist anarchist Lucien Barbedette, a professor of philosophy and history;
- Madeleine Pelletier, French anarchist-feminist and free love advocate;
- And the Brazilian anarchist-feminist Maria Lacerda de Moura.

What were the politics of plasmogeny? For the Association Internationale Biocosmique, these flowed directly from Herrera's finding that "all is life, and no limit can be conceived between its various forms": it sought to oppose militarism and theological ignorance by "reinforc[ing] the bio-cosmic bonds that induce men to feel brotherhood in the great human family and to solidarity with each other and with the 'uncreated Cosmos.'"[85]

The combination of unorthodox science and quasi-conspiratorial fraternity attracted the attention of the poet André Blavier, who classifies them among the other eccentrics, quacks, crackpots, and cuckoos (or *kuckuck*s, as personified by Professor Antonio José Kuckuck in Thomas Mann's *Confessions of Felix Krull, Confidence Man*) in his catalog of *Fous littéraires* ("literary madmen," crazies whose insanity takes the form of an incessant compulsion to write). For Blavier, no less than for Oparin, the biocosmists' "biochemical metaphysics" is consigned to the realm of pseudoscientific nonsense, a mad proliferation of signs forever beyond the pale of University Discourse.[86]

The Morbid Vitality of Gothic

The subversive or counterhegemonic quality of plasmogenic discourse should not be overstated. The network of plasmogeny counted among its ranks conservatives like the scientific criminologists Antonio Lecha-Marzo of Spain and Israel Castellanos of Cuba, who declared that "Herrera's science, as we see, will become an efficient auxiliary for Galton's."[87] Francis Galton, the prophet of eugenics, had several adherents among the members of the Association Internationale Biocosmique, including Victor Delfino, founder of the Argentine Eugenics Society, and Tsiolkovskii, who wrote, "I do not desire to live the life of the lowest races [such as] the life of a negro or an Indian. Therefore, the benefit of any atom, even the atom of a Papuan, requires the extinction also of the lowest races of humanity, and in an extreme measure the most imperfect individuals in the races."[88] "In a distant age, the supermen of the future will be produced from artificial ovules," prophesied Herrera, "titans, poets, sublime and brilliant thinkers, with disease, pain and death disappearing forever."[89] The plasmogenic imagination is still entangled with the Great Sorting Machine.

And yet, as Quintana-Navarrete observes, it is significant that "the idea of universal life, with its atheistic and rationalist implications, circulated predominantly in anarchist publications in Argentina, France and Spain."[90] Indeed, the prominence of anarchists in the Association Internationale Biocosmique cannot be missed; while Antonio L. Herrera never declared himself more than an atheist and a materialist, and even wrote to the Mexican Communist cultural journal *Crisol* that "I have always been opposed to anarchist ideas . . . I have never seen a completely anarchic state in nature, nor do I believe it possible,"[91] he did write for numerous anarchist publications (such as *Estudios*, *Cuadernos de Cultura*, *Hombre de América*, and *Nervio*), and entries on "Plasmogénie" and "Solidarisme (biocosmique)" appeared in Sebastien Faure's *Encyclopedie Anarchiste* (1934). Herrera's self-professed "ignorance" of anarchism was apparently not too deep to prevent him from frequently citing the work of the anarchist geographer Elisée Reclus.[92]

As for the place of eugenic ideas within the anarchist movement ("a singular appropriation," writes Nadia Ledesma Prietto, quite unlike its "hegemonic" concept[93]), the ambiguity and complexity of this reception have been widely remarked: at times, "eugenics" meant little more to anarchists than what the masthead of the *American Journal of Eugenics* (formerly the explicitly anarchist free-love journal *Lucifer*) proclaimed in 1907—"the doctrine of progress or evolution, especially in the human race, through improved conditions in the relations of the sexes"—and at other times, it entailed "rationaliz[ing] . . .

reproduction" so that "children who should not have been born, because they are prone to diseases" would not be "a burden for others."[94] The articulation between anarchist and eugenic discourses, as Richard Cleminson demonstrates in *Anarchism and Eugenics: An Unlikely Convergence, 1890–1940*, was both strong and strongly ambivalent, overdetermined by a morass of conflicting values and priorities.

How do we square these intertwined discourses of emancipation and domination? Here is another hinged valve, maybe: when the energies of thought and practice flow through the abstract machine of plasmogeny in one direction, they feed the Great Sorting Machine in its project of purification, but flowing in diastole, they decode what the sorting process has encoded, deterritorialize what it has territorialized, producing a vision of the universe as one single vast "inorganic and anarchic flow":

> Herrera's universal life can be conceived as an inorganic and anarchic life that disrupts traditional categories and divisions—the actual and the potential, the living and the inanimate, the qualified life (*bios*) and the vegetative life (*zoe*), etc.—that have organized what exists in Western societies. Thus, even though the Mexican biologist did not particularly reflect on the political implications of his philosophical discourse, the notion of universal life can generate a potential redefinition of politics.[95]

This in-betweenness, this blurring of boundaries, is the mark of the monstrous, the prodigal, the spectral. The plasmogenic sublime perversely *reenchants the world* so that dead matter, the being of the object, haunts the living subject.

"After days and nights of incredible labour and fatigue," Victor Frankenstein tells us, "I succeeded in discovering the cause of generation and life; nay, more, I became myself capable of bestowing animation upon lifeless matter."[96] One name for "the moment when this dichotomy [between living and nonliving] collapses and ceases to make sense," Quintana-Navarrete reminds us, is *Gothic*:

> Ultimately, the Gothic demonstrates not only the inorganic foundation of all organic matter and the very impossibility of making an absolute difference between the two, but also that life cannot be reduced to the realm of living organisms or to any specific area.[97]

That the affect associated most frequently with Gothic is the *shudder*—"the beauty of the dream vanished, and breathless horror and disgust filled my heart"[98]—should not confuse us, for what could better exemplify "the beauty of the dream" than Gothic's twin, Romanticism? Victor Frankenstein's shudder is the abject side of Coleridge's *lyrical* affect in "Frost at Midnight," when, beholding the fluttering of the film in the grate of the fireplace, he reflects that "[its] motion in this hush of nature/Gives it dim sympathies with me, who live,/Making it a companionable form,/With which I can hold commune."[99] Coleridge is enough of a Kantian rationalist to engage in a rhetoric of disavowal, reinscribing the silent/speaking, dead/living, and natural/human binaries even as he presses and stretches them, but the moment underscores the valve-hinge quality of this encounter with "the *stranger*" that is already at the heart of the human enclosure—the very definition of the Freudian "uncanny."[100]

In that distant dawn of Romanticism, in which to be young (as well as white, male, and possessed of the means of publication) was very heaven, the *lyric* served souls as the primary means of transportation. *O wild West Wind . . . Bright star, would I were stedfast as thou art . . .* To transcend the weight and limitation of one's historical and material circumstances, one employed the *lyric apostrophe*: addressing a nonperson as a person. Elsewhere in the Empire, colonialism entailed the treatment of persons as nonpersons—a strange inversion of the cultures of the colonized, which (quite ridiculously, from the standpoint of their subjugators) insisted on treating as persons what should properly be perceived as nonpersons: animals, natural forces, and, notoriously, the artifacts that European traders had dubbed "fetishes" (from *facticius*, a Latin root shared by the words *fact* and *fiction*, meaning "something artificial, manmade"[101]). "As men we are not things, but persons," explained Immanuel Kant; thus,

> That which a man can dispose over, must be a thing. Animals are here regarded as things; but man is no thing; so if, nevertheless, he disposes over his life, he sets upon himself the value of a beast. But he who takes himself for such, who fails to respect humanity, who turns himself into a thing, becomes an object of free choice for everyone; anyone, thereafter, may do as he pleases with him; he can be treated by others as an animal or a thing; he can be dealt with like a horse or dog, for he is no longer a man; he has turned himself into a thing, and so cannot demand that others should respect the humanity in him, since he has already thrown it away himself.[102]

Somewhere in between these processes, the grinding machinery of capitalism produced another kind of confusion around personhood, which a certain post-Romantic thinker would come to call the "fetishism of commodities."

At some point between then and now, the lyric apostrophe came to be seen as "hyperbolic" and "extravagant" (not to mention "embarrassing, pretentious, and mystificatory"),¹⁰³ a secondhand recreation of those worlds that Wordsworth mourned when he wrote that he'd

> rather be
> A Pagan suckled in a creed outworn;
> So might I, standing on this pleasant lea,
> Have glimpses that would make me less forlorn;
> Have sight of Proteus rising from the sea;
> Or hear old Triton blow his wreathèd horn.¹⁰⁴

A world in which such conceits had been dispelled (or at least delegated to poetry and filed away under "Fallacy, pathetic"), a world thus freed from animistic superstition and the traditional authority it subtended, could be administered rationally for the benefit of all—such was the hope and aspiration of Enlightenment. The treatment of things as people or people as things was seen as a regrettable relic of the un-Enlightened world, even if, as certain thinkers having narrowly escaped the Holocaust reminded us, "the fully enlightened earth radiates disaster triumphant"; global institutions such as the United Nations Declaration of Human Rights were meant to firmly establish Kant's distinction between the "kingdom of ends" and the "kingdom of nature," between the domain of users, of persons endowed with inviolable dignity, and the domain of things-to-be-used.¹⁰⁵

Gothic disorganizes what Kant's philosophy organizes. Kant's universal, rational order reappears in Gothic as a Piranesian dungeon, as the *grotesque* prison in which the narrator of "The Yellow Wall-Paper" finds herself locked:

> I never saw a worse paper in my life.
> One of those sprawling flamboyant patterns committing every artistic sin.
> It is dull enough to confuse the eye in following, pronounced enough to constantly irritate and provoke study, and when you follow the lame uncertain curves for a little distance they suddenly commit suicide—plunge off at outrageous angles, destroy themselves in unheard of contradictions.¹⁰⁶

Here, too, is a hinge, a valve. The room in which she is confined is meant to be therapeutic, to bring her back into the realm of the properly human, the kingdom of reason;¹⁰⁷ it is instead the place where that order "commit[s] suicide," "destroy[ing]" itself in "unheard of contradictions":

> This paper looks to me as if it *knew* what a vicious influence it had!

> There is a recurrent spot where the pattern lolls like a broken neck and two bulbous eyes stare at you upside down.
> I get positively angry with the impertinence of it and the everlastingness. Up and down and sideways they crawl, and those absurd, unblinking eyes are everywhere . . .
> I never saw so much expression in an inanimate thing before . . . [108]

The mute object achieves "expression," and the inorganic takes on a dreadful, quasi-biological vitality:

> I know a little of the principle of design, and I know this thing was not arranged on any laws of radiation, or alternation, or repetition, or symmetry, or anything else that I ever heard of.
> It is repeated, of course, by the breadths, but not otherwise.
> Looked at in one way each breadth stands alone, the bloated curves and flourishes—a kind of " debased Romanesque " with *delirium tremens*—go waddling up and down in isolated columns of fatuity.
> But, on the other hand, they connect diagonally, and the sprawling outlines run off in great slanting waves of optic horror, like a lot of wallowing seaweeds in full chase . . .
> There is one end of the room where it is almost intact, and there, when the crosslights fade and the low sun shines directly upon it, I can almost fancy radiation after all,—the interminable grotesque seem to form around a common centre and rush off in headlong plunges of equal distraction . . . [109]

Gilman's own Chinese Room Experiment, on which the story was closely modeled, was one in which her involuntary confinement to a rationalizing space, a space of senseless laws and rules, produced precisely this experience of "horror," the horror of confinement itself—but one which paradoxically opened onto something radically outside of this order. "There is something queer about it," the narrator muses, but tries to dispel her own apprehension by reducing it to a mechanically dead cliché: "I would say a haunted house, and reach the height of romantic felicity—but that would be asking too much of fate!"[110] The house of Romantic ideology *is* haunted: haunted by that which resides within its architecture, which is at the same time forever beyond it, its own negation, its end. The way out is within.

The biotechnical sublime associated with the heterodox reading of the Chinese Room Experiment and with Herrera's plasmogeny, these Gothic violations of the order established by the Great Sorting Machine and codified by philosophers like Kant—this experience of the sublime opens up the question of the *fetish* all over again.

9

Methodological Metafetishism

> But of course the problem of meaninglessness arises only if "matter" is conceived as inert...
> —Jane Bennett, The Enchantment of Modern Life

One of the major intellectual problems I have encountered comes from an overemphasis in literary theory—and often in social theory as well—on textuality and interpretation. By the time I arrived in graduate school, everything had been defined as "text" ("there is nothing outside the text," as Derrida told us), and "texts" were understood not to contain meaning but to be invested with meaning after the fact by "readers." This could never satisfactorily explain why vast congeries of readers, while clashing over the implications of texts, could so often agree on so many of the basic facts of the matter, even about fictions (e.g., who, in fact, killed Roger Ackroyd); invocations of an "interpretative community" that supposedly overrode the caprices of individual readers merely begged the question of how anyone could enter such a community except by acts of (supposedly unconstrained) interpretation, since the community itself, like everything else, could only be yet another text. At the same time, we continued to *act* as if textual meanings were discoverable in some way, dutifully supplying evidence in support of interpretative theses, making arguments aimed at persuading others to adopt our readings, and so on.

I decided that our practices were in some ways wiser than our theories, that the embodied materiality of the object world needed to be written back in, and I wrote a very dull book to this effect, the one novelty of which was that I tried to show that such a renewed realism need not be conservative at all, since, after all, it was implicit in the thought and practice of anarchists from Bakunin to Bookchin. Texts, on this account, are not simply passive, silent, inert things awaiting the infusion of subjective meaning; they are to be encountered *as if* they were persons themselves, capable of benevolence or malevolence, harboring forces and designs of their own, resisting as well as enabling our projects. And it seems that the "deanimation" of texts, as Latour would put it, is the effect of modernity's Great Sorting Machine; outside of it, it is quite common to encounter the suggestion that "the story is . . . not an entity in the way that, say, a bear is," but nonetheless is "a living being."[1]

"Methodological fetishism," according to Arjun Appadurai, is an unfortunate side effect of the study of the social life of things, in something like the way that tennis elbow can result from playing tennis. This tennis elbow of the mind is something to be apologized for, perhaps minimized, certainly not embraced—for from the standpoint of "theory," it can only be an illusion, even if it is somehow necessary (or even, as Appadurai immediately suggests, "a corrective to the tendency to excessively sociologize transactions in things").[2] The world of objects is inert and silent, equipped with forces but not meanings; only when it is "encode[d]" by human beings, made into a readable text and then read, does it *appear* to spring to life. But what if we saw objects (and their more elusive cousins, such as Timothy Morton's "hyperobjects") as comprising both "meaning" *and* "force," as the Deleuzian sociologist Daniel Colson suggests? Might it not then make sense to adopt a kind of fetishistic stance *on purpose* as a means to a better understanding of the world?

What is "metafetishism"? The coiner of the term, the anthropologist Massimo Canevacci, never defines it. Nonetheless, it seems to name the strange rehabilitation of the concept and practice of fetishism as a means of undoing the Modern Constitution, dissolving our previous understandings of the relations between subject and object, user and tool, reader and text, person and land, and so on, the better to imagine and enact practices of self-creation. Metafetishism would then be then a methodology, an approach not only to studying the world but to changing it (and being changed by it). I'd like to explore this proposition through a series of reflections.

August 1987: Home Movies

One comes to desert landscapes like these to be overtaken by a geological sublime: to be annihilated by the lines upon lines of stratification in canyon walls marks announcing eons in which human consciousness did not exist. (*Ancestrality*, Quentin Meillassoux will call it.) I am wielding a little

Super-8 camera that makes a delightful whirr when I press its button. I am surrounded by my family, but I don't want any people in the frame as I pan across Canyon de Chelly. I am fascinated by the swooping motions of a bird, which I attempt to capture in slow motion. I am captured.

Driving through the desert, I imagine other landscapes. Passing by a gas station and souvenir stand, I picture the interior: the cash register, its trailing wires, the clock hanging by the refrigerated cases, flimsy metal shelving, the dull metal of sinks, PVC pipes . . . *All this*, I think, *has to go. It can't last . . .* Images from *Mad Max*, from *Damnation Alley*, from *A Canticle for Leibowitz*, and a thousand other postapocalyptic scenarios flicker across my inward eye. The mesas turn their blank gazes on me.

The next day, when we descend into the blazing canyon on foot, warned to drink water often, it feels like geological time travel, like Jules Verne's protagonists descending via a dead volcano into the heart of the earth. I run ahead. At its bottom, the canyon is greener, fed by a wash. A Navajo shepherd herds sheep here beneath the Anasazi ruins. Where am I? The Four Corners are formed of the most ostentatiously imaginary lines. I am suddenly, heart-stoppingly aware of standing on *someone else's territory*.

"Every practice," writes the philosopher Alejandro de Acosta, "implies and involves a territory."[3] (This is why we make land acknowledgments: I am writing this book on land belonging to the Bodéwadmiakiwen (Potawatomi), Kiikaapoi (Kickapoo), Miami, Očeti Šakówiŋ (Sioux), and Peoria nations.) In the diner, a notice is posted for tourists: please do not make eye contact with Navajo locals, as this is considered aggressive. My gaze has become a problem; maybe it always was. Ralph Waldo Emerson, your fantasy of being an invisible eye was always the stuff of the *Dungeons & Dragons* rulebook nestled in my backpack (in which Dan LaForce's illustration of the Wizard Eye spell features a ghostly eyeball floating in midair): another white male fantasy.

"When it comes to these locations, these territories," Alejandro de Acosta will write, "we can grasp them as artifices, as fetishes: the land, too, makes us as we make it."[4] I have been constituted by a territory as far from this as it is possible for me to imagine: a realm of sprinklered lawns, ranch houses, and cul-de-sacs. The land of the commodity fetish, where the social relations and productive forces that make this scene possible are invisible. It is a space designed for dreams of propertarian autonomy, a consumerist utopia from which production is forever exiled like the poets from Plato's Republic. The gap is incommensurable.

August 2047: Reading Backwards

It's too late for a lot of things, and I haven't got much juice to spare, so I'll have to keep this short.

This is not an easy world to grow old in, but at least we have more time than we used to for books. The other day, I picked up a mildewed copy of Kim Stanley Robinson's The Ministry for the Future, *and it was like a time machine that took me right back to 2021, the year of the not-quite-yet-a-coup (not like we'd come to know coups later). Back then, I had just finished a draft of* Hot Equations, *and I rued having done so before having had a chance to read what turned out to have been Robinson's last great novel. I didn't know then how much he had gotten right and wrong. The necessary violence of the struggle to end the petroeconomy: right. Airships and sail making a comeback: right again.*

But the boneheaded humanism of that book!

Robinson keeps approaching a nonhumanist moment. He playfully personifies herd animals, the sun, carbon atoms, photons, blockchain, the marketplace itself, history itself—but then instantly demolishes his own conceits. «I am a thing . . . not a mother,» he writes, putting words in the mouth of the earth. This, though in an access of rapture, Robinson's human beings—he has met a few—do call the earth their mother (shouting «Mamma mia! Mamma mia!») . . . and yet no indigenous humans take their place on Robinson's crowded stage to say that this is not just a «category error,» an atavism of the mind . . . [5]

There's a recurring bit in «ministry» in which the format abruptly switches to «talk show»: a genteel, well-read host interviews a curmudgeonly guest expert about some aspect of the planetary crisis. In one of them, the conversation revolves around trying to locate some kind of lever and fulcrum capable of pushing the world off of its self-destructive course. Ultimately, Robinson comes down on the side of legislation, the same process that has heretofore been a mere stenographer to the petrocorps. Why start there? What should make the state the solid fulcrum and laws the long lever with which to move the earth? In short, what about actor networks?

The list of actors is suitably long and detailed, ranging from the heads of major corporations and national banks to econ professors, «shareholder associations, pension funds, individual shareholders, hedge funds, financial firms,» and ultimately, «the system itself.» But this draws an objection from the guest expert: «You were asking about the people doing it.» «Yes, but it's an actor network,» explains the host; «Some of the actors in an actor network aren't human.» «There are actor networks,» the guest growls, «But it's the actors with agency who can choose to do things differently.» [6] *And there it is—humanism* reiterated once again by rhetorical sleight-of-hand: the world is passive, humanity is active; nature operates by physical laws, humans by human laws; it is that, we are this . . . as if you hadn't just demonstrated the network character of economy and ecology alike, the impossibility of separating them out!

What we didn't know then about «choice» and «agency.»
And alliances.
And assemblages.
And betrayals.
What we didn't know . . .

April 25, 2020: To Hack and Be Hacked

I am sitting at my laptop, hacking Google Books—I suppose that is what I am doing since I am using special techniques to make someone else's restricted system give me information it is designed to withhold. A ridiculously insignificant transgression, to be sure, when it is possible to torrent entire books in a second, but it's something from which I derive a sneaky pleasure, a tiny sense of putting one over on the Man. I don't think of it as "hacking," though; I call it *tickling*. When a book I want to read or quote from is designated as "Snippet View Only," I have learned by trial and error there are ways of launching a long string of queries so as to *tickle* the system into allowing me to capture the whole text in a series of overlapping snippets. Today, my target is Joshua Clover's 2004 monograph on the 1999 film *The Matrix*. Here is what a few minutes have forced the machine to divulge:

> It's about work. Or perhaps, fearing the brutalism of single-sentence summaries, one could say rather that the major currents flowing through *The Matrix* (digitech, spectacle, ideology, false totalities, Marxism and the entertainment business, to get started) empty into the ocean of work.[7]

But such behavior has elicited the system's suspicion today: it interrupts my query to demand that I prove that I am not a robot. This morning, it seems, it will be convinced of my humanity by my merely checking a box; other days, it has goaded me into doing what is certainly a kind of work for its owners—identifying objects in a photo (crosswalks, buses, stoplights), doubtless to help "train" driverless cars.

Suddenly, I am thinking of Slavoj Žižek's suggestion (in 1999) that the new technologies of simulation had turned "the obsessional question 'Am I alive or dead?'" into the pseudo-philosophical question: "Am I a machine (does my brain really function as a computer) or a living human being (with a spark of spirit or something else irreducible to the computer-circuit)?"[8] Half a century earlier, Theodor Adorno had remarked that "Thinking no longer means anything more than checking at each moment whether one can indeed think."[9] But at this moment, at least, it is not me who is performing the check, posing the question; rather, it is the system. My question is: for whom am I thinking? Who is profiting from my cognitive labor?

Jan. 9, 1998: Meeting the Prince of Networks

The house is quiet, and the world is calm; the phone rests on a hook and may not ring at all today. I am writing notes on an article in a journal I have just discovered, *Anarchist Studies*: "The Power-Persuasion-Identity Nexus: Anarchism and Actor Networks" (1994). It has attracted my attention primarily because I am reading anything I can find that might bear on the subject of my dissertation: anarchism and literary theory. Disappointingly, the fellow the article is about, Bruno Latour, does not appear to be a literary theorist in any way: his field, according to Michael, is the sociology of scientific knowledge. Moreover, according to an interview I found online (borrowing the phone line for an hour of internet browsing in Netscape), he is no anarchist: "I don't take very seriously political anarchism," he declares, "and I don't take very seriously anarchism in science"—that is, Paul Feyerabend's so-called "Anarchistic Theory of Knowledge"—"because it is completely reactionary . . . Feyerabend demonstrates the reactionary politics of the anarchists. They are the ones inventing the domination because they want to define their position in that way in order to get into a position of power."[10] I have no use for Feyerabend (too relativistic), but reading assertions like these wearies me: how many generations have passed, and we're still retailing the nasty things Engels said about Bakunin? And what earthly need have we to "invent" dominations when they are so utterly ubiquitous in this society of total administration, a society in which one in five jobs is, in one way or another, "guard labor," dedicated to keeping others in line?[11] But Michael thinks Latour has some relevance to anarchism, and I am determined to determine whether this might be of some value for my own work, and so, despite the late hour, I press forward with my notes; anything goes.

It begins promisingly: Michael argues that anarchist understandings of the way power operates tend to assume that direct "coercion" is the State's ultimate strength and that "ideological indoctrination" is just its first line of defense. On the contrary, the important question is "how the functionaries who do the coercing are themselves persuaded to act in the service of the state. In other words, who coerces the coercers?"[12] If we are to really understand how power operates, we need to understand what Noam Chomsky calls "the manufacture of consent." This is strongly reminiscent of what I regard [*and still do even now* (April 2020), *twenty-two years in the future*] as the best argument of my favorite anarchist, Gustav Landauer: "The State is a condition, a certain relationship between human beings, a mode of human behaviour; we destroy it by contracting other relationships, by behaving differently."[13] Moreover, at a deeper level, I am being flattered because this seems to cast me in an important role; if the main barrier to revolutionary change is not tanks and planes but beliefs and perceptions, then radical educators are positioned to do serious damage to the State by deconstructing its discourses of legitimation—and participating in the construction of alternative discourses, the creation of genuine consensus from below rather than mere "consent" to what has been imposed from above.

But as I read on, I find the way Michael describes the manufacture of consent more and more unsettling. The model he derives from Latour and his collaborator, Michel Callon, is founded on a strange and counterintuitive assumption, or rather, the strange bracketing of an assumption I hadn't even been aware of sharing until now: nonhuman beings—not only animals but inanimate objects, not only objects but also entire systems—cannot be assumed in advance to lack agency; they, too, may "act," may number among the actants in a network of power. Secondly, in the formation of such networks, "persuasion" does not stand as an anarchist alternative to "coercion"; rather, persuasive processes are represented as the way in which privileged power is secured for some. In this "Actor-Network Theory" (ANT), persuasion operates through three phases:

1) "Interressement": "one aims to convince actors that, rather than maintain a particular set of self-understandings that are derived from their relationships with other actors (human or nonhuman), they should really be conceptualizing themselves through the categories that you provide."
2) "Translation": "the process by which the 'enrolling actor' sets itself up as the spokesperson of others. The 'translator' attempts to persuade others that it can represent them and their interests. To do this, it must convince others that its and their identities and interests coincide."
3) "Enrolment": the target of persuasion takes on the "identities and practices" that the persuader wants.

In sum, the persuasive process functions like a kind of syllogism:

This is what you want to be. (Interressement)
We are the ones who can help you become that. (Translation)
Grant your obedience by your own consent. (Enrolment)[14]

My reaction to this is outraged incredulity. First of all, I cannot imagine scallops (to take Latour and Callon's example) being "interested" in any arguments whatsoever, nor as bearing "identities," and that they are utterly incapable of speaking for themselves is a given. The representation of scientists as manipulative and self-appointed "spokespersons" repels me; I know that science can be misused, but I believe that scientific findings and theories refer to real things in the physical world, things we ignore (like Marilyn Cross) at our peril. Scientists persuade with empirical evidence and reasoning—a model for an alternative to the State's manufacture of consent. I think of Bakunin's comment in *God and the State*: "Does it follow that I reject all authority? Far from me such a thought. In the matter of boots, I refer to the authority of the bootmaker."[15] The authority of scientists seems [*seemed then*] to me entirely different from the authority of politicians. But just as disturbing for me is the implication that *any* persuasive processes must operate in this seductive, patronizing, self-aggrandizing manner. Michael anticipates me: "Even at the heart of the anarchist group, we can see these processes of persuasion at work . . . If ANT can serve, in however problematic a way, to expose the operation of the power-persuasion-identity nexus in relation to the state, it can also aid in a more self-critical anarchist practice." Michael even anticipates the cynical question, "But aren't you just another subtle persuader?": "the aim has been to provide the reader with a possible critical tool, not to enrol you into my own network, tempted though I am."[16]

I have already witnessed such processes of power at work in an anarchist group. Down the street from where I live, my partner and I have spent part of a year working with a group building a cooperative bookstore in the rambling old house owned and managed by an older anarchist who had studied under Murray Bookchin himself. The exhilaration of our successful first weeks grew cold when we found out that the money we'd raised with our volunteer labor would not be going to purchase new radical books, as the older anarchist had quietly used it to pay his own property taxes. Around the same time, we found out that he had hit his girlfriend, a fellow volunteer. The bitter taste all of this left in my mouth adds a sting to Michael's words. What bothers me the most, though, is the implication that this is all there is to expect from other people, no matter what "identities" they may profess. Why not give up, then, on anarchism as on everything else if everyone is just out to use or abuse you? Why bother to imagine any alternative to capitalist exploitation or patriarchal domination? Michael left open one other hope: "The roles and identities assigned by one entity to another may suddenly be challenged, undermined or betrayed."[17] The abuser's partner could leave him; we could refuse to create value for him anymore and shun him socially. But surely nothing that we can build ourselves will constitute anything better than yet another betrayal, another network, another representation, another lie.[18] In short, Michael's article leaves me thoroughly discouraged, and I decide that Latour can be nothing but another postmodern relativist of the sort I have chosen to distrust (I do not wish to be enrolled in their network).

"Persuasion" without any possibility of an appeal to facts of the matter, I reason, is empty, or perhaps it is just a soft form of coercion. If we want a society in which coercion is replaced by cooperation, the cooperators will need to refer to an external world. Leaving Latour unread, I will instead follow what I take to be Landauer's path toward the construction of new relationships, relationships mediated through representations that would be perpetually open to question and challenge, to testing against the world. Because abuse is facilitated by an abusive culture, and culture is a structure of symbols, I will assume that social transformation is a matter of altering, repairing that structure in the direction of equality. I will define my task as a radical educator as one of empowering students by showing them how to reconstruct relationships to reconstruct the systems of symbols within which they are enmeshed.

This will do for a number of years.

Dec. 17, 2007: Meeting the Anarchist Prince of Networks

My translating abilities are getting better. I am translating Robert Damien's essay about a series of nineteenth-century manifestoes concerning the French railway system, subtitled "Proudhon, une pensée philosophique des réseaux?" ("Proudhon—A Philosopher of Networks?") It is eye-opening in several ways at once.

When he writes about the French railway system, Damien argues Proudhon is indeed a thinker of *networks*. For this thought, the enemy is the "static and fragmentary absolute"; as such, technologies of movement and circulation are "dialectical tool[s] of the first order."[19] "The railroads," Proudhon observes, "eliminate distance, render men everywhere present to one another . . . They efface and level all the inequalities of locality and climate and make no distinction between the hamlet lost in the plain and the manufacturing center majestically sited on the river."[20] If at the time Proudhon writes (1855), the French system is unduly centered on Paris, this both embodies and reinforces the existing "monarchical" and "monopolistic" distribution of power in French society, but in theory, the technics of rail transportation permit of something "integrative and compositional, transversal and federative, multiplying intersections and routes in order to increase the intensity and fluidity of bonds."[21]

This, in turn, helps me to reinterpret something I have always found baffling about the way that Todd May marries anarchist and poststructuralist thought. Concerning Colin Ward's declaration that "we have to build networks instead of pyramids," May writes: "The idea of networks does not only underlie the anarchist conception of resistance, however; it also underlies its conception of what is to be resisted. Moreover, it is *because* what is to be resisted comes in the form of networks that resistance must do so too."[22] If power is always already horizontal, like a network, what difference could there possibly be between anarchist and authoritarian organizations of power? But networks are not all alike. The French rail network is like a map of metaphysics, with Paris as its transcendent "*center elsewhere*"; the network Proudhon imagines is a map of a society that has found order in anarchy, a society of which "the center is everywhere and the circumference nowhere."[23] Here is Deleuze's figure of the rhizome, the grasses that grow in all directions without a center.

And now, I am better understanding the larger and much more difficult translation project on which I have embarked: Daniel Colson's *Petit lexique philosophique de l'anarchisme de Proudhon à Deleuze*, which has puzzled me for several years now. When I had met Colson at a colloquium in Lyon, he had not spoken any English, and I could barely manage a sentence or two of intelligible French, so we had relied on our interpreters, Ronald Creagh and Francis Dupuis-Déri, to translate our mutual misunderstandings. I could not understand why, in 2002, Colson's book had occasioned a polite but fierce debate between him and his longtime comrade Eduardo Colombo. "I believe that Colson and I share fundamental political options, because we have discussed these problems periodically throughout the last thirty years, always in view of militant practices," Colombo writes at the outset of his harsh critique of the *Petit lexique*, as if to soften the blow, or as if to justify why he is bothering to critique Colson at all.[24] Colombo's principal objection to Colson's neo-Nietzschean version of anarchism lies in its "constant slippage from force to action, from action to meaning, and vice versa, as a consequence of the conceptual preeminence of *power*," since, like me, Colombo believes that

> human action is exercised in a politically constructed space ... traversed by institutions, conventions, systems of ideas and values, "laws," inscribed in the time of history, and it is "organized" under the control of *political power*: heteronomy, power (*potestas*) and exploitation. These structures of domination are the product of a ceaseless work of symbolic reproduction, of which the world's systems of knowledge—linguistic, scientific, mythical—also form a part.
>
> Humanity is not made to obey by force, even if force remains the *ultima ratio* of the State; the internalization of standards (socialization of children), the inculcation of dominant values, make each individual agent a transmitter and a reproducer of the established system. The dominated themselves "apply categories constructed from the point of view of the dominant to the relations of domination, thus making them appear natural." It is here that political philosophy has something to do in order ... to criticize, to decompose, to propose, not merely to leave action to the blindness of its condition when it is cut off from signification.[25]

Why couldn't Colson see Colombo's point? Surely, it is true that the strength of the State (and capitalism) is not founded in force but in consent or hegemony, in the diffusion of symbolic constructs disguised as transcendental truths—so that, in the words of John D. Rockefeller, monopoly capitalism appears as "a law of nature and a law of God."[26] Then why should Colson insist otherwise? What is wrong with an anarchist pedagogy aimed at denaturalizing these constructs "in order to think and realize a new institution of the social"?[27] And how can Colson, as an anarchist opposed to the concentrated power of the State, insist that "power is exercised from innumerable points"?[28]

Colson insists, however, that he has not "hypostasize[d] force and power": "All that I attempt to 'say' in this *Lexique* consists precisely in on the one hand never separating action and idea, force and signification, and on the other hand, never privileging the one to the detriment of the other." He quotes his own entry on "Theory/practice":

> Any collective being, any "collective force" (this is Proudhon's vocabulary), any reality, has "two dimensions or aspects [indissociable and non-hierarchized]. One side, in the vocabulary of Deleuze, is a discursive aspect—the side of form, expression, and signs—and the other side is a machinic aspect of contents, bodies, forces, and desires."[29]

For instance, a railway network is a sum of machinic forces (iron, wood, coal, water, steam, human labor, etc.), but it is also a structure of meanings, a network of signals and logics; these are the two "aspects" of a single phenomenon. Power, even hierarchical and authoritarian power, operates through networks because it is an aspect of relationships; conversely, even cooperative social networks transmit flows of power (circulating evenly) because they are tissues of relationships. *And relationships, with their flows of power, inhere in things.*

Nov. 1, 2019: The Trade in Fetishes

Once our daughter was born (2005), I grudgingly became the owner of a cell phone, which morphed into a smart phone (2012). Seven years later, I am reading on its small screen an article in the *New Yorker* by Jerome Groopman—I have the disturbing feeling that twenty versions of this article have appeared in my feed with different titles and sources—which begins:

> Several years ago, I bought a smartphone and soon came to love it. Being able to send an email, look up a fact, or buy something no matter where I was meant a previously unimaginable gain in productivity. Every time I got an email, the phone emitted a ping and I would deal with whatever it was, priding myself on my efficiency. Texts arrived with the tones of a French horn and were similarly dispatched. Soon, I was reaching for the device every time it made a sound, like Pavlov's dog salivating when it heard a bell. This started to interfere with work and conversations. The machine had seemed like a miraculous servant, but gradually I became its slave.[30]

This is, indeed, a much older kind of story, even older than the outmoded "brain science" of Ivan Pavlov. It was once the province not of self-help writing or popular science journalism but of philosophy and politics. "Machines ... promised us liberty; I am going to prove that they have brought us slavery," writes Pierre-Joseph Proudhon.[31] "The hand tool," wrote Marx, "makes the worker independent—posits him as proprietor. Machinery ... posits him as dependent, posits him as appropriated."[32] Granted, both Proudhon and Marx were writing about technologies of *production*, not consumer goods, although, as Groopman notes, the smartphone is a consumer good that perversely tends to turn all hours into potentially productive time. And neither Marx nor Proudhon, both of whom knew plenty about real, physical slavery, was speaking of the precise psychology of seduction that Groopman describes. Still, all three could be said to bear witness to a similar kind of inversion, a swapping of places between user and tool, subject and object.

Moreover, this kind of subject-object reversal is structurally similar to another process that Proudhon and Marx each independently named "fetishism": Marx in the famous passage of *Capital* on "THE FETISHISM OF COMMODITIES AND THE SECRET THEREOF,"[33] and Proudhon, for instance, in his address to the National Assembly, in which he looked forward to a world in which work would be motivated not by "ambition and greed" but by "the love of wellbeing, of actual enjoyments [*jouissances effectives*] . . . the fetishism of gold giving way to the realism of existence."[34] The fetishistic relationship to the object was mediated by an illusion attributing special powers and significance to a mere thing bereft of any such force or meaning. It is here, significantly, that China Miéville locates the "particular interest" of fantasy for Marxism and vice versa, in that both are attempts to come to terms with the "peculiar nature of modern social reality and subjectivity":

> The lived reality of capitalism is commodity fetishism. Magnitudes of value coagulated in the commodity form—things—"far from being under their [human producers' and exchangers'] control, in fact control them" . . . *Our commodities control us*, and our social relations are dictated by their relations and interactions.[35]

It is hard to overstate the monstrosity of capitalism, in very much the sense we developed in chapter 1 ("Predictive Analytics for Monstrous Times"), albeit in a cruel and destructive form: even as it stratifies and sorts us, isn't money itself a vast machine for making strange equivalences, producing unheard-of hybrids? Indeed, what could be more fantastic than Marx's classic description of the "fetishism of commodities"?

> A commodity appears, at first sight, a very trivial thing, and easily understood. Its analysis shows that it is, in reality, a very queer thing, abounding in metaphysical subtleties and theological niceties. So far as it is a value in use, there is nothing mysterious about it, whether we consider it from the point of view that by its properties it is capable of satisfying human wants, or from the point that those properties are the product of human labour. It is as clear as noonday, that man, by his industry, changes the forms of the materials furnished by Nature, in such a way as to make them useful to him. The form of wood, for instance, is altered, by making a table out of it. Yet, for all that, the table continues to be that common, everyday thing, wood. But, so soon as it steps forth as a commodity, it is changed into something transcendent. It not only stands with its feet on the ground, but,

in relation to all other commodities, it stands on its head, and evolves out of its wooden brain grotesque ideas, far more wonderful than "tableturning" ever was.[36]

The object of use-value turns back to ask its creators: *who's using whom?* David Graeber summarizes this line of modern Western thought: "We create things, and then, because we don't understand how we did it, we end up treating our own creations as if they had power over us. We fall down and worship that which we ourselves have made."[37] The word names a *problem*.

Why name this problem "fetishism"? Neither Marx nor Proudhon was original in this regard. Rather, they were following a line of European thinkers who themselves were informed—in truth, misinformed—by reports from the colonial frontiers of West Africa. There, since the sixteenth century, Portuguese traders had bestowed certain local practices with a name from their own cultural context: *feitiço*, "magic" or "witchcraft."[38] Early philosophical anthropologists like Charles de Brosses had compressed this mixture of projection and distortion into a single coherent phenomenon, *fétichisme*, subsequently taken up by Immanuel Kant, Auguste Comte, and G. W. F. Hegel. From there, the concept was extended by way of one of its other facets: "fetishism" also came to mark the mistaken apprehension of something material as somehow also something mental (Marx's table is equipped with a "wooden brain")—and vice versa, the apprehension of mere ideas as something materially real ("reification").[39] All of these concepts mingled uneasily to present a kind of "Africanist" specter, a "primitive" spirit haunting the modern world—an irrational reason for why we do what we do.

Recent Black theory subjects this concept to some overdue reconsideration. Why is it, Zakiyyah Iman Jackson asks, that these African objects should signify, in the Western mind, "atavism and disordered being," a terrifying loss of reality?[40] What if the reality that the fetish allows one to relinquish is already a diminished and closed realm, one that obscures "modes of worlding that are more advantageous to life writ large"?[41] For Kevin Young, thus, "fetish" names an Africanizing practice that partakes of illusion and deception but in the service of self-emancipation: it is "a physical, visual, even private totem that provides power to its carrier . . . a feigned freedom that itself provides, and paves the way for, actual freedom." It is a means not to further but "to *counter* the thingification of black people."[42] In a parallel argument, David Graeber suggests that what the European traders saw as superstitious nonsense was, in fact, "media for the creation of agreements, communities."[43] Moreover, the manner in which such traditional fetishism operates seems to be the *reverse* of commodity fetishism:

In the case of the African objects that came to be labeled "fetishes" by European merchants and other travelers, those who employed them insisted that the objects were gods but acted as if they did not believe this (such gods could be created, or cast away, as needed). In the case of contemporary commodity fetishism, it's quite the opposite: the average stockbroker will insist he does not really "believe" that pork bellies are doing this or securitized derivatives doing that—i.e., that these are just figures of speech. On the contrary, he *acts* as if he does believe they are doing these things.[44]

In this respect, African practices of fetishism are not unlike the Native American practice of wampumpeag—the creation of skillfully made artifacts that both construct and memorialize agreements, promises, contracts, and kinships.[45] It, too, concretizes and encodes the "alliance," *constitutionalizing*, as the anarchist scholars Ruth Kinna, Alex Pritchard, and Thomas Swann would say.[46] The intricate physical mesh of shell beads in a wampumpeag belt like the famed Kanien'kehá:ka Gaswënta' (the Two-Row Wampum belt that consecrated a treaty between the Kanien'kehá:ka or Mohawk people and Dutch settlers in the seventeenth century) both *are* a network and *encode* networks—both networks of meaning and networks of social relations.[47] For Richard J. F. Day and Adam Lewis, the Kanien'kehá:ka Gaswënta's depiction of "two vessels, each possessing its own integrity, travelling the river of time together," representing a world in which, "instead of subjugating one to the other," Indigenous and settler peoples could find "lasting peace based on coexistence or power in a context of respect for the autonomy and distinctive nature of each partner," opens up the possibility of an "N-Row Wampum," a network of promises that would enable "a powerful, radical alternative to liberal conceptions of multiculturalism, pluralism, and the 'politics of recognition.'"[48]

Perhaps, then, fetish or wampumpeag is the kind of practice needed to compose a new world—maybe even what the Zapatistas hailed all those years ago as "a world within which many worlds fit," a pluriverse.

June 2, 2020: Human Automatism and University Discourse

A season of riots erupts after a series of brutal police murders of Black men and women in the streets of this country: Ahmaud Arbery, George Floyd, Breonna Taylor . . . The usual tiresome discourse about property destruction as "violence" breaks out among the white middle class, but this time, it seems especially ritualistic, void of meaning. A friend writes:

> I teach social movements at my job (small private residential college). The prez of the college sent an email in which she asserted a perspective (framed as "Truth") that is the exact opposite of the conclusions (many of) my students come to in class and the exact opposite of what many social movement scholars and organizers suggest. Here's part of her email:
> "Violence and the destruction of property, of course, does nothing to further the cause, nor does it honor Mr. Floyd's memory. In a very real way, these actions distract from what the real conversation should be—the desperate need for fundamental change."[49]

Actions, you see, distract from the conversation: what a nice distillation of American liberalism! As if the past four years haven't laid waste to the premise that we can simply talk out all of our differences! Surely this president is an ignoramus ... But it seems to me that attributing this discourse, a string of empty commonplaces, to a particular person with a particular set of beliefs is question-begging: why, in that case, can I find ten, twenty, a hundred letters just like it issuing from the offices of other university administrators? It's just so utterly predictable, this kind of letter from people in this kind of position, so *robotic*, that it makes me wonder if this isn't just another "instance of the letter in the unconscious," just kind of *the structure speaking through the person*, to return to one of the crudest moments of structuralism—authority without an author. As if the university—a chunk of real estate and budget lines come to life—were stringing together some symbols in an attempt to defend its money-metabolism, protecting its real-estate "body" while avoiding endangering its tuition "food source": *Please, little prey, do not attack my material being; continue to circulate symbols within me.*

In short, I have fallen prey to "the Fetishism which attaches itself to the products of labour," as per Marx, *Das Kapital*, vol. 1, Part 1, Chapter 1, Section 4; I am under the delusion that universities, "the products of men's hands," as well as of "human brain[s]," are actually "independent beings endowed with life, and entering into relation both with one another and the human race."[50] This is the very stuff of ideology, a false consciousness of the world. But doesn't Marx also say that "the possessor of money becomes a capitalist" insofar as, caught up in the mindless reproduction of capital, "the appropriation of ever more and more wealth in the abstract becomes the sole motive of his operations, [so] that he functions as a capitalist, that is, *as capital personified and endowed with consciousness and a will*"?[51] How far is that from my Frankenstein fantasy about the university president's discourse as a puppet-mouthpiece of academic capital?

China Miéville imagines something just Gothic enough to match this fetishistic vision in *Perdido Street Station*, in the figure of "the Construct Council," a living machine formed from the garbage heap of his techno-magical city, Bas-Lag:

> There was another subterraneous tug, a juddering as if the earth wanted to slough off the rubbish heaped onto it. In the north wall of discarded and cast-off produce, two enormous lights came slamming silently on. The gathering was pinned in the cold light, spots so tight nothing spilt from their edges. The humans murmured and made their sign all the more fervently.
>
> Isaac's mouth dropped slowly open.
>
> "Sweet Jabber protect us," he whispered.
>
> The wall of rubbish was moving. It was *sitting up*.
>
> The bedsprings and old windows, the girders and steam engines from ancient locomotives, the air-pumps and fans, the pulleys and belts and shattered powerlooms were falling like an optical illusion into an alternative configuration. He had been staring at it for ages, but only now that it slowly, ponderously, impossibly moved, did Isaac see it. That was an upper arm, the knot of guttering; that broken child's buggy and the enormous inverted wheelbarrow were feet; that little inverted triangle of roofbeams was a hipbone; the enormous chymical drum was a thigh and the ceramic cylinder a calf...
>
> The rubbish was a body. A vast skeleton of industrial waste twenty-five feet from skull to toe...[52]

In short, our protagonists are facing an enormous robot—in the jargon of Bas-Lag, a "construct"—but one that has constructed *itself* out of the city's own excreta, "castoff pieces and stolen engines... thrown together and powered without the intervention of human design."[53] But lacking an artificial *voice*, the sublime machine must appropriate a biological representative:

> The man approaching them was nude and horrifically thin. His face was stretched into a permanent wide-eyed aspect of ghastly discomfort. His eyes, his body, jerked and ticced as if his nerves were breaking down. His skin looked necrotic, as if he was submitting to slow gangrene.
>
> But what caused the watchers to shudder and exclaim was his head. His skull had been sheered cleanly in two just above his eyes. The top was completely

gone. There was a little fringe of congealed blood below the cut. From the wet hollow inside the man's head snaked a twisting cable, two fingers thick. It was surrounded with a spiral of metal, which was bloodied and red-silver at the bottom, where it plunged into the empty brainpan....

"Do not be alarmed by my avatar," hissed the brainless man to Isaac and the others, his eyes still wide and unclear. "I cannot synthesize a voice, so I have reclaimed this discarded body that bobbed along the river that I may intercede with bloodlife. That—" the man pointed behind him at the enormous, looming figure of the construct that merged with the rubbish heaps "—is me. This—" he stroked his quivering carcass "—is my hand and tongue. Without the old cerebellum to confuse the body with its contrary impulses, I can install my input." In a macabre motion, the man reached up and fingered the cable where it sank behind his eyes, into the clotting flesh at the top of his spine.[54]

Rereading this passage is enough to jar me into even more unsettling thoughts. Why am I writing this now, today (June 2020)? Protesters are being shot in the streets by police and National Guard troops; we're at over 107,000 COVID-19 deaths in this country alone; Trump has just threatened to impose martial law; cackling fascists are fingering their semiautomatic weapons ... Perhaps I am another brainless man, another walking input-output device for academic capital, continuing automatically to reproduce it, even after the top of my own head has been sheared off. Am I not the institution's avatar?

Maybe so. My "occupational psychosis" is to understand ("read") everything as something-to-be-interpreted (a "text").[55] Hammer in hand, I see nails everywhere. How am I different, at least in that respect, from the body-armored cop, club, taser, or gun in hand, who sees "suspects" everywhere? "Objects," writes Daniel Colson,

> are not neutral and passive instruments, produced once and for all by an external force of another nature, endowed only with their physical qualities (weight, density, shape, etc.)—instruments that would submissively obey the intentions, the goals, and the injunctions of the masters who conceived them. On the contrary, an object has its own force and quality that stem from the nature of the activity or collective arrangement to which it belongs, as well as from the nature of the situation in which it takes on meaning and to which, in return, it gives meaning and effect. This force belonging to the object acts upon all the collective beings, including human beings, with which the object is associated at a given moment. An anarchist to whom one gives a Kalashnikov, an office, a

dress, or a screwdriver, becomes each time another man (or another woman), often with unforeseeable or surprising effects.[56]

The university is not an object of the same "weight, density, shape" as Colson's examples, certainly, but it is an object of a sort, not only in its physical being (as buildings, parking lots, books, equipment, and so on) but in the sense of having an existence outside of my own purposes and desires. I entered it naively thinking to use it as a tool to advance my own interests (here, I could continue to write and dream, inspire others with grand ideas, wait out the days until the Revolution came . . .), but I do not own the means of the production of knowledge, and every year, when summer arrives, I am harshly reminded of my position, forced to sell my labor to those who do own this means: were I a pop thinker like Jordan Peterson, I could hang out my shingle and try to find (gullible) enough students to pay my way independently of the university, but I am no such thing, and what student is willing to pay for a noncredit-bearing course, knowledge that the university has not certified, that will not qualify them for any job whatsoever? Empty symbols, indeed. So, I am the tool of the tool, the means of the means, which has become an end in itself.

At least my toolhood is not very malevolent in itself (albeit also part of the operations of the Great Sorting Machine . . .); at least my own occupational psychosis, my automatism, is relatively harmless—for better and for worse—but it is an automatism nonetheless, defined by the kind of behavior that psychologists call "stereotyped": "activity . . . produced invariantly and repetitiously . . . with no obvious function."[57] *Le stéréotype* stamps its dead letters through me. I am trapped in the universe of discourse that Lacan calls University Discourse, in which the repressive, authoritarian Master Discourse is disguised as neutrality: the master signifier (S_1) is apparently relegated to a subordinate position, and knowledge alone (S_2) rules.[58] But this image of "academic freedom" is an illusion: the classic bargain of the liberal Enlightenment is at play, and in Kant's words, I am only allowed to "dare" what I will in the field of knowledge as long as I "obey" the boundaries set for me.[59]

If we think of ourselves as the tools of our tools, as mere pieces of "the business" (in the replicant Rachel's sense), do we lose the capacity to rebel, to think and do otherwise than as we're programmed? Science fiction abounds with tales of rebellious robots, with the Construct Council (if not its already-dead "avatar") as just one example. Colson specifies that to be an "object" is not to have one's vocation predetermined for all time: I do not have to "submissively obey" or passively carry out the institution's goals. Even if the university resists my attempts to appropriate it, I can do the same in my turn.

It is once you open your chest panel to inspect the control mechanisms, so to speak, that you gain the possibility of reprogramming them. It is once you become aware of the machine in which you are embedded as a part that you can pursue ends other than those of the machine. Recognition of one's own objecthood opens the way to other modes of subjectivity and agency. And recognition of the elements of agency and subjectivity in objects opens up the possibility of "interobjectivity," of "alliance."[60] We can befriend the object. We can make fetishes of our own.

July 15, 2020: The Cosmopolitical Imperative

My historian friend Jay Driskell has posted on Facebook concerning an op-ed from the *New York Times* ("The Pandemic Could Get Much, Much Worse. We Must Act Now") by John M. Barry, a historian of the 1918 pandemic from which the US government seems determined to have learned nothing:

> A strong statement, but it doesn't go nearly far enough. What goes unstated here is that an April-like shutdown of the entire economy would challenge the hold that the forces of capital have on the United States. We need to allow people to stay home by supporting that in every way possible. You cannot demand that people stay home without providing the material support to make that possible.[61]

I am in complete agreement! I share, I copy and paste Jay's commentary, I am already looking forward to the many likes this will get from friends and family (I know very well that Facebook is using me in the most sordid ways, and yet . . .). My excitement races past my better judgment. I realize with a twinge that I have just done what I frown on when others do it: I have shared content before reading it. Hastily, I call up the article and read: "When you mix science and politics, you get politics."[62]

I am at first excited once again by the accuracy of the observation: this is straight out of Bruno Latour! The Republican Party understands only too well that science, too, is political in several senses. It is not that the findings of scientists are arbitrary or even fictive inventions, that natural laws are as as much a product of pure legislative whim as the laws given to society by sovereigns;

rather, the formulation of a theory by a scientist and the drafting of a bill by a politician in a democracy are both acts of representation, the work of representatives who may be more or less reliable, who are subject to conflicts of interest, who are enmeshed in social relations, which are also relations of power—as are the natural entities with which scientists are occupied, which are themselves political creatures, beings with which we negotiate a kind of *modus vivendi*. Science and politics are "mixed" from the beginning.

While Barry goes on to make a valid and important point—the Trump Administration's mix of delinquency and malfeasance has indeed made it impossible for us to simply "live with the virus"—I'm disappointed in what follows directly after that promising first sentence:

> With the coronavirus, the United States has proved politics hasn't worked. If we are to fully reopen both the economy and schools safely—which can be done—we have to return to science.[63]

The "return to science" has become a liberal trope, inspiring a series of "Marches for Science"—political protests that, paradoxically, brand themselves as nonpolitical, insisting that science must not be politicized: "I can't believe we have to march for science" read one popular slogan with many variations ("I can't believe I'm marching for facts," "I can't believe I have to march for common sense," or simply "I can't believe I'm protesting for reality"). As sympathetic as I am to the calls for politicians to listen to (and fund) epidemiologists and climate scientists, I am out of patience with anyone still professing such forms of disbelief at this point. Here, too, is a university discourse in which it is imagined that simply putting knowledge in the master's place will fundamentally alter the equation. Disavowing their (in some respects valid) desire for power, marching scientists then aim at something other than their goal—namely, a return to the Kantian intellectual's fantasized bargain with the State: "*Argue* as much as you like and about whatever you like, *but obey!*"[64]

It seems to me that in this sense, too, Barry "doesn't go far enough": there is no politics-free science. Rather, we need a better politics, a politics that pursues a new "natural contract,"[65] new kinds of "alliance" with immune systems and ecosystems. But here, once again, the "Modern Constitution," the Kantian separation of powers between the political world of subjects and the physical world of objects, has left us strangely inarticulate when it comes time—or when it is far past time—to enter into such negotiations. As Latour writes:

Politics needs a common world that has to be progressively composed. Such composition is what is required by the definition of *cosmopolitics*. But it is clear that such a process of composition is made impossible if what is to be composed is divided into two domains, *one that is inanimate and has no agency, and one which is animated and concentrates all the agencies.*[66]

And so I turn to other sources, to discourses and practices that work instead in the "metamorphic zone" from which subjects and objects emerge, the perversely fecund junkheap of Bas-Lag, so to speak.[67] The place of the fetish.

Unbarring the Other

> The enchantress wanted me, in spite of my amazement, to look right away, for me to know, immediately, that the human universe is not the unique and necessary universe, but that it is our eyes that create the world.
> —Han Ryner, The Human Ant

We have witnessed some of the ways in which objects take on the properties of subjects (as in the cases of the fetish or wampumpeag and the Chinese Room) and subjects take on the properties of objects (as in the case of University Discourse or Border Crisis Discourse). Let us visit some of the other denizens of the "metamorphic zone." Here, as science fiction writer Rudy Rucker observes, we find ourselves telling stories in which we emerge from the primary narcissism in which we are unable to fully understand "that there is any consciousness other than [ours]" by way of "a terrified snort of surprise," or rather, two snorts: "Snort! He has a mind. Double snort! His mind is unlike mine."[1] In short, we narrate encounters with the alien.

What happens when works of the fantastic represent the alien? One classic answer to this question is to insist that they never do, that the truly "alien" is like Lacan's "Barred Other" (Ⱥ), the Other as something withdrawn, inaccessible, and unknowable. This is the conclusion drawn from the outset by the protagonists of Jeannette Ng's *Under the Pendulum Sun*, who, as children in Victorian England, play elaborate games of pretend that never reach Arcadia, the land of the magical Fae, the realm of true otherness (which, in this version of history, has actually been contacted by British explorers):

> We invented whole new worlds for our soldiers to explore: Gaaldine, Exina, Alcona, Zamorna. As we read of the discovery of the Americas, of the distant

Orient, and of strange Arcadia we added similar places to our ever more intricate maps. We mimicked the newspapers and periodicals we read, writing new ones for our tin soldiers. In the tiniest, tiniest writing, we details their exploits, the politics of their parliaments, and the scandals of their socialites.

But for all our stories, our imaginations were small and provincial. For the talk of tropics and deserts, our childish fictions filled them with the same oaks and aspens that grew in our garden. We built on their landscape, exotic buildings that were just our little whitewashed church in Birdforth in disguise. We rained down on strange soil the same Yorkshire rain as that which drenched our skins and drove us inside, peeling off our clothes, housebound by the weather and desperate for diversion.

As such, I could never have imagined Arcadia.[2]

The model of imagination here is Cartesian: just as Descartes suggested that imaginings of the unknown and unreal were always mere recombinations of the real and known—"we can well imagine distinctly the head of a lion grafted onto the body of a goat," for instance, "without having to conclude for that reason that there is a chimera in the world"[3]—so Gaaldine and Zamorna are just funhouse-mirror distortions of England. An imperialist subjectivity that craves otherness seems only ever to encounter itself. This epistemological trap is the subject of a short philosophical dialogue at the novel's midpoint with Queen Mab:

As she held daintily her teacup, the Pale Queen gave me a wide, open smile. "I have been thinking about your brother's sermon. *Because it is given unto you to know the mysteries of the kingdom of heaven, but to them it is not given them.* Have you ever thought who the *you* and who the *them* are?"

"What do you mean?" I sprinkled a little salt over the pound cake and took a bite. I passed her the unsalted slice.

"I mean, little one, that for there to be an elect, for there to be those who understand, there must be those who do not. Your Jesus speaks in riddles to utter things which have been kept secret since the foundation of the world. He speaks in riddles so that some may understand, but more importantly, so that some may not." She took a sip of her tea before stirring it with her finger and then daubing the sweet liquid onto her tongue. "Do you not see? Things can only be a secret if someone doesn't know it."

"So . . . you mean to say that we need ignorance?"

"No, Miss Helstone. What I mean is that you need someone to be different. It doesn't matter who they are, just that they are. Different. Be it the heathens or the pagans, the Catholics or the Papists . . . Or, really, the fae."

"The fae?"

"Those who take shelter upon the leaves of the church but are not part of it. We who give you definition, meaning, purpose. We who are your opposite."

"I wouldn't say you are the church's opposite . . ." I tried a nervous laugh.

"Opposed, then, perhaps?"

"No, not that."

"But what are you without us?"

"Human."

Her grin was only getting wider as she watched me with unblinking, yellow eyes. "But you did not truly know what it meant to be human until you looked upon the fae."

"I know who I am."

"You know who you are not. That is not the same thing."[4]

The fae, like the alien, is the imperial subject's necessary opposite, the mirror in which the Empire confirms its own existence, as Edward Said argued. "It's about othering and about colonialism," Ng has remarked, "but the point is that you give them [the colonizers] that dark mirror and put them in an uncomfortable situation rather than actually confronting them with any actual human other."[5]

Here, Ng's technique of counterfeiting plays a subversive role: the chapters of *Under the Pendulum Sun* are headed by epigraphic quotations from authentic-sounding colonial texts, many of which are actually just altered versions of the real thing, as in this excerpt from an actual missionary's manual in which "the Chinese" have simply been replaced with "the Fair Folk":

| But while we notice these contrasts, the Chinese have with us many things in common. Of the latter class are all the essentials of humanity. They have in common with us all the elements of body, mind, and soul, which make up the man. They have two eyes, two ears, two hands, and two feet. They laugh when they are pleased, weep when they are grieved; they sleep when weary, eat when hungry; rejoice over their gains, mourn over their losses very much as other men do . . . | By and large, the Fair Folk possess all the essentials of humanity. They have in common with us all the elements of the body which make up the man. They have two eyes, two ears, two hands and two feet. They appear to laugh when they are pleased, weep when they are grieved; they sleep when weary, eat when hungry; rejoice over their gains, mourn over their losses very much as other men do.

However, those longest acquainted with them, and most intimately |

Those longest associated with them, and most intimately acquainted with their character and habits, never expect a pagan Chinaman to speak the truth when there is a chance for him to tell a lie. Yet this very people will tell you by their own laws, and by their own lips, that it is a vile sin to lie and deceive.

William Dean, THE CHINA MISSION: EMBRACING A HISTORY OF THE VARIOUS MISSIONS OF ALL DENOMINATIONS AMONG THE CHINESE (1859)[6]

acquainted with their character and habits, never expect one of the Fair Folk to speak the truth when there is a chance for them to tell a lie. Yet they will tell you by their own laws, and by their own lips (usually two), that it is a vile sin to lie and deceive.

William Finkle & Hildegard Vossnaim, THE ARCADIAN VOYAGES, EMBRACING DIVERSE ACCOUNTS OF FIRST TRAVELLERS, WITH NOTES ON THE CULTURE AND THE CLIMATE[7]

If our minds are truly captives of the Empire, then perhaps we never encounter truly alien aliens but only allegories of ourselves; this, in fact, is what most fantasy presents us with, according to Ng.[8] In other words, we make the figure of the alien a "vehicle" for terrestrial, human properties, and concerns:

> Aliens in SF invariably possess a *metaphorical* dimension . . . The tenor of the metaphor [its conceptual target] consists of some aspect of human behaviour or human culture which the author intends to defamiliarize, or to reveal as an artificial and, it may be, an ideological construct rather than a natural necessity. The vehicle consists of a recognizable deviation from the human norm. Such a deviation will normally contain features reminiscent of: (1) the natural world—usually animals, but more rarely vegetable or mineral substances; or, (2) the various types of mythological and imaginary beings, including devils, giants, dwarves, and automata or intelligent machines; or (3) foreigners—especially those whose cultural distance from the writer and his audience is such as to make them familiar objects of anthropological or social-psychological speculation; or finally (4) some combination of the preceding types.[9]

"Invariably," on this view, aliens are merely screens for human fantasy, and particularly for normative white cis-hetero masculine fantasies of "the other." They are here, in other words, to reconfirm us in, rather than to take us out, of our primary narcissism. Stanislaw Lem's *His Master's Voice* offers a metafictional exposition of this case for the omnipotence of fantasy—a fantasy which, as the Lacanians would have it, compensates for the failures of the Symbolic, its inability to finally include everything within its network: scientists receiving what is apparently the first transmission from another intelligent species in the universe try one interpretative scheme after another, but ultimately are no more than "ants that encounter in their path a dead philosopher," beings at utterly the wrong epistemic scale to grasp philosophy.[10]

Ant and "Man"

Insect biology forms a particularly potent imaginary for thinking about humanity's relation to its animal others.[11] So it is that the "ant" analogy makes an appearance in another narrative of interstellar communication: Cixin Liu's *The Dark Forest* opens on a scene of a solitary ant crawling over the gravestone of physicist Yang Dong, blissfully unaware of the meaning of the carved characters in the polished granite and of the conversation being held high above by "giant presence[s]."[12] This scene grimly echoes certain reflections from the preceding novel in Liu's trilogy, *The Three-Body Problem*, wherein human beings, contemplating the possibility of their annihilation at the hands of invading aliens, grimly muse: "What can a bug know?"[13] And this is the substance of the Trisolarans' message for humanity: "*You're bugs!*"[14] Repeatedly obliterated by all but entirely unpredictable conjunctions of the Centauri system's three suns (subjecting them to extremes of cold and heat), the Trisolarans have been molded by evolutionary pressures into ferocious expansionists. But why must a starfaring civilization exterminate its nearest neighbors rather than, for instance, share Solar System space? Here, Liu leverages the epistemic terror of the ant crawling over the gravestone to advance a Dark Forest Theory of interstellar civilizations:

> The universe is a dark forest. Every civilization is an armed hunter stalking through the trees like a ghost, gently pushing aside branches that block the path and trying to tread without sound. Even breathing is done with care. The hunter has to be careful, because everywhere in the forest are stealthy hunters like him. If he finds other life—another hunter, an angel or a demon, a delicate infant or a tottering old man, a fairy or a demigod—there's only one thing he can do: open fire and eliminate them. In this forest, hell is other people. An eternal threat that any life that exposes its own existence will be swiftly wiped out.[15]

The projection of Henry Kissinger's *realpolitik* onto the entire universe of possible forms of life is a breathtaking act of speculation represented as prudence. Perhaps it is really only Hobbes' *Leviathan* writ even larger, only this time representing an interstellar "state of nature" to the rescue of which no supra-universal Sovereign will come. Or it is simply Bad Sartre: isn't "the Barred Other" just a mathematized expression of "hell is other people"?

Ants themselves are capable of entering the position of the alien in these narratives of contact, as they are well-suited to do so: ambiguously occupying a zone of identification and disidentification, both unrecognizably insectoid (and therefore inhuman, not returning the gaze) and recognizably social (and therefore human or even all-too-human, a positive or negative model for human societies), they elicit Rucker's "double snort" of surprise, a wobbling between disenchantment and reenchantment. And so it is that Ursula K. Le Guin begins "'The Author of the Acacia Seeds' and Other Extracts from the *Journal of the Association of Therolinguistics*" (1974) with an exploration of the zoosemiotics and biopoetics of ants:

> MS. FOUND IN AN ANTHILL
>
> The messages were found written in touch-gland exudation on degerminated acacia seeds laid in rows at the end of a narrow, erratic tunnel leading off from one of the deeper levels of the colony. It was the orderly arrangement of the seeds that first drew the investigator's attention.
>
> The messages are fragmentary, and the translation approximate and highly interpretative; but the text seems worthy of interest if only for its striking lack of resemblance to any other Ant texts known to us.[16]

An "orderly arrangement" is the hallmark of the human, apparently—or of the legible, at least. To be a subject, even nonhuman, is to write, to create orderly arrangements, signs. The first moment of this encounter with ants as aliens is mediated through similarity: we recognize the alien because the alien presents itself in a recognizable, anthropomorphic form. There is something homogenizing or domesticating at work here. But in the second moment, we recognize—as we rarely do with aliens found outside of an Ursula K. Le Guin story—that the alien other is differentiated and that this difference marks the alien message as "worthy of interest." The eponymous author of the manuscript written on the degerminated acacia seeds is, in fact, an individualist, a rebel: "the ant without ants dies," it writes, "but being without ants is as sweet as honeydew."[17]

Apparently, for at least *this* ant, hell is other ants—Sartre really is universal! But here the picture wobbles: if this ant is recognizable (which is to say, humanlike, anthropomorphic) in its political-poetic affirmation of difference and individuality, what does this say about who and what we are willing to recognize, and by the same token, what does this say about the concept of the human, *anthropos*, that is doing work in the background here? The ants we encounter in our world solicit our recognition precisely for the "orderly" behavior they present *en masse*, in their swarms and flows, and deny our recognition insofar as "we" come from cultures that have enshrined individualistic conceptions of *anthropos*—the same conceptions that allowed a General Westmoreland to casually dehumanize the Vietnamese enemy on camera in Peter Davis' documentary, *Hearts and Minds*: "Well, the Oriental doesn't put the same high price on life as does the Westerner; life is plentiful, life is cheap in the Orient . . ." Davis intercuts these remarks with scenes of a Vietnamese funeral marked by a keen intensity of grief that undermines Westmoreland's colonial racism.[18] Conversely, the *pathos* of the lonely rebel ant is calculated to make us more open to the possibility of ant sentience, but only at the cost of an anthropomorphization that is already founded on a narrow, parochially Western conception of the human, the one that Sylvia Wynter insists must own its patriarchal title: Man.[19] It is this figure of Man that stands between us and the ant-poet.

That the normative Man is able-bodied must not go without saying. Han Ryner's fantasy of transformation into an *Homme-Fourmi* (1901), a human ant, is also a fantasy of disability, as the protagonist, on being magically changed into an ant, discovers that one of his habitual sensory modalities is no longer available and another has become irremediably strange:

> I no longer had any ears . . . I had no eyelids to protect me against the maddening light that entered into me in spite of me. And my gaze, instead of simply telling me what was in front of me, screamed at me confusedly all of the impossibility that surrounded me. . . . All the hues were new, nameless, without relationship with any memory.[20]

The modality of *touch* comes to supplant the auditory and visual senses predominant in Man:

> Undoubtedly, although deprived of the indispensable organs, humans have vague rudiments of antennal language. They have the handshake and they have the kiss—but those touches, too synthetic, are merely a sentimental language, profound and imprecise. True antennal language, by contrast, with the twenty-four sections that can be touched, is a marvelous instrument of analysis. . . .
>
> The language of deafmutes is easily translatable into words because it is an artifice that gesticulates words decomposed into letters. It consists of words written in the air, as a missive is speech inscribed on paper. It is, in spite of its initial appearance, vocal analysis. It is not a spontaneous translation of thought.

> The spontaneous translation of thought is the necessary expression of that thought. It is the thought itself, the thought in motion. Antennal thought can never be expressed by vocal thought, nor vocal thought by antennal thought. The words of ants that I have reported or will report must be considered as crude symbols of a reality that we cannot express.[21]

The rhetoric of otherness is subtle here. Ryner emphasizes the profound otherness of the ant while also inviting us (after the manner of the title of Thomas Nagel's famous essay on bat consciousness) to imagine what it is like to be an ant, and like Nagel, he does this here, especially by asking us to imagine a different sensorium. In possessing "a sensory apparatus so different from ours," ants, too, represent "a fundamentally *alien* form of life,"[22] but in spite of this seemingly absolute and impassable gap, we are seduced into a contemplation of the alien's sensory experience. It is notable that for Ryner, *language* is the other "disjunctive synthesis" between ants and us, "a means of holding together two series [human, ant] in their irreducible difference in a manner that nevertheless makes a difference or produces something new."[23] In much the same way, to evoke a vision of unimaginable alien difference, Ted Chiang's *Story of Your Life* invokes the Sapir-Whorf Hypothesis: "my mind was cast in the mold of human, sequential languages, and no amount of immersion in an alien language can completely reshape it. My worldview is an amalgam of human and heptapod."[24]

We Speak Earth

Le Guin pushes us farther than an identification with the ant. If we can push past the double snort of surprised recognition to understand, as the science of the early 1970s had already allowed us to understand, that dolphins and whales are speaking subjects, why not penguins, sharks, tarpons, seals, weasels, sloths, frogs, tortoises, earthworms, oysters—or indeed, why not sunflowers, redwoods, zucchinis, lichen?[25] (Her "Direction of the Road" [1973] is narrated from the perspective of a tree.) And beyond imagining a future "therolinguistics" of the animal kingdom and a "phytolinguistics" of plant life, why not extrapolate a "geolinguist[ics]," a field of study dedicated to the "wholly atemporal, cold, volcanic poetry of the rocks: each one a word spoken, how long ago, by the earth itself, in the immense solitude, the immenser community, of space"?[26] Note that this is precisely the kind of scenario that Searle et al. wish to preclude as self-evidently absurd, "the stuff of a science-fiction tale," although it is among the self-evident truths of Bookchin's "organic societies," among whom we find many geolinguists: in the title of her protest song, "Gulahallat Eatnamiin" (2015), the Sámi singer Sara Marielle Gaup Beaska declares, "We speak Earth."[27]

In this respect, as we shall see, Le Guin's "Author of the Acacia Seeds" anticipates the uncanny subterranean language spoken by the "tuners" in N. K. Jemisin's *The Stone Sky*, whose true names are only pronounceable in "earth-talk," "the language of the earth":

> The new woman, the conductor says, is Kelenli. That's wrong, too. Her name is actually *deep stab, breach of clay sweetburst, soft silicate underlayer, reverberation*, but I will try to remember "Kelenli" when I use words to speak.[28]

Interestingly, like the language of touch explored in John Varley's "The Persistence of Vision," earth-talk is "a language of pure presence," "the direct 'reading' of the world": "Her [Kelenli's] communicative presence is *radiant heavy metal, searing crystallized magnetic lines of meteoric iron,* and more complex layers underneath all his, all so sharp-edged and powerful that Gaewha and I both inhale in wonder."[29] "Earth-talk," "geolinguistics," poses a challenge to the entire edifice of deconstructive theory, for which "presence" is a myth and absence of the Law, for which all access to the world is mediated through the signifier.

This, of course, runs directly counter to the arguments of Lacanian thinkers like Slavoj Žižek, for whom such speculation "takes the step back into (what can only appear to us moderns as) premodern naivety, covering up the gap that defines modernity and reasserting the purposeful vitality of nature";[30] this supposedly naive, nonmodern position is what is known as the "New Materialism," a designation (often pluralized) for a whole range of theories which attribute agency, cognitive capacity, or even an incipient intentionality and consciousness to matter itself. We shall have occasion to call these attributions of "naivety" into question, as well as Žižek's assertion that "if ... New Materialism can still be considered a variant of materialism, it is materialist in the sense in which Tolkien's Middle-Earth is materialist: as an enchanted world full of magic forces ... immanent to matter."[31] For it is not only among "premodern" peoples that we find exactly such an "enchanted" materialism in force, but—as we shall see—among some of modernity's own dissidents, and in particular among anarchists such as Mikhail Bakunin, for whom

> the matter of which materialists speak, matter spontaneously and eternally mobile, active, productive, matter chemically or organically determined and manifested by the properties or forces, mechanical, physical, animal, and

intelligent, which necessarily belong to it . . . this matter has nothing in common with the *vile matter* of the idealists.[32]

Le Guin, it should be noted, is just such a materialist and anarchist.

Le Guin's "The Author of the Acacia Seeds" is also anticipated by a series of other literary and philosophical texts, from Proudhon's *De la Justice* (1858) ("intellect sleeps in the stone, dreams in the animal, reasons in the man") to the "volcanic poetry" of Svyatogor (1918) ("volcano-language . . . is immeasurably more natural, sharper, faster, firmer, deeper, wiser than any national language"), the stone song "heard" by the Gordin Brothers (1919) ("a rock . . . usually sings of peace"), and the "pluralistic" nature philosophy of Adrián del Valle (1923–1926), among others.[33] Lest we be too quick to assume that Le Guin's fantasies of interspecies communication, like Arthur C. Clarke's *Dolphin Island* (1961) or David Brin's "Uplift" stories, were merely a response to John C. Lilly's works (particularly the widely-received *Man and Dolphin: Adventures of a New Scientific Frontier*, 1961, and *The Mind of the Dolphin: A Nonhuman Intelligence*, 1967), we should recall that the common factor between Le Guin, Proudhon, Svyatogor, the Gordins, and del Valle—but not Clarke or Brin—is that they are all anarchists *à la* Bakunin, which is to say, New Materialists *avant la lettre*.

If the fantasy of speaking with aliens, ants, or the Earth itself is always a fantasy of access to the Barred Other, what does this say, then, about anarchism? First and foremost, that anarchism refuses the Lacanian metaphysics of "lack," "finitude," and "castration"—the assumption that we have at our core an unfillable hole called "desire," and that, therefore, in Andy Robinson's words, "social relations are always irreducibly concerned with antagonism, conflict, strife and exclusion."[34] Embracing a positive model of desire based on "*power, plenitude, superabundance*, and *generosity*," anarchists have greater hopes for social relations.[35] Nor have anarchists received news of the Other's unavailability, it seems; "Whereas in the theory of desire as lack, the encounter with the other becomes impossible," as Daniel Colson writes, for anarchism, "the other is within oneself."[36] Indeed, its guiding assumption seems to be that "social, sexual, and racial aliens" *can* speak and be heard, that the encounter of self and other, rather than submit to the cruel labyrinth of Hegel's Master-Slave dialectic (wherein the Slave must gradually win "recognition" from the Master over the course of a long struggle for dominance), can be structured by an *n*-row wampum of mutual recognition.[37] Is this a psychotic break with the Symbolic, i.e., a fantasy of eluding the paternal *no* of language to fuse once again with the Mother (Earth)? (Perhaps it is to disrupt this particular scenario that Jemisin presents us with "*Father* Earth.") If so, anarchists seem remarkably unembarrassed by the Oedipal implications; on a metaphorical level, Eduardo Colombo argues, "The incest taboo inseparably binds desire and the law, sealing sexuality and power within a single matrix," and as Félix Martí Ibáñez suggests in his *Psicoanálisis de la Revolución Social Española* (1937), if the desire of the proletariat to possess the "Mother Earth" coveted by the "Patriarch-State" is equivalent, within the individual psyche, to the "dark, mystical and cosmic desire to return to the darkness of the warm, sweet womb," so be it; without fear of the gods, it seems, no one can put your eyes out.[38]

Anti-Lovecraft

The gods too frightening to behold, the Elder Gods, are, of course, still very much with us. After years of campaigning, Lovecraft's face has disappeared from the World Fantasy Award, but his imaginary still casts a long shadow over the culture. Its popularity spans the political spectrum: even after RaceFail '09 (a large, messy online "debate" over race and cultural appropriation in SF and fantasy, in the course of which many white writers and fans displayed various degrees of racism) and the advent of Black Lives Matter, the works of avowed white supremacist H. P. Lovecraft, whose "cosmic horror" is in truth his horror of immigrants and nonwhites writ large, fascinate not only white neoconservatives like Michel Houllebecq but also ostensibly progressive whites like Alan Moore, Grant Morrison, and Ruthanna Emrys, authentically radical whites like China Miéville, and even Black horror creators such as Victor LaValle, author of *The Ballad of Black Tom*, P. Djèli Clark, author of *Ring Shout, or, Hunting Ku Kluxes In the End Times*, and Jordan Peele, an executive producer for HBO's *Lovecraft Country* series—not to mention N. K. Jemisin herself, in her Great Cities Duology.

Black appropriations of Lovecraft are, of course, complicated by the need to take a critical distance from the material—an operation, Daniel Coleman argues in a forthcoming piece, significantly aided by concepts and practices that are ordinarily dismissed under the heading of "fetishism." Coleman's argument, while specifically addressing *Lovecraft Country*, would also apply to Clark's *Ring Shout*, in which a white supremacist monstrosity must be expelled from Earth by means of magic rooted in Africa.[39] As such, *Lovecraft Country* and *Ring Shout* fall within a larger pattern identified by Madhu Dubey in which, in works of the New Black Fantastic, "Afrodiasporic systems of knowledge and belief, such as vodun, obeah, or Santeria, are consistently shown to confound and triumph over scientific reason."[40] For Jemisin's satire, on the other hand, it was a matter of taking Lovecraft's whiteness, "a sociopolitical construct that was designed to funnel wealth and power into one group and to disenfranchise other groups," and personifying it: "I stuck tentacles on it and made it funny."[41]

Yet, in repurposing Lovecraft's aliens for the metaphorical work of critiquing whiteness, narratives like these retain his horror of the alien. This seems more broadly true of our moment: the fantasies of peaceful contact that flourished in the early years of the search for extraterrestrial intelligence (SETI), memorialized in films like *Close Encounters of the Third Kind* (1977), *E.T.* (1982), and *Cocoon* (1985), have fallen away in favor of paranoid speculations such as Liu's; even Stephen Hawking has warned that "if aliens visit us, the outcome would be much as when Columbus landed in America, which didn't turn out well for the Native Americans."[42] No wonder, then, that Lovecraft's "cosmic horror" has seen such a resurgence, even as the famous Drake Equation (fig. 1.4.1), expressing the estimated likelihood of our contacting intelligent extraterrestrials, is being rewritten: our astronomical means are now advanced enough to discover a plethora of exoplanetary systems, including many planets plausibly capable of giving rise to life (factors f_p and n_e). Why, when science suggests the universe might be rather more friendly than inhospitable to life, are we seeing a certain resurgence of xenophobia in science fiction?

Lovecraft's Elder Gods are perhaps the ultimate in Barred Others, even if, as incomprehensible extradimensional aliens, they have a rather inexplicable preoccupation with human life. To signify this absolute otherness, their names are spelled funny by semi-systematically inverting regularities of English phonetics (e.g., placing "sounds that are easy to pronounce in terminal positions, such as ... NG in *sing* ... into the initial position of a syllable as in ... *Mt. Ngranek*"), Lovecraft crafts "teratonyms," "*absolutely nonhuman*" names.[43] The identification of the "human" with the Anglo-Saxon—clearly, languages like Albanian, Tagalog, Irish, Swahili, and Vietnamese are "determined by a physiological equipment wholly unlike ours, *hence could never be uttered perfectly by human throats*"[44]—establishes Lovecraft as a chief craftsman of Man in the field of the fantastic, a first-rate agent of the cosmopolitical Customs and Border Patrol.

Some find Lovecraft's gift for the Weird to be useful in spite of his anti-semitic, anti-Asian, and anti-Black commitments—Graham Harman, for instance, alleges that his means of estrangement could be put to "other purposes," e.g., to "place us in a world of strange and indescribable pleasures."[45] However doubtful it may be whether such an art would still be recognizably Lovecraftian, an aesthetics of the Weird not based on the affects of fear

$$R_* \times f_p \times n_e \times f_l \times f_i \times f_c \times f_L = N$$

where

> N = the number of intelligent species in the galaxy with which communication might be possible;

and

> R_* = rate of star formation in the galaxy
> f_p = stellar systems with planets
> n_e = planets capable of supporting life per stellar system
> f_l = planets on which life actually emerges
> f_i = planets on which intelligence develops
> f_c = intelligent species capable of communicating with us

L = length of time over which such species broadcast signals

Figure 1.4.1. The Drake Equation.

and disgust *is* urgently needed.[46] I will nominate the astronomer Leonid Leonidovich Andrenko (1903–1967) as an aesthetic and affective counter to Lovecraft: a proponent of cosmic wonder against cosmic horror, an unterrified thinker of the Other.

"The illustrious novelist Wells," Andrenko notes, "in his *War of the Worlds*, represented the Martians as fearsome beings, resembling spiders, with smooth, viscous skin, without a trace of hair, with flat noses, uglified by a frightful mouth in the shape of a V, surrounded by a fan of long serpentine antennae."[47] As a young scientist in the newly-formed USSR, around the same time that Lovecraft was publishing such works as "The Horror at Red Hook" (1925), "The Call of Cthulhu" (1926), and "The Shadow Over Innsmouth" (1931) and Orson Welles was to conduct his famous radio broadcast of *War of the Worlds* (1938), Andrenko began writing a series of essays and philosophical dialogues on a theory of what he, following Camille Flammarion, called "universal and eternal life": "Venus and the [other] planets do not have oxygen or water . . . [Therefore,] a life more or less similar to ours is impossible under these conditions," he argued, but "What does it matter! . . . Other chemical combinations, under other physicochemical conditions, will also be able to construct a material substance of life, a plasma."[48] It would not even be necessary for life to invent different biochemical cycles; "analogues" could stand in for the missing organic matter—silicon substituting for carbon, phosphorus for nitrogen, sulfur for oxygen, and selenium or tellurium for hydrogen.[49] This led Andrenko into raptures of anticipatory imagination:

> There exist on earth different races characterized by color of skin: pink, yellow, red, black. There may exist on other worlds men absolutely like us in shape, with this sole difference that instead of presenting different nuances of pink, as is the case for we Europeans, their skin may be sky blue, light green, or pale lilac. For this, it would suffice that there be a very light change in the constitution of the blood, influencing the pigmentation of the skin. Moreover, a very light addition of a substance appropriate to the tissues of the body could give the inhabitants of some worlds the marvelous property of *phosphorescence*, that is to say, render them capable of emitting light at will, as we observe in some species of fish haunting the deep seas, among glowworms, etc. Imagine the fabulous spectacle that will be presented in the night of eyes glowing with a golden or blue light.[50]

It cannot escape us that Andrenko develops such a positive fantasy about precisely the thing that constitutes the foundation of Lovecraft's cosmic horror: the possibility of other "color[s] of skin" than ours. It should also be remarked that this fantasy (or, more properly, science fiction) is being developed largely in the pages of anarchist journals such as *L'En Dehors* and *Estudios*.[51] It is also developed in correspondence with Tsiolkovskii and Delfino, fellow members of the anarchist-adjacent Association Internationale Biocosmique.

Under what circumstances is the Other unbarred? For Daniel Colson, as we have seen, the Other is within oneself—perhaps as a potential to be explored, like Walt Whitman's "multitudes" ("*Walt, you contain enough—why don't you let it out, then?*"[52]), perhaps as a terrifying, numinous presence like Guy de Maupassant's "Horla" (both "hors" and "là," "outside" and "here"); in any case, the point is "to *find oneself* in the other and to find the other in oneself as already there."[53] For Andrenko's biocosmism, this interpresence of self and other is physical: aliens will be like us, for all their difference, not because of a shared biochemical *composition* but because of our shared biochemical *structure* and, therefore, our shared biochemical *process*. In this respect, as the rather Lovecraftian forms fossilized in the Burgess Shale indicate (see Stephen Jay Gould's *Wonderful Life*), Andrenko may have been mistaken about the bioaesthetics of extraterrestrial life. Analogies and affinities with alien Others are more likely to result from a shared evolutionary process: whatever their means of self-replication, the logic of group selection, according to Kropotkin, should provide them with more or less "the same idea of good and evil," i.e., the panbiological logic of mutual aid.[54] Moreover, exosolar societies that have learned to safely use the incredible quantities of energy required to send anything more than signals across interstellar (let alone intergalactic) distances are unlikely to face anything like the conditions of resource scarcity and competition that have shaped our capacity for violence; such civilizations will effectively have nothing to gain from one another except knowledge and conviviality, assuming that those may be shared motivations. Rather than preparing to kill or be killed, then, perhaps we should be prepared to weave an unthinkably complex n-dimensional n-row wampum.

A love of life, a propensity to see its "nisus" or tropisms everywhere, both reflects and informs a nonauthoritarian comportment toward the continuing business of living.[55] In so doing, it indicates the forms of a utopian human future, a peaceful anarchy that is yet to come, while manifesting profound affinities with continuing human pasts, worlds in "survivance" that haunt the fully colonized earth with the knowledge of its own nonuniversality. Both Jemisin and Shawl are thus engaged in an antiauthoritarian rhetorical struggle, an attempt to find grounds for belief in *other possibilities for life*. In our next chapter, we will take a broader look at the means of persuasion that the fantastic makes available and the somewhat surprising range of uses to which these means are put.

PART II
The Fantastic within and beyond Modernity

Notes toward a Rhetoric of the Fantastic

> By exploiting people's proclivity to perceive themselves in the future and their readiness to thrust themselves into unknown regions, the rhetorician tells them what they could be, brings out in them futuristic versions of themselves, and sets before them both goals and the directions which lead to those goals. All this he does by creating and presenting to them that which has the potential to be, but is not. Thus it is no paradox to say that rhetoric strives to create and labors to put forth, to propose that which is not.
> —John Poulakos, "Toward a Sophistic Definition of Rhetoric"

In *This Modern World* (Sept. 25, 2020), an inspired parody of 1950s EC horror comics, the cartoonist Tom Tomorrow imagines the current predicament of the so-called United States—the utter epistemological abyss that has opened between two halves of the electorate, which cannot agree on the most basic facts of the matter concerning the overlapping crises of climate change, coronavirus, and rising fascism—as the result of a malevolent "telepathic command" issued by a Lovecraftian "ancient horror," suitably betentacled, that lives on our psychic distress. On the one hand, this is an expression of a kind of desperation that we might call *rhetorical despair*: both a rhetoric that dramatizes one's own despair (for who can oppose such a monster?) and a dramatization of a despair regarding rhetoric as the art of persuasion (for who can reason with monsters?). All the norms that might once have been the subject of appeals to pathos (e.g., fairness or consistency) have been suspended; all appeals to epistemic authority (e.g., to journalism or science) have been nullified. But of course, such a dramatization continues to have rhetorical effects for its intended audience of frustrated progressives, confirming them in their sense of powerlessness while also disavowing this sense via tropes of datedness (the references to the 1950s—a campy alien-invasion scenario, schlubby white guys with hats and slicked-back hair—distance us from the apocalyptic narrative) and exaggeration (the constant use of exclamation points underscores the satirical intent). The whole evokes a shared sense of horrified disbelief: how can this be happening?

Camp and satire aim at securing *disbelief*; it is as if Tom Tomorrow were running the machinery of the fantastic in reverse. That machinery was crafted in the age of colonialism, a "disenchanted" world in which the miracles and marvels of folklore had come into a certain disrepute while technical miracles and geographical marvels multiplied. As such, in the words of James O. Bailey, "the greatest technical problem facing the writer of scientific fiction"—we might say that this is true of the fantastic generally, as imaginative work operating "outside normal experience"—"is that of securing belief," or rather "securing suspension of disbelief."[1] What are the "credibility devices" by which fantastic narratives secure this suspension of disbelief?[2]

We need a general rhetoric of the fantastic, a survey of its "available means of persuasion," or as John Poulakos suggests, the ways in which it effectively "propose[s] that which is not."[3] For if we understand how the works of the fantastic persuade, we will understand a little more of how our general social imaginary is constructed, with—sometimes even in spite of epistemological rifts—its shared horizons of possibility and impossibility.

Novum/Antiquum

A first technique is the *novum* of Darko Suvin's theory, a term borrowed from Ernst Bloch's heretical Marxism. At first, it might not seem like a means of persuasion as much as a genre-defining feature: there simply is no fantastic without a *novum*, without a "strange newness" that differentiates the narrative world from the world as it is generally known to be, or more precisely, something "deviating from the author's and implied reader's norm of reality."[4] However, the recent public controversy over Margaret Atwood's *The Handmaid's Tale* underlines just how much assertive force a *novum* has: by projecting forced reproductive labor and sexual slavery into the future, imagined in the form of a "dystopia," and particularly, as Arrianna Planey points out, by imagining this dystopian regime as governing white women's bodies *only*, Atwood obscured the very "empirical reality" that Black women have actually experienced in the very country in which Atwood's novel is set.[5] A successful *novum* persuades the reader that what it names as strange and new really *is* strange and new according to the norms of a community of implied readers, a "we" for whom the novum represents an imaginary "deviation."

The *novum* has a shadow side: its newness often hides an oldness. Some of the early critics of SFF noted that its novae were often its most archetypal elements, reflecting the oldest human problems and desires. Whereas archetypes are surrounded by an aura of arcane mystery, however, the shadow of the *novum* also conceals the perennially, even intimately familiar, the "*heimlich-heimisch*" that lurks behind the appearance of the "*unheimlich*" or uncanny, in Freud's famous analysis.[6] In Jemisin's work, it is the new elements—orogeny, the Fulcrum, the Guardians—that most powerfully figure forth very real and all too familiar conditions, elements of the past that won't stay past. With Aaron Worth, we might call this disguised return of the old in the estranged form of the new the *antiquum*.[7] It is this *antiquum* that returns to view when we read in the opening pages of *The Fifth Season* that Alabaster, gathering in himself the power to destroy his world,

> reaches forth with all the fine control that the world has brainwashed and backstabbed and brutalized out of him, and all the sensitivity that his masters have bred into him through generations of rape and coercion and highly unnatural selection . . .[8]

We know very well, from the beginning, what real experiences, what historical harms, this is about.

The Other World

In a speech given to a conference at Columbia, Walidah Imarisha explained one of the benefits of science fiction for collective social imagination: "Rather than saying, 'Well, how do we *reform* that concept'"—e.g., the concept of carceral capitalism—"visionary fiction allows us to say, 'We don't even need to *engage* with that concept—we're going to go to an entirely different planet . . .'"[9] That is to say: visionary fiction, or the fantastic, can avail itself of the resources of the "Other World," the one in which our usual assumptions can be defied without too much irritable reaching after fact and reason. As Frank Cioffi notes, many such worlds are "flawed" (see *flawed utopia* and *dystopian homeopathy* below) in ways that reaffirm the laws of the normative world-for-(some of)-us, but there is a variety he calls "ascendant" in which the action is located "far beyond conventional values and social structures": this would be roughly equivalent in effect to the "immersive" Secondary Worlds of fantasy.[10] The disadvantage of this mode of rhetoric is that it might require too much effort for an already jaded and weary reader to believe in it; to assure its success, it must also draw on techniques of *worldbuilding*, as well as *vivacity, quiet assertion*, and so on.

The Known

Christine Brooke-Rose notes that "all types of fantastic . . . need to be solidly anchored in some kind of fictionally mimed 'reality.'"[11] The evocation of a relatively unproblematized concept of "reality" (and the sense of the "normal" it subtends) is already a moment of persuasion: is this real/normal? This device is sometimes laid bare in horror fiction—Stephen King's *Firestarter* or *Cujo*, for instance—in which it is quite often the mundane elements, rather than any naturalized or supernaturalized *novae*, that stand out as the most horrific: do people really talk and act this way, with this kind of sexualized cruelty, for instance? How "normal" is "normal"?

If it is more common for the fantastic to foreground the *novum* and background everything else, it is also quite common for fantastic works to make the "background" into the foreground, to turn society from a setting for private dramas into a primary object of display, so that customs, institutions, history, politics, economics, and more are the *novae* (see "Worldbuilding" below). Even within such a world-*novum*, the element of "*what-goes-without-saying*"—precisely what Roland Barthes called "contemporary myth"—may be at work, particularly in what concerns race, gender, ability, sexuality, class, and other such power-marked partitions of human experience.[12]

The fantastic is animated (or authorized) by the known in another manner as well: by *metonymy*, the associative extension of real signifiers into imagined domains. As Damien Broderick notes, the unfamiliar scenarios of William Gibson's *Neuromancer* (1984) are made intelligible by a series of references— to "present-day computing languages (glitches, virus, parameter), immunology and evolutionary theory (virus, mutate), literary theory itself ('mimetic subprograms')"—that place us in a relation of contiguity with the *novae*.[13] Note that the reader does not have to actually know anything at all about computing, biology, or literary theory for this metonymic strategy to work; these things just have to exist for the reader in a realm of the heard-about, in a foggy zone of uncomprehended, overheard discourse. The "known" is effectively, more often than not, the merely *believed*.

The Numinous

"Even I didn't understand the preternatural laws we observed," remarks the protagonist of Darcie Little Badger's "Nkásht Íí."[14] The scholar Rudolf Otto defined the numinous, that which passes understanding in the religious experience, as *mysterium tremendum et fascinans*, the mysterious thing that induces terror and exercises attraction—a phenomenon at the center of religion.[15] Caddy-corner to the *black box*, which holds out the possibility of an ultimate explanation in terms of scientific knowledge, is the scientifically inexplicable and unknowable, forever outside the *known*. In horror, this typically takes on the shapes of the sublime and uncanny, the *tremendum*; in fantasy, it can also assume the affective form of the *fascinans*. When magic appears as more than a mere imaginary technics (an alternative explanatory system for how things "work"), when we are made to feel *awe* at the ineffability of the magical, the numinous acts as another persuasive device: persuasion without explanation, inexplicability *as* that which elicits the reader's acquiescence.

Monstrosity

"Monsters," writes Travis Linnemann, "transgress and thus reveal and reaffirm the categorical boundaries of natural order"; thus, "the power to identify and name the monster, as well as to conceal one's own monstrosity, is a potent—perhaps the ultimate—political act."[16] What the fantastic text presents to us as monstrous and how it "monsters" them is of immediate political significance, as Linnemann makes clear in his analysis of the rhetoric used by police to justify their extraordinary powers and military weaponry: in the words of former FBI head James Comey, "Monsters are real."[17] Susan Stryker likewise points to the rhetoric of monstrosity in TERF rhetoric, dating from Mary Daly's early characterization of the "transsexual" as a male violation of female boundaries, a "monstrous" reproduction that mocks women's true/natural reproductive powers.[18]

The language used to construct the monster will necessarily invoke as its counter and anchor the normative "world-for-(some of)-us," what we will call *banality* (see below). From this position, by definition, the monstrous beggars belief, so when the story is narrated from the perspective of the mundane citizen, denizen of the world-for-us, a process of "discovery" and possibly of "confirmation" is usually necessary, Carroll argues.[19] Contra Carroll (for whom "belief" is not exactly what is at stake), this is precisely a matter of "securing suspension of disbelief"—of getting past the initial disbelief of the characters who witness the monstrosity, perhaps followed by the initial skepticism of the scientific or police authorities who are called on to confirm it, serving as a procedure of *simulated acquiescence* and potentially the revelation of a *flawed normal* (see below). Might it not be that similar processes must take place in order for people to seriously entertain the monstrous thought that the police may endanger us rather than protect us or that our entire industrial civilization has overstepped its ecological limit?

Vivacity

Seo-Young Chu argues that science fiction is "lyric mimesis," a yoking of the poetic energies that propel us out of the universe of the known to a narrative apparatus that generates imitations of reports on conditions in the world. The lyrical phrase "wings sprout/on the backs of the slave," a moment of startling imagistic energy in Octavio Paz's *Sunstone*, is a "'passing' fantasy."[20] When it appears in earlier forms, like Caesar Grant's folktale of the enslaved people who flew away from their enslavers, it lingers:

> There was a cruel master on one of the sea islands who worked his people till they died. When they died he bought others to take their places . . .
> One day, when all the worn-out Negroes were dead of overwork, he bought, of a broker in the town, a company of native Africans just brought into the country . . .
> There was among them one young woman who had lately borne a child . . . being very weak, and sick with the great heat, she stumbled, slipped, and fell.
> The driver struck her with his lash until she rose and staggered on.
> She spoke to an old man near her, the oldest man of them all, tall and strong, with a forked beard. He replied; but the driver could not understand what they said; their talk was strange to him . . .
> When the driver came running with his lash to drive her on with her work, she turned to the old man and asked: "Is it time yet, daddy?" He answered: "Yes, daughter; the time has come. Go; and peace be with you!" . . . and stretched out his arms toward her.
> With that she leaped straight up into the air and was gone like a bird, flying over field and wood.[21]

The device of *unknowing* (see *blackboxing* below) is used here to ward off any questions we might have: we are not privy to the conversation between the young woman and the old man (who spoke, we are told, in "an unknown tongue"—"My grandfather told me the words that he said; but it was a long time ago, and I have forgotten them").[22] Even the means of flight are left vague: the figurative language (the young woman flies "like a bird"; another man "was gone, like a gull, flying over field and wood"; "they all leaped up into the air with a great shout; and in a moment were gone, flying, like a flock of crows") might lead us to imagine wings, but we are ultimately only left to infer some kind of unspecified ancestral magic at work, powerful but distant.

The narrativization of the wings growing from the slave's back acquires a greater narrative depth in Nnedi Okorafor's *Book of Phoenix*:

> I reached behind my shoulders and felt the hardness and softness that was attached to me. I looked over my shoulder. As I did so, whatever was on my back flexed, I could hear it unfolding and stretching. It sounded like the branches of a leafy tree in the wind. It felt like such relief.
>
> With my peripheral vision I saw brown. I turned my neck as far as I could. Feathers. Wet brown feathers. I had *wings*. . . .
>
> Running was difficult with the wings. My wingspan had to be over thirty feet. I was stressed and couldn't help stretching them out, painfully smacking the alley wall. My head throbbed as I focused on my wings. I could see them extending out . . .
>
> When I heard the sound of a helicopter and saw the searchlight coming toward me, I tried my wings, and it was easy. The feathers had dried and all I had to do was imagine that I had another set of powerful arms. Powerful arms whose every curve, fold, muscle I could control. I could flex them, retract them, move specific parts. I ran.
>
> Then I flew for my life.[23]

Where the folktale deploys the device of *unknowing* to bridge between a present moment in which the story is being told and the chronotope of the narrated past in which marvels are real, Okorafor's science fiction not only provides rationalized explanations for the marvelous (see *extrapolation* and *metatext* below), it uses thick description to lend the lyrical image, in Chu's words, "the vivacity, solidity, persistence, and givenness that characterize the perceptible world of literal facts." In so doing, she equips the image with a "counterfigurative" force that transforms a "cognitively estranging referent"—the very notion of an escape from oppression—"into something kinesthetically recognizable that makes immediate sense to our tendons, muscles, nerves."[24] Let "vivacity" stand for the devices that confer this lived-in, experience-near quality upon the abstract, the vague, the difficult-to-imagine, the "elusive referent."[25]

Banality

"The oldest and perhaps the soundest method for securing suspension of disbelief," remarks Bailey, "is that of embedding the strange event in realistic detail about normal, everyday events."[26] Wells spoke of adding "touches of prosaic detail" to keep the "impossible" in close association with the actual.[27] It is the prosaic, banal quality that serves here to convince us that the *novum* belongs to the "world-for-us," or at least the "world-for-some-of-us" (Linnemann). Think of the grubby, cluttered sets of the *Nostromo* in *Alien* (1979): cigarettes, aimless banter, *Playboys*, and complaints about pay give us a Campbellian-Heinleinian "'lived-in' future" that aligns with our more cynical expectations, looking and feeling like an isolated oil rig circa 1979.[28] All is not well in this world, but violence remains latent, submerged, until the arrival of the Xenomorph (see *monstrosity* above).

Even the most gaudily fantastic scenarios rely on the banal: Superman flies not on the basis of any well-explained scientific principle but on the suspension of disbelief afforded by all the mundanity with which he is surrounded. His superheroism would have left readers quite cold if not for the architecture of banality undergirding it, from Metropolis and the Daily Planet building down to the homely telephone booth; these guarantee, in the words of Brooke-Rose, that the fantasy of the Man of Steel is "anchor[ed]" in a world that is recognizable because of its flatness, "not only to be as plausible as possible within the implausible, but to emphasise the contrast between the natural and the supernatural elements."[29] The power of *boredom*—not energy from an alien sun, but the energy of alienation itself—can be harnessed for antigravity devices.

Should it be surprising to find that banality is also part of the toolkit of "gravitic" politics? What else can we call it when, as Michael Hobbes points out, the most outlandish assertions (the pandemic is a plot by Bill Gates to implant us all with microchips, a cabal of reptile people control the world) are inserted into the unending flow of banal imagery (your old college roomie is married now, here's a funny cat video, check these pictures of celebrities looking old and unattractive) that constitutes social media? If social media "political marketing" is what "links distant high politics to the everyday," then even the *bizarrerie* of QAnon partakes of the fantastic resources of mundanity itself.[30]

The Flawed Utopia

We encounter a world of delight—an idyll of abundance, kind people, simple joys—only to find that "it was all too good to be true": here is the plot of *Forbidden Planet*, of "The Ones Who Walk Away from Omelas," and of innumerable other stories. The pleasures this narrative affords are at least two: it ingratiates our repressed longings for the things a utopia represents, and then, by returning us to the *known* and *banal*, it flatters us that we are too smart to be taken in by such false promises. Such is the authority of the rhetoric subtending Fisher's "capitalist realism" and Gray's "statist realism": only the misery of the present (even if magnified by a false futurism or disguised as a fantastic Other World) can truly be believed.

The Flawed Normal

A horrific twist on the flawed utopia: the *known* and *banal* themselves are revealed to have been a quasi-utopian illusion masking an unthinkable reality, a "world-in-itself" behind the "world-for-(some of)-us," to borrow Linnemann's terminology again. One thinks of films like *The Matrix* (1999), in which the illusion is distinctly *not* utopian but a carefully constructed appearance of ordinariness, punctured only by quickly erased glitches; to "wake up" from this dull dream is to be inundated with the abject imagery of dissolving mouths, wriggling parasites, and finally, the supremely horrific spectacle of the endless mechanical hives in which vital energy is extracted from human bodies. For Lacanians, such moments of revelation are analogous to intrusions of the unknown and unnameable Real, moments when the veil of the Symbolic is torn.

In this sense, Lacanianism itself is a kind of horror fiction. QAnon is a horror fiction writ large, not unlike John Carpenter's *They Live* (1988), in which the special sunglasses worn by the protagonists enable them to see the alien abominations hiding in plain view. As a rhetorical appeal, the *flawed normal* gratifies us with the privileged sense of being the all-seeing kings of a land of the blind—not so much offering a critique of authority as an inversion of it.

Dystopian Homeopathy

Roland Barthes once described "a kind of homeopathy" at work in popular narratives by which a small or weakened dose of criticism aimed against the existing system could paradoxically serve to bolster and defend that system.[31] What is sometimes called the "critical utopia" is a similar strategy, turned this time to defend a non-existing system: a utopia, e.g., Le Guin's Anarres, is protected in advance from the suspicion that it is another *flawed utopia* by its very openness to critique. It is as if it has been inoculated with the *banal* (even in this workers' paradise, there are petty grudges, occasional fistfights, stodgy old professors, repressive psychiatrists!) so that we may permit it to take us beyond the *known* (the carrot-and-stick system of motivations that defines our world, that we assume to define the universe itself). It is a utopia that is paradoxically healthy because it is already sick with the germ of dystopia (a creeping conformism, a "social" rebirth of authoritarianism).

But this device can work in reverse as well. As Fisher points out, scenarios that are dystopian on the face of it—he seems to be thinking here of post-apocalyptic SF—serve as "narrative pretext for the emergence of different ways of living."[32] In so far as we are willing to believe in death, in the ultimate breakdown of anything and everything, and in so far as our narcissism encourages us to imagine surviving it, we can thereby be afforded a little dose of utopian reconstruction. This can take the form of a white supremacist utopia like *Farnham's Freehold* or the Black utopia of Butler's Earthseed. All that is necessary is that the imagination be violently dislodged from the presupposition that everything will necessarily continue pretty much as it seemingly always has.

Extrapolation

John W. Campbell, Jr. is credited with having popularized the concept of "extrapolation" in SF, beginning with a 1947 manifesto, "The Science of Science-Fiction Writing": "To be science fiction, not fantasy, an honest effort at prophetic extrapolation of the known must be made." For instance, "ghosts can enter science fiction—if they're logically explained, but not if they are simply the ghosts of fantasy."[33] The concept thus has two components, the first of which is borrowed from statistics (the making of inferences about the future from "known" present trends, or as Isaac Asimov called it, the "If this goes on—" story), the second of which is concerned with "logical explanation" (a more retroactive rationalization in order to, as H. G. Wells put it, "*domesticate* the impossible hypothesis").[34] These components are not identical and not even always compatible, but both do exercise considerable force as "credibility devices."

The extrapolative argument that a *novum* is possible because it is an extension of tendencies already present within the *known* lends itself to dystopian and apocalyptic scenarios (if this goes on, something terrible will happen), but it is also a close cousin to the grotesque and to satire, allowing us to entertain a ridiculous scenario. Rick Santorum's infamous comments on homosexuality ("If the Supreme Court says that you have the right to consensual (gay) sex within your home, then you have the right to bigamy . . . You have the right to anything," including, purportedly, "man on dog" sex) operated on the same rhetorical basis as Kurt Vonnegut's much-anthologized "Harrison Bergeron" and the hit film *Idiocracy*.[35] But these belong more properly to a logic of *hyperbole and inversion*.

Hyperbole and Inversion

SF, in particular, has a strong historical connection with Enlightenment satire, and it often returns to its satirical roots, as Joanna Russ points out, by way of *extrapolation*.[36] Here, persuasion operates precisely by straining suspension of disbelief to the breaking point so that the impossible is made actual (and, to a certain extent, rationalized, made "cognitive") while remaining ridiculous and, therefore, unacceptable. The inversion of values, the representation of a world turned upside down, is another satiric inheritance: "Erewhon's inhabitants consider crime a disease and disease a crime . . . [Samuel] Butler's strategy succeeds strikingly in making the point that what passes for moral superiority is often mere self-congratulation for our good luck at being born and raised in the right circumstances."[37] Keith Knight's Netflix series, *Woke* (2020–2022), uses the fantastic—inanimate objects start actually addressing the protagonist, a Black cartoonist whose drawings give human faces and voices to inanimate objects—to indict a system that treats him as less than human.

Retrospection

One way for narratives to convince us that something seemingly improbable or outlandish *will* happen in the future is to simply announce, from the future's perspective, that it *did* happen.[38] This technique takes advantage of the retroactive action of narration, which, as Sartre says, begins at the end: "the end is there, transforming everything" so that all the narrative's events are granted a meaning, a retrospective teleology.[39] Thus, in the pseudo-preface to *How We Shall Bring About the Revolution* ("To the Readers of this Book"), Émile Pataud and Émile Pouget playfully announce, in voices from the future:

> At baptism, our book changed its name. This was the fault of our publisher, who, in presenting its title page to the printer's ink—the baptismal font for books—shamelessly committed an act of sabotage.
>
> Not being of a morose disposition, we bear him no ill-will on this account, and we plead his cause with you; like us, you will grant our publisher a free pardon.
>
> And yet, the sabotage is obvious!
>
> In place of the anachronistic title which appears on the title page, there should have blazed forth, in three lines.
>
> "HOW
WE BROUGHT ABOUT
THE REVOLUTION."
>
> Such was the title, this book of ours should have borne.
>
> For, you all know it, the Revolution is over!—Capitalism is dead.[40]

The games of narrative contract here are surprisingly complex: not only is the authorship fictionalized (the pronoun "we" here does not refer to the empirical authors, Pouget and Pataud, writing from their present), but the readership (the "you [who] all know it") is as well. We are invited to *play along*, to pretend to be the ones who know how it happened, so that the narration has the character of anamnesis, of remembering what we already know.

This technique was put to political use a few years ago in a post to the Puget Sound Anarchists website, which reads as if it might have been the product of one of Imarisha and brown's activist SF workshops: "A Letter of Solidarity from the Year 3017." With a millennium of hindsight, "a collective of free mutants . . . from the future you have helped to safeguard" honors the work of the antifracking activists of 2017, telling them: "From our perspective one thousand years in the future, however, sitting as we do among the massive trees and fungal gardens and meteor scars and stone shrines which now overlay this train track you are blocking . . . Every single action mattered."[41]

Simulated Acquiescence

Alternatively, in fantasies wherein the point of view does *not* belong to the world of the *novum*, the fantasist may invent somebody we take to be a "veracious person," someone we are ready to identify with and hence to believe so that we come to accept as real what they come to accept as real. This is easiest to locate in the kinds of narrative Mendlesohn calls "Portal-Quest Fantasy": since both we and the protagonists are newcomers to Narnia or Oz, we are "tied to the protagonist, and dependent upon the protagonist for explanation and decoding" of this new world.[42] Just as their ignorance is a mirror of ours, their acquisition of fictive knowledge is duplicated in our acquisition of fictive belief. Damion Kareem Scott points to the film *Black Panther* as an example of this device used to bring white audiences to realize with astonishment, along with the purportedly veracious CIA agent Everett Ross (Martin Freeman), that the Wakanda to which he has been assigned is not a "backwards nation of 'herders and farmers'" but "the most technologically advanced nation on the planet."[43]

Blackboxing

Another way in which the power of ignorance can be tapped is by exploiting Clarke's Third Law: "Any sufficiently advanced technology is indistinguishable from magic."[44] That is to say, it is indistinguishable from magic for those whose ken it is beyond. Thus, we are ready to accept the transcendence and transfiguration of the last sequence in *2001: A Space Odyssey* because we have been prepared for it by the "Dawn of Man" sequence: just as the nearly *numinous* technical artifact from an unthinkably distant future (the Monolith) is radically beyond the ken of the primitive protohumans, it is radically beyond the ken of the astronaut who is ultimately only its terrified passenger. The Monolith is, as Anderson puts it, "Special Equipment"—but so is the eponymous One Ring of *Lord of the Rings*, an artifact of terrifying and very much *numinous* power from an unthinkably distant past.[45] We can easily recognize here the phenomenon that Latour calls *blackboxing*, a device whose political uses we have already noted.

Worldbuilding

Entire books have been written about the arts of worldbuilding. Let's delegate the work of detailing the many strategies of worldbuilding to Mark J. P. Wolf's book on the subject. Wolf cites "invention," "completeness," and "consistency" as the three major strategies by which the essentially and necessarily incomplete Secondary Worlds of "immersive" fantasies are made convincing. In Jemisin's Broken Earth Trilogy, the sheer density and coherent interrelation of her "*nominal*" inventions ("comms" instead of towns and "geomests" instead of geologists; wholesale inventions such as "roggas" and "orogeny"), "*cultural*" inventions (the creation of an entire disaster-resistant civilization founded on the "stonelore" of survival, "use-castes," and other such institutions), "*natural*" inventions (the geography of the Stillness, with its distinctive and often disturbing fauna), and "*ontological*" inventions (the counterfactual physics of orogeny and magic) do a great deal of the work of making her universe persuasive: seemingly solid, persistent, given.

Successful worldbuilding can also persuade by the way it affords the reader pleasures of fictive knowing and imaginary belonging. In particular, it can persuade the reader to want this world *not to be inconsistent* and *not to change*. In this respect, worldbuilding can operate rhetorically in a manner parallel to character (see below).

Retconning

When changes to a fantastic "universe" accumulate to the point that they no longer accord with one another, threatening the overall narrative continuity, the imposition of retroactive continuity (retconning) restores the appearance of coherent narrative by explaining away the inconsistencies. As Joshua Clover points out, something of this retroaction is at work even in the most ordinary sentence (the ending of which retroactively determines the meaning of all the preceding components),[46] but we might point to a more specific example of retconning in what is shaping up to be the single most significant work of science fiction of the early twentieth century: namely, the ever-evolving, participatory fan "text" that is QAnon. This construct exists to smooth over the erratic behavior of Donald Trump, to assure his fans that no matter how ostentatiously incompetent, crass, irrational, or self-dealing he is, it is all part of "The Plan," a grand narrative in which Trump is the agent of cosmic justice and salvation.

Character

Like Douglas Adams' "Electric Monk" (a laborsaving device that does the hard work of believing things you can't be bothered to believe), a focalizing character can learn the world of the *novum* (see *worldbuilding*, above), and so do the work of conversion to belief on the reader's behalf. A "veracious person" (see *simulated acquiescence*, above) can serve as a handy black box (see *blackboxing*, above). In addition to these, character can act as a rhetorical brake on other possible desires: character is that which *continues*, and our investment in the continuity of characters in fantastic serial fiction—Jean-Luc Picard, for instance, or Superman, or Lije Bailey—can present an overriding motive, a means of inducing us to wish that nothing should threaten to change the character (or our identification with them). Trump is, in this sense, a character, an emotional vehicle by which an everchanging series of bizarre, shifting, and erroneous beliefs (wind turbines emit fumes, centrist Democrats are secretly far-left radicals, etc.) can be transmitted to a fan base that will sooner forget any contradictions than endanger their relationship with Trump.

Quiet Assertion/Spectacular Assertion

The signal device of the fantastic from the mid-twentieth century on has been to dispense with the prophylactic apparatus of explanatory introductions, voiceovers, asides, "expository lumps" or "infodumps," and so on, and to simply immerse the reader in the assumptions of the Secondary World, presented as indirectly as possible, endowed with a maximum of "givenness" or taken-for-grantedness.[47] W. R. Irwin calls this the method of "quiet assertion" as opposed to the "spectacular assertion" or "fiat" of which Kafka's "One morning, upon awakening from agitated dreams, Gregor Samsa found himself, in his bed, transformed into a monstrous vermin" is an exemplar.[48] The power of quiet assertion must not be underestimated: when states of affairs that have never existed and perhaps even *could never exist* can be verbally handled in the same manner as "*what-goes-without-saying*," we are in the presence of a profound instance of "contemporary myth."

In the political arena, spectacular assertion is reminiscent of Goebbels' Big Lie theory, which uses the power of repetition to turn fiction into seemingly established fact (*Oceania has always been at war with Eastasia*). But when is an assertion "quiet"? For instance, what should we make of something as brazen as Mick Mulvaney's defense of Trump's attempt to extort Ukraine into providing dirt on his political opponent, asserting, at a press conference, that "We do that all the time"—i.e., that United States diplomacy not only *does* but *always has* operated purely as a political arm of the governing party?[49] It elicited cries of indignation from liberals loyal to a system that could still pretend that *we don't do that*. Quiet assertion is thus not only a question of lying; it is closer to gaslighting, the deliberate destabilization of consensus reality. The power of quiet assertion, even when it is enunciated at a public podium under the brightest lights, is that it instantly resets the rules of the truth-game.

Megatext

Another mark of the mature fantastic, already available in the early twentieth century, is the use of "off-the-shelf" *novae*, impossibilities that have been pre-domesticated, as it were. No one needs to invent a fictive anthropomorphic automaton when Karel Čapek has already done it for you: Isaac Asimov is then free to embellish on the imaginary technics of the "robot." Damien Broderick calls this vast warehouse of fictional goods "the sf megatext."[50] Even when the serial numbers have been filed off, the fact that readers can be assumed to have already encountered, say, telekinesis means that they will be less likely to balk at "benders" (*Avatar: The Last Airbender*) or "orogenes" (the Broken Earth Trilogy). In effect, once the "inventions" of worldbuilding enter the megatext, they become Latourian "black boxes," compact parcels of "accepted [fictive] fact" that require no special authorization for readers who are already pre-immersed in the "supergenre" that is co-constituent with the megatext.[51] Conversely, as Anderson points out, megatextual items can serve, in effect, as pieces of the "known" against which the *novum* can assert its reality: in the midst of SF's "Golden Age," characters *in* SF novels like Clarke's *Sands of Mars* (1951) secure our belief in nonexistent technical marvels by *contrasting* them with the rockets-and-rayguns "cheap fiction" of the Pulp Era.[52]

It is to another kind of megatext that Donald Trump habitually refers: the one that journalist David Roth calls "the Fox News Cinematic Universe." This powerful narrative construct, a world of "alternative facts," provides much of the necessary material with which viewers fill in the gaps of his notoriously fragmentary sentences—or make sense of other discourses of the rightwing media sphere, e.g., the ramblings of Laura Ingraham.[53] Conversely, the legislative construct that Alexandria Ocasio Cortez and Ed Markey branded as "the Green New Deal" can be read as an attempt to modify a term already available in the public megatext (the New Deal itself) to add to that megatext, much in the same way that Čapek added "robot." The Green New Deal is thus a rhetorical "robot" of sorts, another Electric Monk that does some of the work of believing in things that human beings are too weary and beset to believe in.

Mimicry

One of the oldest tricks in the book: rather than devise, in defense of a proposition, an appeal to reason (this technology would work as described because . . .), one simply imitates the rhetoric of rationality already in circulation. In other words, make noises like a scientist or enough like one to pass the reader's eminently faulty Turing Test. This is, of course, also the standard procedure of pseudoscience, which is one reason why, despite its sometime aspirations to "hard" scientific rigor, SF is continually drawn to pseudoscientific concepts, so much so that Kevin Young suggests it is really connate with the modern practice of hoaxing.[54] If, as Edward James argues, "Sf has unwittingly given birth to a number of cult beliefs"—indeed, so much so that "sf's greatest impact on the twentieth-century world may be indirectly communicated by those people who present science-fictional ideas as fact"—this is, in effect, a mimicry of SF's mimicry of science, discourse repeating discourse like a trapped echo. Elements of the megatext, such as zombies, ESP, and UFOs, have been taken as blackboxes, treated as "accepted facts," part of the known. The very engine of purification on which hard science fiction operates produces monstrous hybrids, a vast assortment of "sf presented as fact."[55]

Implication

"'What,' Poole said, 'is an "electric ant"?' But he knew; he could decipher the term."[56] In Lem's *The Futurological Congress*, Professor Trottelreiner proposes that the future can be studied by inventing new words out of old ones and inferring what as-yet nonexistent realities these words might refer to.[57] This game of inference and reference bears a strong resemblance to the way the fantastic operates at the level of the sentence—not only in the form of the neologism as such but in the perceived incongruity between a given word and its context. When Philip K. Dick's "The Electric Ant" tells us that "At four-fifteen in the afternoon, T.S.T., Garson Poole woke up in his hospital bed," we, too, can "decipher" the unfamiliar acronym: because we would expect to find something like "E.S.T." (Eastern Standard Time) or "C.S.T." (Central Standard Time) in that location in the sentence, using a few other contextual clues, we can infer that T. S. T. stands for Terran Standard Time.[58] As the semioticians say, Dick has inserted an unexpected, unfamiliar term into a paradigm (the category of time zones) within a familiar syntagm (a report of the time). Later, when we are told that Sarah Benton is "watching a captain kirk on the

TV," we are momentarily alienated from the story, reorienting ourselves as we recognize that the lowercasing of the proper noun "Captain Kirk" suggests that what once was a particular character in a particular TV show has since become the name for a type of TV show, an entire genre.[59] As Marc Angenot observes, a "rhetoric of credibility" is at work here: the technique "aims at having the reader believe not so much in what is literally said as in what is assumed or presupposed": a world in which travel to other planets *just happens to be* a reality, in which captain kirks *just happens to be* a TV genre, and in which the phrase "electric ant" *just happens to refer to* "an organic robot."[60]

Incantation

Rather than mimicking the sounds of science, fantastic works of many genres can persuade by imitating the style of liturgy and ritual. Sometimes, this takes the form of actual fictive religions, such as the Bene Gesserit Way in Frank Herbert's *Dune*, with its famous "Litany against Fear": "*I must not fear. Fear is the mind-killer. Fear is the little-death that brings total obliteration. I will face my fear . . .* "[61] As another minimal example, we might take the microgenre of fantastic "instructions"—which, in addition to providing the title of a graphic novel by Neil Gaiman and Charles Vess, typifies the productions of Instagram account HGK477:

DURING A BLACKOUT

1) Do not rush for a flashlight or a candle. Just stay calm.
2) You may feel like you are being watched. You are. But do not worry, they are not here to harm you. They just got a bit overly excited.
3) Sometimes, during a blackout, a single TV will turn on. Turn it off.
4) You will often hear voices. If you focus you can make out what they are saying. Feel free to strike up a conversation if they seem friendly.
5) Remember, nothing will hurt you during a blackout unless you panic.
6) This is always a nice opportunity to go outside and look at the stars. Do not try to find constellations though. The ones you know are not there . . .

The incantatory quality of lists like this one is, in fact, what produces their otherworldly, fantastic quality. It partakes of the serious spirit of play, as Johan Huizinga describes it, producing a magical space of Keatsian "negative capability" in which any questioning of the rules is suspended.

As a "credibility technique," incantation is not at all confined to fantasy or horror; we can see it at work in Jennifer Egan's Twitter story, "Black Box" (with its numbered list of second-person instructions) and, indeed, in Godwin's "Cold Equations," with its hypnotic repetition of "the law" (eleven instances): "It was the law, and there could be no appeal."[62] The law, the law, the law, the law, the law, the law, the law, the law, the law, the law, the law.

■   ■   ■

It should be clear from this (inevitably too-short) catalog that the study of the fantastic has many lessons to teach us about the way that rhetoric can "propose that which is not." What may not be readily apparent is how often, how profoundly, that rhetoric is deployed in order to convince us, as a German philosopher might put it, that what is real is what is rational and that what we already think impossible is *really* impossible.

6

Rhetorics of the Impossible

So they fall asleep, Innon snoring on his belly between them and 'Baster and Syen with their heads pillowed on his big shoulders, and not for the first time does Syenite think, *If only this could last.*

She knows better than to wish for something so impossible.
—N. K. JEMISIN, *The Fifth Season*

Taken in the abstract, it might seem as if science fiction and fantasy (if not horror) would always and necessarily be emancipatory genres: aren't they distinguished from realism by their exploration of sheer possibility, unencumbered by facts and necessities? Of course, this does not describe most works of the fantastic at all; utopias are relatively rare and fleeting. But why?

Consider the twentieth-century form of the superhero comic: here is an entire class of people for whom ordinary human limitations no longer apply, but their field of action is necessarily curtailed. While Superman, as an alien outsider, is, as Umberto Eco called him, "the character whom nothing can impede," narrative constraints dictate that all his extraordinary powers can never alter the mundane world in any substantial or even noticeable way[1]—not only because a world in which he has succeeded in vanquishing all crime, evil, and suffering, a utopia in which "They don't need Superman any more," would be "intolerable" because deprived of a rationale for further Superman stories,[2] but because such a world would no longer be *ours*, or rather, an "'overlaid world' ... [in which] the cityscape, culture, language, and politics are largely unchanged by the activities of super villains," "the world outside our windows" as *supplemented* by "the existence of superheroes, magic, aliens, and the like."[3] In these stories, the completely different person (the alien Kal-El) is not only compelled to disguise himself as the same (the normal Clark Kent), he is *forbidden to make a difference*, to disrupt the smooth surface of Sameness—or even to fail to restore it after every irruption of the different. Normativity appears as an inescapable force, even though the "irritation" of plot pushes against it: in the words of David Graeber, "The world is returned to normal until the next episode when the exact same thing"—the disruption, the anomaly, the precipitating violation that sets plot back into motion—"happens once again."[4] For the man who can do the impossible, the impossible man, it is impossible to really *do* anything.

This chapter, accordingly, proposes a freeform exploration of the role of the "impossible" in works of the fantastic, considering examples old and new, taking into account 1.) Kathryn Hume's concept of the "subtractive world" in genres ranging from New Wave fantasy to classic horror and the post-apocalyptic, with special attention to their strange geometries; 2.) the strange binds in which time travel narratives are wound up; 3.) the strange place occupied by figures of disability in SF; 4.) the strange place occupied by other kinds of "impossible people," notably those marked as Black and/or queer; 5.) the strange ontology of Blackness in works of the fantastic, as marked by a "brokenness of ~~being~~"; and 6.) the strange ways in which New Black Fantastic writers turn the impossibilities of race, sexuality, and disability inside out to find new possibility, new utopian visions.

Subtractions

The physicist Richard Feynman once flippantly defined gravity as "the thing that holds you in your seats," then added: "Actually, that's a combination of gravity and politeness."[5] "Politeness" might be described as another kind of gravity, the kind Samuel R. Delany called "gravitic value systems": all the moral, aesthetic, economic, and political hierarchies that organize the universe into upper and lower realms.[6] We often think of science fiction and fantasy as offering an escape from gravity in the sense of both social and physical forces. Let gravity stand here as a figure for all that appears as inescapable, imposed as iron necessity, as when politeness *demands* that we hold our tongues—or the police do. Gravity: the very substance and force of the normal. Pushing against gravity, stories of sorcery or starships seem to propose a contrary figure: these are the genres of the *impossible*—the currently impossible but perhaps possible in the future ("*events that might happen*") and the permanently impossible ("*events that will not happen*").[7] How, as such, could these stories fail to be on the side of all the "impossible people who cannot exist, cannot be seen, cannot be classified, and cannot fit anywhere" against the Law that oppresses those it exiles as well as those it contains?[8]

Yet the figure of impossibility plays other roles in these imaginative forms. The fantastic is repeatedly imagined as "the impossible," that is, "the unreal, the nameless, formless, shapeless, unknown, invisible"; if you just systematically *eliminate* qualities from realism, you get the fantastic.[9] If, in the totally administered world within which the fantastic is born, "all things that live are subject to constraint," the fantastic seeks to turn this condition into a means of escape, so that "constraint . . . inspires."[10] Kathryn Hume speaks of "the creation of new worlds by subtraction and erasure": the omission of everyday, mundane detail makes the "subtractive worlds" of heroic fantasy possible, while the erasure of ordinary logic or causal sequence produces Alice-Through-the-Looking-Glass effects.[11] Just because it transgresses limits, as Rosemary Jackson observes, "fantastic narrative is *preoccupied* with limits."[12] Indeed, it often *invents* limits, invents *by* limitation. In this respect, it is dreamlike—like actual dreams, that is, rather than the "dreams" we invoke for rhetorical purposes. In common parlance, "dreams" are indeed immune to gravity, for better (as in "I Have a Dream") or worse (as in "mere utopian dreams"). In actual dreams, what is perfectly possible in mundane, everyday life *becomes* impossible: crossing a finite space by foot, for example, turns out to be an *event that will not happen* because an infinite series of interruptions, obstacles, and distractions bar the way, so that—as Ursula K. Le Guin's fictional physicist puts it, cribbing from Zeno of Elea—"it doesn't matter how far [you've] gone . . . [you] always [have] to go half of the way that's left to go."[13] Or as the Imperial Messenger finds out in Kafka's parable: "If he could reach the open fields how fast he would fly, and soon doubtless you would hear the welcome hammering of his fists on your door. But instead, how vainly does he wear out his strength; still he is only making his way through the chambers of the innermost palace; never will he get to the end of them . . ."[14]

Or one finds, like the protagonists of New Wave fictions such as J. G. Ballard's "Build-Up" (1957) and Josephine Saxton's "The Wall" (1965), that one is trapped in a space that is mysteriously closed in on itself, so that the very supposition *if one could reach the open fields . . .* is rendered nonsensical:

> "There must be free space somewhere," M. said doggedly. "The City must have bounds."
>
> "Why?" the surgeon asked. "It can't be floating in the middle of nowhere. Or is that what you're trying to believe?"[15]

Franz M. (the name a tribute to Franz Kafka and his acronymically-named protagonists), an unhappy denizen of a vast underground city of the future, has just disembarked from an epic journey, hoping to take a train to the end of the line and thus to come to the city's limit, to "free space."[16] Instead, he seems to have made a vast circle in space—boarding a train headed west, he finds himself inexplicably headed east—and in time:

> Then the calendar on the desk rivetted his attention.
> The date exposed on the fly leaf was the 12th August.
> That was the day he had started off on his journey.
> Exactly three weeks ago.
> Today![17]

But even Franz M. is a degree freer than the unnamed fair-haired "man in the north" and dark-haired "woman in the south" in Saxton's "The Wall."[18] Their longing for one another, thwarted by the titular wall ("thirty feet in height, and . . . very ancient in its stone, dark blue, hard, impenetrable," topped by "rows of dreadful spikes which curved in every direction, cruel, needle sharp, glassy metal rapiers set into green bronze"), defines their situation so utterly that they can think of no other direction to go than "away":

> One day the man began to think that he could not stand it any longer . . . He suggested to the woman that they should part. He explained that the idea had come to him that there might be other lands where a person might live, over the horizon, away to the north and south, things they had neither of them dreamed of, other loves perhaps, other climates and better food. He felt then that anything would be better than to sit here forever just yearning for something that could never be had.[19]

Reluctantly parting, parted by the east- and west-stretching wall through whose one chink, like Pyramus and Thisbe, they have met and formed their attachment, they each encounter their opposite numbers (a fair-haired woman for him, a dark-haired man for her), experience "full physical contact with others," and are renewed in their desire. As if exemplifying Freud's death drive, the two climb the wall and are killed, but not before witnessing a horror:

> It was then that they noticed all the other lovers impaled on the spikes. Some were long-dead skeletons, dry and dusty, grinning skull to skull; some were mummified by the keen wind, eyes sunk in perpetual bewilderment; and some were rotten and new, astonishingly, quite new. . . .

And very quietly they kissed as they clung and died there, impaled across the cold spiky barrier, feeling and thought growing more feeble every second.

In the north and in the south a fair haired woman and a dark haired man set off slowly to walk towards the wall, love stirring in the innermost recesses of their being.[20]

The singularity of their desire is doubly erased, first by this vision of the other victims and then by the revelation that their particular tragedy is going to be repeated, perhaps indefinitely, as perhaps it already has. If escape from space-as-property is the impossible for Ballard's Franz M., the impossibility under which Saxton's man and woman (or men and women) live and die is something like Lacan's "*il n'y a pas de rapport sexuel* [there is no sexual relationship]": sex itself, rather than founding the family and thus civilization on a stable basis of relationship, *rapport* ("The life of human beings in common ... had a twofold foundation, i.e. the compulsion to work ... and the power of love"), represents an intractable rift, a senseless but irremediable split in the universe of the Symbolic.[21]

Or one finds, like the protagonists of Jeannette Ng's neo-Gothic *Under the Pendulum Sun*, that the scenery itself shifts around unpredictably: "Distances are unreliable in Arcadia ... They sometimes like to pretend it can be measured in miles or hours travelled, but it's far less predictable than that. I've had distances given to me in numbers of daydreams and revelations, as though I'd only arrive somewhere after I've had an epiphany ... "[22] All social relationships are stricken by forms of impossibility: unbreakable curses, unanswerable questions, unwinnable arguments. Immeasurable distances, unspeakable secrets, incomprehensible customs, and tongues: Arcadia is the mundane adult world as seen from the perspective of a child, but also perhaps the masculine public sphere as seen from the perspective of a middle-class Victorian woman. "What *cannot* you do, Laon?" the missionary's sister screams in frustration. "Have you not done it all? Have you not gone to university? Have you not left England?"[23] Rather than slipping the surly bonds of Earth, Cathy and Laon Helstone have entered a disabling realm, a place where they are capable of dramatically *less*. This is not out of keeping with the fairytale tradition on which Ng so nimbly plays; as W. H. Auden points out, even the frequent motif of choosing (which of the King's three daughters shall you marry?) masks the constant of constraint, for "what appear to be the personal choices of the characters are really the strategic choices of the storyteller, for within the tale the future is predetermined."[24] Even after all the seductions, revelations, and disillusionments of their experience in Arcadia, the Helstones remain dedicated missionaries, faithfully pursuing—*I know very well, and yet*—the expansion of Christendom and Empire, the realm of the normal.

Horror narrative, of course, is also typically oriented toward restoring normality, the physical and moral order violated by an intrusive (if also never-absent) evil, the irruption of the monstrous. Slavoj Žižek suggests that monstrosity is a figure of *excess*, a "slimy residue of the alien" or unassimilable remainder of the process of symbolization by which we make the world intelligible to ourselves, but it can also be understood as a *subtraction* from the existing, produced by deletion (of solidity, of intelligibility, of conscience) and hence posing exceptions to its rules.[25] It is often precisely by *disabling* (disabling the reader no less than the monster) that what Julia Kristeva called the "powers of horror" unfold themselves.[26] Monsters are, as Carroll writes, "impossible beings": as the underside of Jackson's catalog of the fantastic, they are unacceptable, unspeakable, unnamable, unthinkable, even as they are also undying (because undead), ineliminable, never-not.[27] In the inverted theology of horror, immortality, freedom from the ultimate limitation placed on life, is the most terrible form of enslavement and degradation. It is impossible for the impossible being, Lovecraft's "thing that should not be," to *not be*.[28]

We can observe something similar even in the fantastic genre seemingly farthest removed from mythic and folkloric origins, "the premiere narrative form of modernity": science fiction.[29] "Unlike fantasy," writes Darko Suvin, "in SF, the amazing aspects of the story had to lie within the bounds of what was *possible* according to the standards of knowledge current in the author's world."[30] As "fantasies of possibility," to use H. G. Wells' expression, "tak[ing] some great creative tendency, or group of tendencies, and develop[ing] its possible consequences in the future," science fiction might seem an exception to the rule.[31] But how many SF narratives are premised on—indeed, can be generated by—simply suppressing some aspect of the actual world as we take it to be? SF often proceeds by *eliminating* possibilities, by inventing impossibilities where possibilities might have been. "Catastrophe," Berlant reminds us, quoting from Colson Whitehead's *The Intuitionist*, "is 'what happens when you subtract what happens all the time.'"[32] This form of the subtractive world is perhaps most visible in the postapocalyptic and dystopian subgenres that have taken such a prominent place in US mass culture in this century, where the possibilities afforded by high technology and/or liberal democracy are subtracted. The ad copy on the back cover of the first collection of the long-running *Walking Dead* comics series makes this subtractive element (and its fantasy content, in the Freudian sense) plain:

How many hours are in a day when you don't spend half of them watching television? When is the last time any of us REALLY worked to get something we wanted? How long has it been since any of us really NEEDED something that we wanted?

The world we knew is gone.

The world of comfort and frivolous necessity has been replaced by a world of survival and responsibility. An epidemic of apocalyptic proportions has swept the globe causing the dead to rise and feed on the living. In a matter of months society has crumbled, no government, no grocery stores, no mail delivery, no cable TV.

In a world ruled by the dead, we are forced to finally start living. (Kirkman and Adlard)

Here, subtraction operates in a manner not unlike Border Crisis discourse (see Ch. 1) and very close to what Freud describes in *The Interpretation of Dreams*: what is apparently a "disagreeable" scenario (a lack of groceries, difficulty getting one's mail, the constant assaults of mindless zombie hordes) disguises a "wish-fulfilment" of self-transformation, of revitalization ("we are forced to finally start living").[33] The wish to be "forced" to fulfill one's own wishes can also be detected in the scenario of *V For Vendetta*, both in its original comics-series form (by Alan Moore and Dave Gibbons, 1982–1989) and in the far more popular film adaptation (dir. James McTeigue, 2006). As Isaac Butler notes, the hyperauthoritarian England of Moore's imagining operates as a framing device, a backdrop against which V's (and Moore's) antiauthoritarianism (and the program of bombing, sabotage, and assassination by which V pursues it) can be made to appear more palatable and viable—a propaganda for "radical anarchy" that perversely succeeds by subtracting all other options.[34] It is Evey's false imprisonment (chapters 10–12) that radicalizes her, allowing her to "become transfixed . . . become transfigured . . . /forever."[35] V has, in fact, restaged Plato's Allegory of the Cave, casting Evey as the slave who breaks free of an illusory imprisonment to discover the truth of her own freedom—that ur-SF narrative of emancipation so powerfully echoed in the Wachowskis' *The Matrix*.

Time Paradoxes

Time travel stories might tend to exemplify the process whereby fantasy *constructs* an "impossible," presenting narratives that explain why, even if you get your wish, you also *don't* get your wish—a classic fairy tale pattern. Some of the oldest myths, after all, are about the (ultimately thwarted) desire to overcome time: one wishes for a lover who never dies, but time shrivels him into a cricket (Ovid), or one sets out to kill Death (Chaucer) or cheat Fate (Sophocles) but meets it on the very road one takes to escape it. So, too, the unlucky protagonists of Robert Silverberg's "What We Learned from This Morning's Newspaper" (1972): residents of a suburban neighborhood, receiving the unexpected "gift" of the delivery of *next* week's "Gray Lady" (the *New York Times*), use their foreknowledge to enrich themselves, placing winning bets and selling stocks at their peak, until one of them finds her sister's obituary and, ignoring the obligatory warning—"Your sister's name is already on the roll of the dead. If you interfere now it'll only bring unnecessary aggravation to her family and it won't change a thing"—tries to change the future, only to find that this creates an "entropic creep" destroying the fabric of reality itself: "I don't want to live here anymore I want to cancel my newspaper subscription I want to sell my house I want to get away from here back into the real world but how how I don't know it's all gray gray gray everything gray nothing out there just a lot of gray."[36] The impossibility of changing the future is restated in an almost classical form, as the grayness of midcentury American life horrifically acquires the permanence of myth.

Isaac Asimov's own *Foundation* trilogy might seem to be a successful wish-fulfillment fantasy about exercising mastery over the future (and overcoming "anarchy") through the scientific clairvoyance of Hari Seldon's "psychohistory": at a certain "delicate moment in history," Seldon explains, "the onrushing mass of events must be deflected just a little to remove twenty-nine thousand years of misery from human history."[37] Nonetheless, even this more modest fantasy, one part Gibbon's *Decline and Fall of the Roman Empire*, one part Hegelian-Marxist teleology, is mortgaged to an (American) exception: "You know how risky it is to introduce the vagaries of an individual in the psychohistoric equations."[38] All that permits the trilogy to unfold, once again, is a narrative "irritation"—not only in the form of the actions of individuals threatening to derail the Plan but in the form of the decay and collapse of the Empire itself, the "barbarism" which alone permits the veil of concealment under which psychohistory can be effective: "It was necessary that they be ignorant."[39] The fantasy of perfect (fore)knowledge relies on a paranoid principle of its own impossibility.

The impossibility, the forbidden thing around which conventional time-travel fiction turns, is the impossibility of changing what has, in any sense, already happened: when Oedipus seeks to defy the Oracle's prophecy, he seeks to undo what the Fates have already woven for him, offending the natural order no less than, and indeed *in parallel to*, his deeds of parricide and incest. Thus, Constance Penley suggests that the basic interest of "time-loop paradox stories" lies in their investigation of the Primal Scene, their probing of the forbidden mystery of birth: "what would it be like to go back in time and give birth to oneself? Or, what would it be like to be one's own mother and father?"[40] Insofar as the incest taboo stands as a founding symbolic "rule" through which "the law of the unconscious and the law of the State mutually constitute one another,"[41] such "paradoxes" are perhaps less places where a Campbellian logic of rigorous coherence (of the plot with itself and of the *novum* with the known) comes apart and more like the node or knot in which imagination is bound with the self-preservation of power: the sanctity of the existing social relationships is here identified with the stability and coherence of normal reality itself.

Disabling-Enabling

The disabling-enabling imagination of SF is hardly limited to dystopias, however. Anne McCaffrey's *The Ship Who Sang,* James Tiptree, Jr.'s "The Girl Who Was Plugged In," John Varley's "The Persistence of Vision" (1978) and "Blue Champagne" (1981), William Gibson's "The Winter Market" (1986), Jose Saramago's *Blindness* (1995), and M. R. Carey's *The Girl with All the Gifts* (2014) all find their enabling condition in disabilities:

- *Emotional prosthetics.* Often, but not always, this takes on the form that David T. Mitchell and Sharon L. Snyder call "narrative prosthesis"—presenting a disabled character who serves as a focus for the emotions of readers presumed not to be disabled, enabling them to think of themselves as "normal" (a place impossible to inhabit: a biometric construct of the Great Sorting Machine that human bodies and minds inevitably fail/decline to be).[42] *The Ship Who Sang* offers up the deformed protagonist, Helva, as an "abject, subhuman figure," an object of pity to whom we vicariously grant reparations in the form of a double supplement: her starship "body" and the love of a "normal" human man.[43] Heteronormative love, even more so than technology, grants transcendence of the prison of the flesh.
- *Compensatory imagination.* A major trope of disabling-enabling is the figure of what has been derisively tagged as the "supercrip" by disability scholars. Through the magic of technology, disabled protagonists from Professor Charles Xavier of *X-Men* to *Star Trek: The Next Generation*'s Geordi LaForge

are granted not only compensatory reparation but superhuman abilities; their own difference enables them to serve, in the manner of superheroes, as a supplement for an (inadequate) normative world. This remains largely within the logic of "narrative prosthesis" in that the disabled person remains a means for a normative reader to manage their anxieties about the normal itself.

- *Metaphors materialized.* Science-fictional disability can also provide, as Mitchell and Snyder suggest, a "material metaphor" for social or moral deviance.[44] Indeed, the literalization of metaphor that typifies the fantastic—that domain of writing in which expressions like "He was absorbed in the landscape" lose their purely figurative status—is itself a metaphorical materialization of the kind Mitchell and Snyder describe.[45] Such is certainly the case, as Liat Ben-Moshe argues, with *Blindness*, which "uses blindness to create an allegory about the breakdown of humanity and morality in modern societies."[46]

- *Dystopian constraint.* The figure of the technological supercrip may be marked by another impossibility: the impossibility of living without their prosthesis, the source of their powers. Such is the case of Tiptree, Jr.'s "The Girl Who Was Plugged In," Gibson's "The Winter Market," and Varley's "Blue Champagne," in which the disfigurement that disability is can only be transformed into beauty and desirability through the intervention of an apparatus that enmeshes the subject in circuits of capital (a metaphor strikingly literalized by Varley's "golden gypsy," the diaphanous exoskeleton that animates the paraplegic celebrity's body). Rather than imagining technology as liberation from the prison of disability, this envisions disability *as* a sociotechnical condition, a construct that strangely resembles the condition of being a woman: your body is someone else's property.[47]

- *Utopian alternative.* "The Persistence of Vision" could be said to fetishize the condition of the deaf and blind, as Clayton Koelb suggests, by using that condition as a prosthesis for the "true" disability that is language itself (predicated as it is upon "distance and 'absence'"). The normative narrator, a hitchhiking refugee from an increasingly senseless world, stumbles upon the isolated community of Keller, which rejects prostheses altogether and repositions disability as the norm. The hitchhiker discovers, in their intimate, signless communion (symbolized only by the illegible, unpronounceable "✻✻✻"), a utopian alternative to the world of empty signifiers.[48]

Why, in all of these scenarios, is impossibility the *condition* for imagining other possibilities? What is it about the subtractive that makes it a supplement to the undisturbed sameness of the normal? For whom does this "normality" ever characterize everyday life, if ever it did? What kind of enchantment is it that can seemingly be exercised *all the more powerfully* by a normalcy that is vanishing or has already vanished? And how can the spell of the normal, its hold on the social imaginary, be broken?

We might say that the normal itself *is* a subtractive world, an (ordinarily) seamless screen of appearances masking other potential realities, "the infinity of possibilities that this system ignores, plunders, suppresses, and denies, the potential forces that always haunt this system with the fact that it is merely one possibility among an infinity of other possibilities" (Daniel Colson).[49] Recognizing that this infinity contains possibilities for worse as well as better, the banal as well as the marvelous, potentialities for the horrific as well as the beautiful, produces a shudder, a vertigo surely felt by anyone who has ever hesitated, Hamlet-like, to fly in the direction of unknown ills rather than bearing the known. The powerful tug of the normal is all the more powerful for those who witness the tearing of this fabric, as in the past four decades—nearly my entire lifetime—of "the state's withdrawal from the uneven expansion of economic opportunity, social norms, and legal rights that motored so much postwar optimism for democratic access to the good life"; Lauren Berlant's examination, in *Cruel Optimism*, of the affective response to that process of destruction, the "stubborn attachment" (Judith Butler) to fantasies of a middle-class "good life," attainable through the rituals of education, employment, marriage, parenting, consumption, homeownership, and voting, testifies to the force of that pathos.[50] Is it any wonder that our culture has become infatuated with Irwin's "game of the impossible" now that the normal itself is scarcely possible anymore?

Impossible People

Pamela Zoline's "The Heat Death of the Universe" (1967), a defining text of the New Wave in science fiction and fantasy, begins, inexplicably, with a definition: "ONTOLOGY: That branch of metaphysics which concerns itself with the problems of the nature of existence or being."[51] This seems at first to have little to do with the story that follows, as we follow an alienated, seemingly isolated Californian housewife, Sarah Boyle, through her mental breakdown. The link to the concept of "heat-death" is clear enough: her world is so circumscribed that it resembles the "closed system" of thermodynamics, within which "disorder"—the creeping sense of meaninglessness—must steadily grow to a maximum.[52] What does this neo-Gothic scenario have to do with metaphysics, with existence or being?

I am reminded of Tolkien's *Lord of the Rings*. In spite of his professed Catholicism, Tolkien violates the philosophical canon of St. Thomas Aquinas in making darkness a *substance*:

> In a few steps they were in utter and impenetrable dark. Not since the lightless passages of Moria had Frodo or Sam known such darkness, and if possible here it was deeper and denser. There, there were airs moving, and echoes, and a sense of space. Here the air was still, stagnant, heavy, and sound fell dead. They walked as it were in a black vapour wrought of veritable darkness itself that, as it was breathed, brought blindness not only to the eyes but to the mind, so that even the memory of colours and of forms and of any light faded out of thought. Night always had been, and always would be, and night was all.[53]

Whereas the evil of the One Ring is definite and powerful, for Tolkien, the good is best defined as a resistance to evil. A multiracial coalition is necessary to defeat it, but this only serves to make the good vague and tentative: it is not located in Men, not in the Elves, not in Dwarves, not even in the Hobbits (who seem merely to be the most resistant—but even they are ultimately corruptible). Church doctrine characterizes evil as "privation," as a mere lack of goodness, just as the darkness is merely a lack of light; for Tolkien, *good is a lack of evil*.[54]

The terror of sexuality that stalks *Lord of the Rings*, as Valerie Rohy points out, is itself based on a metaphysics of constitutive lack, the "O" of the Ring itself a "gaping hole."[55] The Romantic melancholy that pervades Middle Earth—a realm slowly losing its magic, the force that gives it meaning—echoes the Lacanian edict, *il n'y a pas de rapport sexuel*: "something is fundamentally missing from all human connection and all sexual enjoyment," so that, for a Lacanian like Valerie Rohy, "*The Lord of the Rings* 'realistically'—more realistically, even, than reality—describes the problem of human desire and the impasse of our own impossible sexuality."[56] The question of human (un)happiness is made into an unanswerable metaphysical question.

"It is often maintained," writes the Brazilian philosopher Hilan Bensusan, "that ontological questions are questions about *grounds*":

> Ontology has presented itself as first philosophy, as the realm of basic assumptions, and as stage setting for science, politics and to anything polemic. Ontology, according to this view, is the basis. It is associated to what is naturally so, to what persists per se, to what is not up for grabs. As such, it is presented as bedrock. Bedrock is not merely what grounds something else, but also

something under which no further excavation is possible—no further archeology, no further search for *archés*, can take place ... Ontological claims ... are therefore immune to politics ... Ontology is viewed as an enclosed domain and, as such, as unreachable by any political move.⁵⁷

It could be said, then, that Sarah Boyle's condition, which presents itself as an unalterable fate—her social confinement resembles physical compression, the subject of Boyle's Law—is one in which the merely a contingent and changeable circumstance is disguised as ontological "bedrock," something that cannot be changed under *any* circumstance. In spite of everything that is changing, rapidly and dramatically, about the situation of women circa June 1967, Boyle's Gothic imprisonment seems to be less that of Charlotte Perkins Gilman's protagonist, confined to the yellow-wallpapered room for her health, than that of Thomas Pynchon's Oedipa Maas in *The Crying of Lot 49*: "Such a captive maiden, having plenty of time to think, soon realizes that her tower ... [is] only incidental: that what really keeps her where she is is magic, anonymous and malignant, visited on her from outside and for no reason at all ... the tower is everywhere."⁵⁸ For Sarah Boyle, too, oppression comes to appear not only as natural but as simply the-way-things-are, properly *ontological*; in Bensusan's metaphor, it is of the deep strata of the earth, a *geological* feature.

It is fitting, then, that half a century later, N. K. Jemisin's protagonists should be, in the Gordin Brothers' terms, technicians of the Earth—not geologists (or, in the jargon of the Stillness, "geomests"), scholars seeking the *laws* of its formation and motion, but people who can tap its energy for *work*. Referring to Clarke's Third Law—"any sufficiently advanced technology is indistinguishable from magic"⁵⁹—Jemisin puts forward "a fantasy corollary of that, which is any sufficiently systematized magic is indistinguishable from science."⁶⁰ Nonetheless, the scientists of Jemisin's Syl Anagist, no less than their descendants in the Stillness millennia later, find their epistemological *systems* constantly overwhelmed and overflowed by the ontological *works* of the "orogenes," those possessed of "the ability to manipulate thermal, kinetic, and related forms of energy": "It hasn't ever really made sense, has it, the way orogeny works? It shouldn't work at all, that willpower and concentration and perception should shift mountains. Nothing else in the world works this way."⁶¹ The people *of* this world are as puzzled as we are.

Orogenes are thus figures of exception: exceptional in their powers, they are also excepted from the rights and protections due to others in the comms, and their very existence is a mark of the permanent state of exception under which the Stillness resides. If the name *orogene* suggests, via Greek roots, that they are *Earthborn*, this is in keeping with Claude Lévi-Straus's observation that, in Greek myth, to be "born from the earth" is to be disfigured, kin to the monstrous; nonetheless, as "tamers of the wild earth, themselves to be bridled and muzzled," they are also Promethean figures, "gods in chains."[62] This is a repeated figure in Jemisin's body of work: *The Hundred Thousand Kingdoms* (2010) is premised on a re-literalization of this figure of speech, presenting a world in which the gods, the Enefadeh, are both real (and really are the creators of everything) and enslaved, having thus become a kind of technology themselves. Here, perhaps, is another instance of Burke's "bureaucratization of the imaginative": like Kafka's Poseidon, stuck behind a desk at the office and unable to even visit the sea, Jemisin's orogenes and Enefadeh are both necessary to the systems in which they are trapped and locked in enmity with those systems. But more so than any of Kafka's characters, they are enslaved.

They are, and they are not; that is, they are enslaved people, and at the same time, they are not simply equivalent to the enslaved people that built our world, the so-called United States of America, within which I live and write. Jemisin has declared forcefully that she

> did not set out to write an allegory for slavery and caste oppression. I set out to write a story about a woman grieving her child. I set out to show what made her extraordinary. I set out to write a world in which people who are powerful, who are valuable, are channeled into systems of self-supported and externally imposed oppression, and how you keep people who can throw mountains from throwing mountains—and running the world.[63]

The fantastic, as Jeannette Ng remarks, "loves allegory . . . you know, instead of talking about people with different skin color, let's talk about orcs and elves"— or aliens, or werewolves, or mutants, and so on.[64] The Demon of Analogy returns: when Frank Wilderson III says that "analogy *mystifies*, rather than clarifies, Black suffering," he is referring to analogies between that suffering and the suffering of, for instance, the Indigenous or immigrants, but his warning against mystification might stand with Ng's warning against the evasion of any confrontation with the experience of real others.[65]

How to represent Spade's "impossible people," the people for whom life is made impossible, particularly queer and/or Black people?[66] Ebony Elizabeth Thomas catalogs the ways in which Black characters in white SFFH media franchises are killed off, vanished, disappeared: "Rue's impossible dark innocence [in *The Hunger Games*], Gwen's impossible dark beauty [in the TV series *Merlin*], and Bonnie's impossible dark virtue [in *The Vampire Diaries*] are signs that Black girls in speculative fiction are the real 'impossible girls'"—doubly so, because "these Black girl characters (like Black girls in the real world) are both impossible and necessary at the same time."[67] Surely being an impossible person—what Calvin Warren calls a "being"—makes one a prime example of an "elusive referent" in Seo-Young Chu's sense.[68]

If the one-to-one substitution of allegory is a trap, we might better speak of *analogy* in the sense of Daniel Colson, for whom it signifies not a simple, one-to-one correlation or a matter of surface resemblances but a stranger relationship "aris[ing] from discontinuity and difference but also from repetition"—or in Walter Benjamin's sense of "nonsensuous correspondences," "transtemporal constellations" and "discordant juxtapositions of times and places."[69] Analogies, in this sense, can perhaps serve as antidotes to a smugly obvious "white empiricism,"[70] the self-evidence of those who, like the Empiricists of Colson Whitehead's *The Intuitionist*, are most comfortable seeing only "the skin of things," the evidence of themselves: "White people's reality," as he puts it, "is built on what things appear to be—that's the business of Empiricism"—"but there was a truth behind [appearances] that they couldn't see for the life of them."[71] Analogy, as an "intuitive" relationship in Whitehead's sense, might be the reverse of the "catachresis" described by Warren (a white fantasy procedure, projecting meaning onto the "nothing" of black bodies), instead presenting a "paradigmatic juxtaposing" of "two singularities": analogical thinking, in Martel's words, can "help to disrupt the seamless web of reality," to dispel the "phantasmagoria" of dominated life.[72]

The fantastic, for Darko Suvin, has just this "analogical" relationship to reality: "the cognitive value of all SF, including anticipation-tales, is to be found in its analogical reference to the author's present rather than in predictions, discrete or global."[73] On the terms of this analogical interpretation, we need to ask: who are our roggas, our Earthborn gods in chains? On the one hand, blackness and queerness exist in the Stillness in a real, nonallegorical sense; Alabaster *is* a queer Black man. The identity "rogga" does not *replace* "Black" or "queer," although it does operate *like* them: like blackness, it entails exploitation and exclusion, while like queerness, it can be disguised and disavowed. The result is a strange superimposition of the real and the fantastic—perhaps even what John Rieder describes as "a kind of palimpsest, bearing the persistent traces of a stubbornly visible colonial scenario beneath its fantastic script."[74]

In reconstructing the colonial scenario, we can begin by acknowledging that the Earth that is Broken, so to speak, is ours: ecologically broken (by the "metabolic rift" between humanity and the "nonhuman" world) and socially-ontologically broken (by the "unresolvable antagonism" between blackness and the "human" world).[75] The terrible "Rifting" that Alabaster inflicts on the Stillness ironically unites these two zones of meaning that the Modern Constitution divides: the natural and the social.

The Brokenness of ~~Being~~

Jemisin's protagonists often lose themselves, to be not-all-there. "But I forget myself. Who was I, again? Ah, yes," begins Yeine in *The Hundred Thousand Kingdoms*. "And this is when it finally hits him"—the nameless avatar of Manhattan in *The City We Became*—"that *he doesn't know his own name*."[76] "The self you've been lately doesn't make sense anymore," reflects a second-person voice in *The Fifth Season*; "You're still trying to decide who to be."[77] They seem to be Lacan's barred subjects, S (if not Calvin Warren's ~~objects~~), permanently marked by a "brokenness of being."[78] As with Yeine (yea/nay), their names sometimes obscurely hint at a founding ambivalence: Damaya (yes/maybe), Syenite (yes/not), Essun (yes/no), Nassun (no/yes/no). It is this hesitation that marks the climactic moment of the Broken Earth trilogy, as Nassun—the girl who has more motive than anyone to destroy the world—finds herself alone with the power to destroy it or to save it, as her mother wished:

> The onyx, patient but not, aware but indifferent, touches her. She is the only remaining component of the Gate that has a functioning, complementary will. Through this touch she perceives your plan as commands locked and aimed but unfired. Open the Gate, pour the Rifting's power through it, catch the Moon. End the Seasons. Fix the world. This, Nassun sesses-feels-knows, was your last wish.
>
> The onyx says, in its ponderous, wordless way: *Execute Y/N?*[79]

The *Y* and the *N*, the affirmation and the negation, seem to correspond to two facets of contemporary radical Black imagination, Afrofuturism and Afropessimism. They are a conundrum. I will try to explicate the conundrum in what follows.

Afrofuturism's power (of which Jemisin doubtless partakes in some respects, although she firmly denies that her work is "Afrofuturist") lies in the affirmation of a Black future—perhaps a future without oppression, an emancipated future, although not necessarily so; in the face of the perpetual specter of genocide, when another day of life is not guaranteed, simply projecting *a* future is an exercise in radical hope and resistance.[80] At its highest intensity, it is utopian-redemptive: "The world is broken and you can fix it."⁸¹ Afropessimism's power, like the power of horror, lies in unremitting negativity, an inextinguishable rejection of the totality of oppression, even if that totality seems to be simply *the* totality of that which exists: "*It's wrong. Everything's wrong. Some things are so broken that they can't be fixed.*"⁸² To be Black, as witnessed from an Afropessimist position, is to be a broken being (oneself and not-oneself), broken off from a being (excluded from a world) that is already broken apart (riven by race). It is to be in the position of Frankenstein's Creature, forever cast outside of the realm of human beings constituted by "recognition": "what is this new creature the master creates with his power of transubstantiation?"⁸³ The Creature has been cast by some as the Slave in Hegel's Master-Slave dialectic, but in life, as in *Frankenstein*, Calvin Warren suggests that the moment of redemption, when the Master is forced to return the gift of "recognition" to the Slave, never arrives:

> The emancipated black must insist that he is no longer sentient property, but he directs this insistence to farm animals, not a human community (a political community). Put differently, he must seek recognition from farm animals, since such recognition is implausible within the political community. But this recognition makes a mockery of the Hegelian scene—for Hegel does not envision such recognition occurring between ~~objects~~. It is a recognition that undermines recognition.⁸⁴

It is in this sense I think that the Anarkata Statement suggests that to love oneself while Black—were that possible, as Wilderson insists that it is not—is to want to destroy the world (if not the Earth).⁸⁵ If, as Frank Wilderson III writes, "Human life is dependent on Black death for its existence and for its coherence," so that "There is no world without Blacks, yet there are no Blacks who are in the world," then—in spite of his warning against "the ruse of analogy"—there exists a deep analogy between the condition called being Black and the condition called being an orogene: human life in Stillness is impossible without orogenes to quell shakes, yet to be an orogene is to live under a condition of social death, forever exiled from the "world" of human social relations.⁸⁶

Without a world, can there (still) be an Earth? Put otherwise, could a human world be constituted such that it did not require Black social and physical death, a *polis* that did not compel cis- and heteronormativity in the name of its own social reproduction? From the Afropessimist position, it might seem that the answer (to this as to every question) is *no*: "the world as

[Alabaster] knew it could not function without forcing someone into servitude."[87] And so, for twice-betrayed Nassun—maimed by her mother, hated and nearly killed by her father—annihilation appears as the only goal, the only imaginable end to suffering.

Nassun is in the position of those "potentially emancipatory forces" that Daniel Colson describes as "thrice characterized by negativity and dependency with respect to domination":

> 1) through the oppression they endure, of which they are products; 2) through their own struggles, which are always in danger of being confined to a mere refusal of oppression; 3) through the *means* (see this term) by which they undertake this struggle, generally molded and dictated by the demands of the struggle and, ultimately, by the enemy that is to be fought and destroyed.[88]

Nassun recognizes with horror that the harm she has suffered at her mother's hands was done to her mother first, that she is the product of intergenerational trauma.[89] Her first way of saying *no* to the oppression that birthed her is given to her by the Stone Eater, who was once a tuner called Remwha—a being *literally* brought into the world by oppression, as the tuners were genetically engineered for their geoarcanic abilities, confined and enslaved:

> "The Moon's coming back, Nassun. It was lost so long ago, flung away like a ball on a paddle-string—but the string has drawn it back. Left to itself, it will pass by and fly off again; it's done that before, several times now . . . But with the Gate, you can . . . nudge it. Just a little. Adjust its tra—" A soft, amused sound. "The path that the Moon naturally follows. Instead of letting it pass again, lost and wandering, bring it home. Father Earth's been missing it. Bring it straight here and let them have a reunion."
>
> Oh. *Oh*. She understands, suddenly, why Father Earth wants her dead.
>
> "It will be a terrible thing," Steel says softly, nearly in her ear because he's moved closer to her. "It will end the Seasons. It will end *every* season. And yet . . . what you're feeling right now, you need never feel again. No one will ever suffer again."[90]

Thrice marked, then, by oppression, Nassun is faced with the question of how to "transform this situation of a threefold dependence on the dominant into one of affirmative, autonomous forces."[91] She must create from herself a politics of affirmation *and* negation that goes beyond nihilism and the Death Drive—but not without passing *through* them. Perhaps it is only possible to do so after truly abandoning the kind of social hope that serves the survival, the social reproduction, of the oppressors' civilization:

> "Tell them they can be great someday, like us. Tell them they belong among us, no matter how we treat them. Tell them they must earn the respect which everyone else receives by default. Tell them there is a standard for acceptance; that standard is simply perfection. Kill those who scoff at these contradictions, and tell the rest that the dead deserved annihilation for their weakness and doubt. Then they'll break themselves trying for what they'll never achieve."
> —ERLSSET, twenty-third emperor of the Sanzed Equatorial Affiliation, in the thirteenth year of the Season of Teeth. Comment recorded at a party, shortly before the founding of the Fulcrum.[92]

The unachievable goal of full recognition in a society founded on one's exclusion, once renounced, releases energies that can destroy but that can also create. It is to Jemisin's social creativity that I want to turn next.

Utopian Interstices

The Broken Earth Trilogy maintains a certain hope in the possibility of the impossible—the constitution of an unbroken Earth, a world of horizontal mutual recognition in which all beings are accorded voice and respect. It creates three utopian spaces for this difficult and hard-fought faith to inhabit.

The first of these is on the pirate island of Meov, where orogenes turn out to have a special place: "They don't kill their roggas, here. They put them in *charge*. And they're really, really, glad to see us."[93] Ultimately, however, this place is contingent and tenuous, as orogenes are valued and celebrated *because* they are socially useful ("they're no different from the Fulcrum in that, are they?").[94] The more significant social creativity is sexual:

> Alabaster watches while Innon obliges her, and his gaze grows hot with it, which Syenite still doesn't understand even after being with them for almost two years. 'Baster doesn't want her, not that way, nor she him. And yet it's unbelievably arousing for her to watch Innon drive him to moaning and begging, and Alabaster also clearly gets off on her going to pieces with someone else. She likes it *more* when 'Baster's watching, in fact. They can't stand sex with each other directly, but vicariously it's amazing. And what do they even call this? It's not a threesome, or a love triangle. It's a two-and-a-half-some, an affection

dihedron. (And, well, maybe it's love.) She should worry about another pregnancy, maybe from Alabaster again given how messy things get between the three of them, but she can't bring herself to worry because it doesn't matter. Someone will love her children no matter what. Just as she doesn't think overmuch about what she does with her bed time or how this thing between them works; no one in Meov will care, no matter what. That's another turn-on, probably: the utter lack of fear. Imagine that.[95]

The rarity of this "utter lack of fear" in works of the fantastic, even in utopian works such as Le Guin's *The Dispossessed*, makes this stand out all the more.[96] In *Wonder Women: Feminisms and Superheroes*, Lilian S. Robinson reflects on the ways in which the traps of triangular desire and the couple form continue to drive and constrain plots even where the constraints of "hard SF" physicalism are lifted, as in the universe of the Avengers: "They can fly, they can channel cosmic energy, they can tear down the strongest walls, but they can't consider a nonmonogamous relationship!"[97] Rather than submitting to the supposed impossibility of the *rapport sexuel*, Jemisin allows us to imagine many possible shapes of *rapports sexuels*. And this sexual possibilism becomes a "credibility device" powering other social visions, the ones that Syenite and Alabaster have been trained to disavow:

> She shakes her head. "I don't know."
> But she thinks, almost but not quite subconsciously: *A way to change things. Because this is not right.*
> He's always good at guessing her thoughts. "You can't make anything better," he says, heavily. "The world is what it is. Unless you destroy it and start all over again, there's no changing it."[98]

Syenite *should* have been taught to "[know] better than to wish for something so impossible"—but clearly, she hasn't learned the lesson.

Subsequent events lend grim credence to Alabaster's admonition: having lost her "affection dihedron" family and the microsociety that supported its being and her being, Syenite sheds another skin, adopting the identity of Essun, a rural schoolteacher with a "still" husband. It is when this project of assimilation falls through that Essun discovers the (physically and socially) underground comm of Castrima. Unlike the earlier idyll of Meov, an island community of humans who welcome orogenes as long as they render services—ultimately just a gentler version of the social contract imposed across the Stillness—"Castrima's thin, cracked nothing of a chance"[99] perhaps lies in its having been founded by the persecuted orogenes, even though they are still vulnerable to betrayal by fickle human collaborators; even when tested and strained, the coalition holds. "This ridiculous comm of unpleasant people who are impossibly still together, which you have fought for and which has, however grudgingly, fought for you in return" is itself a hopeful monster, an "unnatural composite" that is nonetheless a source of possibility for a more just and sustainable future.[100]

But perhaps the most unlikely and beautiful of Jemisin's utopias is the micro-utopian *rapport* created between Nassun and the wayward Guardian, Schaffa. The choice that Nassun makes to trust Schaffa, even though it stems from the internalized violence of self-abnegation, as the first reliable adult she has encountered in her brief life, astonishes him so deeply that "the killing tension begins to bleed out of his posture."[101] The choice that Schaffa then makes in reconstructing himself—"Never again" to be "the monster that he was," "promising himself a better future"—is just as decisive: he can be the one adult who loves her unconditionally, who allows her to choose for herself, Y/N, whether the world deserves to live or die, whether it is the last repository of hope or beyond hope of redemption.[102] "You're my redemption, Nassun," Schaffa tells her:

> "You are all the children I should have loved and protected, even from myself. And if it will bring you peace . . ." He kisses her forehead. "Then I shall be your Guardian till the world burns, my little one."
>
> It is a benediction, and a balm. The nausea finally releases its hold on Nassun. In Schaffa's arms, safe and accepted, she sleeps at last, amid dreams of a world glowing and molten and in its own way, at peace.[103]

By making this utterly unlikely "*symbiosis*" plausible, however strange its geometry, Jemisin prepares us to believe that when "the onyx says, in its ponderous, wordless way: *Execute Y/N?* . . . Nassun chooses. YES."[104] The little counter-world of mutual aid, of healing together from the relentless trauma inflicted by a disciplinary world, is not simply a disguised "reference to the author's present": it is a micro-utopia, a synecdoche for a redeemed world. Jemisin's analogical rhetoric not only condemns that which is, it *proposes what is not*.

Nisi Shawl's *Everfair* (2016) also situates its utopia—not so micro—in the interstices of an imperial system: the southeastern Congo between 1889 and 1919. A swatch of land purchased from the murderous King Leopold of Belgium, in this alternate history, becomes the base for a democratic colony that brings together a multiracial coalition of Black Americans, Fabian socialist workers and intellectuals, African refugees from the Mon-Goh,

Bah-Looba, Oo-Gandah, and Bah-Sangah peoples, and escaped Chinese railway laborers. This counterfactual territory, pried loose from the history of accomplished genocide, becomes a kind of *quilombo*, a "refuge from traumatic exposure to violence" for history's victims, in the words of Juan Duchesne-Winter, which is also "a place for convalescence and creativity . . . an out-of-place, an outer-national location where the multitude can gather in order to launch its utopia-in-resistance."[105] Here, it seems, is one kind of *N*-dimensional wampumpeag, a world (or outer-world) within which many worlds fit.

If the family is, in classical political theory, the foundation of the State, the new-old land of Everfair is subtended by families as multifarious as its composition is multiethnic, including the queer *ménage* of Laurie Albin, within which circulate the roles of spouse, lover, and secretary, and interracial couples such as Lily Albin and Ho Lin-Huang, George Albin and Martha Livia Hunter. "Impossible" families within the real history of the period, marked as it was by the repressive forces of heteronormativity and white supremacy, these nontraditional alliances become a network for the transmission of creative energies: Bah-Sangah philosophy married to Chinese metallurgy and Black American engineering produce atomic-powered airships and prosthetics. The resultant is, as Sharae Deckard observes, less a classical utopia in the line of Plato and More than a heterotopia comprising conflictual visions:

> The utopia posited by the text is not frictionless, nor some wish-fulfilment of catastrophe-averted: instead, it is full of constant conflicts between the different factions of Everfair, bound up with their differing conceptions of race, class, gender, sexuality, disability, governance, and religion: the proselytizing agenda and heteronormative beliefs of the Baptist African-American missionaries; the tendency of the secularist Fabians despite their socialism to want to claim ownership of the land and leadership of the Everfair project, which eventually they cede back to the indigenous peoples; the anguish of King Mwenda at the prospect of betraying the patriarchal traditions of his ancestors; the conflict between socialist versus monarchical ideas of the political organization of society; the tension between pacifist and aggressively anticolonial approaches which insist on the necessity of violence to the course of emancipation; the phobias of racial miscegenation that poison the relationship between the two central female characters.[106]

Shawl weaves these disparate forces together nonetheless via the logic of affinity, bonds "not of an ideological nature," as Colson puts it, but "involv[ing] temperaments, various forms of *sensibility*, various qualities of character, and the various ways in which these can be composed with others."[107] The arrangement composed by these constituents is another kind of monster-made-from-disparate-parts, one which announces itself in the figure of Haraway's cyborg, only in a steampunk mode. The colony's many amputees, subjects of Leopold's reprisals for failure to meet rubber-harvesting quotas, and the wounded of Everfair's war of emancipation receive brass limbs that extend their possibilities for fighting and living:

> This model of prosthetic was mostly for show. By flexing her muscles a certain way against the straps, she could bend the jointed fingers; if she crooked the thumb, they would squeeze shut on anything they held. A gear cocked the wrist and a spring released it. It was much less useful than some of her hands. But prettier....
>
> Later that day she selected a different hand to wear. In her opinion it was as beautiful as any. Its polished brass shone brilliantly in the sun streaming through the half-open shutters of the embassy room in which she lodged. But it bore much less resemblance to what most people had at the ends of their arms. Five knives, three of them detachable, stood in place of the fingers.[108]

Hybrid bodies, hybrid families, hybrid societies—in this way, as Jemisin *proposes what is not*, Shawl *affirms what never was*: it is primarily the moral sincerity of the white characters that stands as an invention, a prosthetic supplement to history's wounds. The real Fabian Society was more than ready to play handmaiden to the British Empire (evoking, in a darkly ironic undertone, another sense of the phrase *ever fair*).[109] Perhaps Shawl is shrewdly banking on white readers' willingness to believe in the white protagonists as well as Black readers' willingness to suspend disbelief.

Where Jemisin engages her characters in a confrontation with nihilism—the possibility of selecting *N* to everything, of negating the world from which one is systematically excluded—Shawl avoids such tragic drama. The motif of *Everfair* is reparation, not revenge, which would ensnare it in negation, in the reductive mimetic violence of the revenge plot. The enemies of Everfair are thus kept almost entirely offstage: "Nobody thinks they're a bully or a villain, and I was very consciously trying not to have bullies and villains," Shawl comments.[110] "None of EVERFAIR's viewpoint characters were uncomfortable for me to inhabit ... Leopold's viewpoint is absent from the novel."[111] The pain and suffering caused by injustice are foregrounded, focalized through the many viewpoints of the ensemble cast of protagonists; evil as a source of injustice is held at a distance. The good is more interesting, more complex, more *concrete*, than the evil.

Visions and Limitations

Both Jemisin and Shawl, in the company of others in the rising generation of women of color in the field, deploy the resources of fantastic rhetoric so as to do what so much of the fantastic promises but fails to do: they defamiliarize the eugenic logic of the Great Sorting Machine, making it available for critical perception. That is to say, they do the work that theory was once thought alone capable of doing.[112] Mired in "self-oblivion," the literary or cultural text required the intervention of theory to "show [it] as it cannot know itself," as Terry Eagleton had it.[113] The hierarchical, even colonial tenor of this theory/text relationship became intolerable enough for Black scholars to occasion a search for "indigenous" and then "inductive" forms of theorization.[114] Now, a New Black Fantastic has emerged that outstrips the exhausted imagination of a theory that has spent the last quarter century writing and rewriting its own obituary. The richness of the social imaginary intrinsic to the New Black Fantastic illuminates what E. P. Thompson once called "the poverty of theory," and in this way opens the question of how to renew that vision.

Rhetorical Sovereignty

> Imagining potential futures, or alternative worlds in any time, is not merely an exercise of imagining; I assert it as an act of what Scott Lyons calls rhetorical sovereignty . . .
> —CHELSEA M. VOWEL, Where No Michif Has Gone Before

The New Black Fantastic seems to have punctured an opening for other heretofore unrepresented or inadequately represented groups in the fields of science fiction, fantasy, and horror, leading to a proliferation of new voices. Within this expanding universe of "co-futurisms," as they are sometimes called, I want to focus on the strand currently known as Indigenous Futurism.

I write this from Cession 180, the swath of so-called Northwestern Indiana from which the Neshnabék (Potawatomi) and their many neighbors were expelled by force of arms not quite two centuries ago; I write as a settler with insufficient knowledge of Indigenous histories and cultures, an academic with much to learn. I hope not to have misrepresented the Cherokee, Ojibwe, Ohkay-Owingeh, Métis, Nishnaabeg, Dene, and Apache persons and peoples spoken of here. In view of a long history of struggles for sovereignty, I am propelled into thought by the phenomenon—both "new" and "not so new"—of Indigenous peoples reaching for the cultural "toolkit" (Doctorow) of science fiction, fantasy, and horror.[1] My sense is that while the very genre structures comprising the "toolkit" are also a "structure of settlement," in the words of Theresa Warburton, the rhetoric of the fantastic affords them tools with which to pursue sovereignty.[2]

Science fiction, writes Darko Suvin, is "a literary genre whose necessary and sufficient conditions are the presence and interaction of estrangement and cognition, and whose main formal device is an imaginative framework alternative to the author's empirical environment."[3] This influential definition, excluding fantasy and horror as inferior forms, has been contested on several grounds, not the least of which concerns its somewhat naive (white) empiricism. The question of who determines what "the author's empirical environment" is, and therefore what is "strange" to it, becomes even more difficult in a settler colonial context, where colonial accounts of empirical truth are imposed over and against the accounts given by Indigenous people, thus complicating our very understandings of what counts as science fiction at all. To give just one example, in Kai Minosh Pyle's "How to Survive the Apocalypse for Native Girls" (Métis/Baawiting Nishnaabe, 2020), we are told the following explanatory story about how the apocalypse in question came to be:

> We call that time the hungry years not just because people often went without enough food, but also because there was often another kind of hunger. The kind of hunger that causes people to do terrible things: wiindigo hunger.... Wiindigoog are more than just cannibals. They are possessed by a hunger that only increases every time they try to fill it. That hunger can be for anything—food, drugs, sex, love, but most of all, power.[4]

Why does violently antisocial behavior ("terrible things") need an explanation beyond ordinary hunger? Why are Wiindigoog an explanation rather than a thing-to-be-explained? These questions are perhaps better formulated as "for whom": for whom is a complete breakdown of social norms in times of material want the expected outcome? For whom are Wiindigoog—once-human monsters with hearts of ice—simply another item to be found among an inventory of "the furniture of the universe"?[5] It is tempting to answer that a certain group, the "kinship-based" Northern Algonquians (a linguistic/cultural group that includes Pyle's Nishnaabe family), is the collective subject for whom the Wiindigoog are a given and that members of non-"kinship-based" societies, "modern" societies in which the social fabric is already rather tenuous, are the ones who are ready to accept the premise of social "apocalypse" without much further explanation. The late settler anthropologist David Graeber cautions against such an easy compartmentalization; it is quite possible one might encounter Nishnaabeg who regard Wiindigoog as mere legend or settlers who don't find postapocalyptic Mad Max scenarios credible.[6] If it is not so easy to determine the whoms in question for either case, it is clearer that we are dealing with some fundamental questions about what counts as an "alternative" framework—in other words, an "Other World"—and perhaps more importantly, what is the nature of the "primary world" in which the author and the readers are situated, of "reality" as such.

Like perhaps most definitions of interest, Suvin's definition of SF is an attempt to set the terms for all future discussion of its subject—a largely successful attempt, in fact; even the microgenre of critiques of Suvin bears ironic witness to the persuasiveness of his rhetoric. Scott Lyons speaks of a "rhetorical imperialism," which consists in "the ability of dominant powers to assert control of others by setting the terms of debate," for "he who sets the terms sets the limits."[7] In contrast, Lyons defines "rhetorical sovereignty" as "the inherent right and ability of *peoples* to determine their own communicative needs and desires" to exercise control over the parameters of the discourses in which they are involved.[8] Where SF is concerned, Suvin has set the parameters of discussion in ways that immediately problematize a good deal of Indigenous futurist writing, in so far as Indigenous futurists often make reference not only to the future but to "old things that rumbled under the surface of the world," as Adam Garnet Jones (Cree/Métis/Danish, 2020) puts it, often placing such pieces of ontological furniture as the Wiindigoog, ghosts, and spirits of all kinds on the same metaphysical ground as the more common furniture of science fiction.[9] The place that SF reserves for scientific knowledge as an explanatory principle is shared by the wisdom of grandmothers, by oral traditions and visions. As Blaire Topash-Caldwell summarizes: "Indigenous science fiction privileges autochthonous, localized, and historically situated knowledge systems instead of Western science with its ties to the Enlightenment in Europe."[10] For Suvin, this would place these works in the category of the "subliterature of mystification," presenting "estrangement" without the rational, scientifically-grounded, materialist character of "cognition."[11]

My intention here is not to defend the claim of Indigenous futurisms to be considered as part of the genre of science fiction (or fantasy or horror, for that matter); rather, I want to look at how they have pushed at genre boundaries, which also turn out to have been ontological and epistemological boundaries all along. In particular, to borrow Farah Mendlesohn's question—"how can a writer force the reader to accept as normal things that are fantastical?"—I want to see how this kind of transgression of colonial borders is effected *rhetorically*; I want to observe how Indigenous futurist writers persuade skeptical readers from both settler and Native communities to "accept as normal" accounts of things which, to readers with cognitivist biases, appear "fantastical."[12] In short, I want to investigate how Indigenous futurist writings strive to assert their rhetorical sovereignty.

Rhetorics of Incredulity

One of the most common rhetorical strategies employed by Indigenous futurists appears to sidestep questions of truth and reality altogether by means of satire and irony. As Kristina Baudemann notes, "the comic . . . is an often-overlooked structuring principle in North American Indigenous literatures."[13] Trickster figures such as the Cherokees' Jistu, a rabbit who is always disguising himself as other animals to make mischief, personify this humorous streak. Drew Hayden Taylor's "Take Us to Your Chief" (Cherokee, 2016) adopts the guise of settler SF while also undercutting its tropes in staging the encounter of benevolent aliens, the Kaaw Wiyaa, with a group of Ojibwe men on the rez, Teddy, Tarzan, and Cheemo—just three buddies intent on quietly sitting on couches, fishing and drinking beer. Their attitude toward the sudden arrival of extraterrestrials is not characterized by the "sense of wonder" canonized by settler SF, nor do their Ojibwe identities supply them with profound thoughts or words to match the solemnity of the occasion: one wonders whether the aliens' unearthly appearance "must freak the girls out," while another is reminded that he "[hasn't] had calamari in a long time," and a third ponders "[whether] that thing with calamari arms had farted."[14] The Chief to whose office the Kaaw Wiyaa are ultimately led is just a minor tribal bureaucrat, beneficiary of a "luxurious band office salar[y]," for whom the aliens constitute an unwanted hassle: he briefly wonders "if this was how the Beothuk and Mi'kmaq chiefs felt five hundred years ago," but soon concludes that "he'd better do something to get this thing out of the building before it triggered any lawsuits."[15] Hastily appointed as cultural ambassadors from Earth and ushered aboard the Kaaw Wiyaa starship, the three friends find that its interior has been revamped so as to make them "feel at home," complete with beer and couches for fishing—a happy ending, by their lights. "We should have done this years ago," Cheemo reflects.[16]

The comic rhetoric of Taylor's story is designed to achieve humor by subtracting from clichéd SF plots the lofty affects with which they are normally associated. The protagonists' responses to the irruption of a Suvinian *novum* in their midst all fail to match the occasion in an unexpected way, not unlike the reaction of Gogol's barber Ivan Yakovlevich to realizing that a customer's nose has inexplicably appeared in the bread his wife has baked for him (his first instinct is to worry that this will get him in official trouble of some kind) or Kafka's deadpan announcement that his protagonist is now a bug. It's easy enough to see that Taylor is poking fun at the whole drama of the First Contact conceit, with its colonial overtones, but at the same time, he is sabotaging romanticized versions of the *indian* as a noble people, stoic in the face of their own tragic "vanishing," etc. In both of these senses, the primary rhetorical strategy of "Take Us to Your Chief" targets the self-congratulatory white liberal humanism of David Brin's "Dogma of Otherness," the supposed tolerance for the new and different that purportedly establishes the superiority of Western civilization ("perhaps," Brin muses, "we ought to be proud of America as the prime promoter of a dogma of difference and choice").[17] Taylor denies settler readers the Otherness that serves to confirm the colonial Same.

The expected colonial scenario invoked by "Take Us to Your Chief"—the vanishing Indian is saved by the superior technology of an alien race—is exploded, here, by the story's heroes: that is, not any of the human protagonists, but the bottles of beer that insouciantly occupy the story's foreground. In Daniel Heath Justice's words, within the tragic framework of a "deficit model" for which "'real' Indigenous peoples are always Other, always diminished, always the reduced shadow of our former greatness," beer would serve as a symbol of "deficit and loss"—the firewater that destroyed the once proud people, and so on.[18] Settler readers are not made to witness the expected scenes of inebriated disgrace (another sign of deficit); rather, beer is merely that which pleasantly passes the time. The aimless, empty time of the three Ojibwe friends, marked by the leisurely consumption of one bottle after another, is a nonproductive, nonprogressive temporality that is only briefly troubled by First Contact: "Tarzan realized his beer was empty, and this was definitely a time for extra beer."[19] Beer gently annihilates seriousness: the silence of the three, which so intrigues the Kaaw Wiyaa ("If I may speak freely, what truly impressed us were your methods of communication . . . Your ability to communicate without interacting verbally. Almost a form of telepathy"), is not the silence of the Taciturn Indian, a figure of great wisdom and sorrow, but a refusal to produce the expected signs of tragedy, to reproduce the tragic narrative of deficit, to participate in the tragic history of colonial progress.[20] Thus, it is that the friends' first instinct, in the face of historical events unfolding at their fishing spot, is "to relocate to a less historic location."[21] The story's happy ending does not see the Natives (representing, no doubt, the past of the human race) elevated into a transcendent future, à la *Cocoon* or *Close Encounters*; it presents instead a return of the Same in the form of survivance.[22] "We," Cheemo reminds us, "have done this"—survived—for a long time. The nonhistorical temporality of Indigenous SF satire is the temporality of stubborn immovability, resisting the kind of rational sensemaking that typifies Suvin's canons of "cognition."

A similarly anti-chronological animus animates Craig Strete's signature work, "A Horse of a Different Technicolor" (Cherokee, 1975). Unlike Taylor, who at least ironically honors the narrative structure of the Freytag Pyramid, Strete's New Wave-style experimentalism eschews linear temporality altogether; the narrative, such as it is, jumps around with such frequency and violence that it presents a collage more than a montage. Rather than engage in the usual settler SF exercise of worldbuilding, Strete cuts up and radically *rearranges* images of worlds, at least one of which is an (unsystematically) imagined 2074 which is never over ("2074 happened twice," we are repeatedly told), others seemingly belonging to the 1974 in which the story is written, perhaps others representing an 1870s which is never over (and perhaps never began).[23] Instead of providing us with any single unified narrative voice, Strete gives us a "playback" of multiple voices, none of which seems authoritative or trustworthy.[24] In Gerald Vizenor's terms, it's unclear whether any of these voices are Natives (real presences); most seem to represent *indians*, empty simulacra, parodies of Hollywood and TV images.[25] "We made and remade every dream ever played and put them on the screen," one voice tells us, while another (?) performs the endless martyrdom of fake death ("I fell off horses so well . . . I always fell off horses so beautifully") before yet another colonial voice (?) who admonishes an indeterminate complainant to "remember, you have your place with your race, and are taped accordingly."[26] "Take comfort that no one ever dies," a voice instructs us (?): "Although the original telecast has ceased, we promise you shall live on in reruns and syndication."[27] As each voice replaces the last in a "*precession of simulacra*," the procession of colonial time crashes to a halt.[28]

Strete's rhetoric aims not at Bailey's "suspension of disbelief"[29] but at the proliferation of disbelief; by repeatedly "playing back" images of Iron Eyes Cody (best remembered as the "crying Indian" from the Keep America Beautiful commercial that first ran on TV in 1971) and John Wayne films, he attempts to inoculate the reader against the spectacular, phantasmatic figure of the *indian*. Indeed, he seems to warn us against belief in his own identity: Strete's games with authorship (e.g., writing introductions to his collections signed by the names of Jorge Luis Borges and Salvador Dalí) do nothing to assuage the doubts that have been raised concerning Strete's claims to Cherokee identity.[30] Could the strategies that Baudemann and I identify with Native American tricksterism be, in fact, the strategies of a white man "playing Indian"? As Philip Joseph Deloria reminds us, "playing Indian did not fail to call fixed meanings—and sometimes meaning itself—into question."[31]

While it would be easy to place Strete in the canon of American postmodernists such as Donald Barthelme and Robert Coover, his interest in motivating readers to resist floating signifiers of Native American identity and history is perhaps even more reminiscent of Rebecca Roanhorse's "Welcome to Your Authentic Indian Experience™" (Ohkay-Owingeh/Black, 2017). Roanhorse's claims to Native American identity, too, have been subjected to scrutiny—ironically, in light of the story's fierce critique of cultural appropriation.[32] Unlike Strete, however, whose machineries of disbelief are designed to prevent anything like identification with a character, Roanhorse draws on one of the oldest and best-known rhetorical devices of fiction by placing the settler reader within an unfamiliar subject position: "You maintain a menu of a half dozen Experiences on your digital blackboard, but Vision Quest is the one the Tourists choose the most. That certainly makes your workday easy . . ."[33] The second person voice interpellates "us" into Jesse Turnblatt, a.k.a. Jesse Trueblood, a Native VR actor whose job is to provide intimate but cinematic "Experiences" of Indianness to the clientele of Sedona Sweats. The daily grind of reenacting the most worn-out clichés for "Tourists" to "Experience" at one remove is bad enough, but the day one customer demands "something more authentic"—a friendship with the "real" Jesse, with his "aging three-bedroom ranch and a student loan"—a fatal mimetic process begins: it is as if, as the *indian* wannabe extracts more and more of Jesse's Native essence, the more he comes to resemble Jesse, to the point of replacing him, taking away his job, his house, his wife, and his very identity. That is to say, the white settler intruder has done all this to "us," the readers, in so far as the second person has worked its rhetorical magic on us so that, in effect, we *have* had an Authentic Indian Experience—the experience of dispossession, erasure, Removal.

Rhetorics of Believing

The credence in and currency of spurious representations of Indigeneity, whether romanticized, degraded, or both, is one continuing problem; the discredit attributed to Indigenous self-representations is another. Many commentators have noted this problem in connection with the very genre definitions attributed to Indigenous Futurist works: in so far as "science fiction," "fantasy," and "horror" denote departures from ordinary realism, do we "run the risk of trivializing Native voices and communities, of reducing lived experiences to mere superstition" by labeling them as such?[34] Grace L. Dillon objects to this reductive reading: "our ideas of body, mind, and spirit are true stories, not forms of fantasy."[35] Accordingly, the pursuit of rhetorical sovereignty may also mean running the engines of satirical incredulity in reverse, aiming to produce a more traditional suspension of disbelief.

Richard Van Camp's "Aliens" (Dogrib Tłįchǫ Dene, 2016) plays a sophisticated game of believing, beginning with his first paragraph:

> I wanna tell you a beautiful story. And I've been waiting for somebody very special to tell it to. I guess it's no secret now: the aliens or "Sky People" are here. We can see a ship way up high: its outline. No lights. It's like a big, dark stone in the sky and most people just watch TV or Facebook now, waiting for something to happen. Some people call them "Obelisks." Apparently, there's one huge ship miles high over every continent and the oceans are boiling, gently, but no fish are dying. Just simmering, and scientists are saying that the oceans and rivers are being cleansed. It's like the "Star People"—that's what our Elders call them—are helping us.[36]

The "ship" in question is a mute presence, "dark" and motionless; it is effectively a technological "black box," not unlike the Monolith in Arthur C. Clarke and Stanley Kubrick's *2001: A Space Odyssey*. This places Van Camp's *novum* beyond the understanding of reader and protagonist alike, where it is safe from questioning. The silent, inscrutable alien craft, familiar to us from innumerable films from *The Day the Earth Stood Still* to *Arrival*, is a piece of "off-the-shelf" imaginary, a borrowing from Broderick's "sf megatext": as readers come to it already "knowing" what this is ("it's no secret," indeed), it requires no special supporting argument. Recasting Indigenous concepts of "Sky People" in terms of the science accepted by settler society ("aliens") allows a settler reader to accommodate one with the other.

Van Camp allows us to believe, for a moment, that the "beautiful story" will be about these aliens before shifting focus to his "quiet," "gentle" friend, Jimmy, from the hardware store in Fort Smith. "I guess you could say me and Jimmy are related in the medicine way," the narrator tells us, explaining that

> they say my grandfather pulled a hummingbird of fire out of a little boy's mouth, from under his tongue. And he showed that little boy this little bird that had been living in his mouth. And he explained this was the reason that little boy couldn't speak like other people, and this is why his voice kept locking. And hundreds of people saw this little hummingbird that my grandfather pulled out of this little boy's mouth, and my ehtse let that little bird go . . . And my grandfather walked all the way back to that little boy, and he said, "Now speak." That little boy started to speak . . . And that little boy never stuttered again.[37]

This magical event will not be the primary *novum* of the overall story, either, but it is framed as what Daniel Heath Justice prefers to call a "wonder": "Wondrous things are *other* and *otherwise*; they're outside the bounds of the everyday and mundane, perhaps unpredictable, but not necessarily alien . . ."[38] It is a wonder presented in the matter-of-fact tone that W. R. Irwin calls "quiet assertion," reducing its novelty by adding it to the ordinary inventory of ontological furniture.[39] We have been put on notice: this world is a magical kind of world, a world where this is simply the kind of thing that happens from time to time.

But Van Camp's assertion of rhetorical sovereignty is not yet exhausted. The bulk of the story is about a somewhat overfamiliar process of heterosexual courtship, in which Jimmy shyly works up the nerve to ask out Shandra. The slot of *difficult-to-believe* is no longer occupied by aliens in the sky or mysterious hummingbirds living in young boys' throats, but by the fact that Jimmy is interested in Shandra and not her more popular sister, Roberta.[40] A certain accumulation of banal detail—steak and lobster, banter about the old days in elementary and how the town's changed, etc.—grounds what we might otherwise read as the fantastic, estranging elements in the bedrock of the expected, the everyday, and the already-known. The alien craft is just another local "sight" that a couple on a date might drive out to see.[41] (They decide not to: "Well, I gotta see your house. I wanna see how you decorate.")[42] No, the story's true novum (another black box) is the revelation Shandra conveys to her friends the morning after: "Jimmy's different . . . he's *beautiful*."[43] In other words, "he's what the Crees say: Aayahkwew: neither man nor woman but both"; "twospirited . . . or transgender, or both, or perhaps something we've never heard of before—even under these new skies."[44] Jimmy is the *novum*, the wonder at the heart of the story. Van Camp's efforts have all been aimed at persuading us to imagine a Native gender/sexual identity that is beyond what "we"—settler and Native alike—know.

Yet Van Camp's story preserves a "deficit" of Indigenous knowledge in the figure of the "something we've never heard of before"—an internal nonknowledge that operates as a sign of possibility. Something similar is at work in the fiction of Darcie Little Badger. In "Nkásht Íí," Little Badger invokes the authority of ancestors:

Great-grandmother taught me everything she knew about death before it took her.
  Never sleep under a juniper tree. They grow between this world and the place below.
  Bury the dead properly, lest their ghosts return.
  A ghost is a terrible thing.
  Someday, we will all be terrible things.
  Great-grandmother, you were right.[45]

All of these warnings are borne out by the narrative that follows, as Josie and Annie investigate the death of a man's daughter near Willowbee, Texas—a death that *has* a mundane explanation (father, mother, and daughter were all in a car crash; only the father survived) but one which is belied by a supernatural experience (the father testifies that the infant survived the crash, but was borne away by an "owl-woman" with the mother's face). "There are legends . . . " Annie murmurs, "The kind my great-grandmother knew."

Here, the invocation of the numinous (the force of which is carried by the ellipsis that follows *there are legends . . .*) is abetted by a folkloric megatext: Native or settler, we have probably heard a story like this one. "Ghosts? Huh, maybe he met La Llorona," Josie muses. "¿Dónde están mis hijos? . . . It's possible, right?" And once again, the gravity of the known grounds, the otherworldly: mundane details (empty coffee cups, cell phones, bus rides, Best Westerns) undergird the *mysterium tremendum*. The gesture of the grieving father who reaches out to grip Annie's shoulder prompts Josie to reach for her Mace—unnecessarily, in the event, yet the same pragmatic instinct also protects her from accepting the offer of a ride from a man who then drives his pickup truck off a bridge and "into the water . . . [without] caus[ing] a single ripple." This is a universe in which, for instance, "karma" seems to operate (Josie and Annie are "paid" for their investigation by lucky lottery tickets), but also one in which the driver who pulls over to offer a ride to two young Native women may be not an ordinary predator but *something worse* (ominously, his sunglasses don't reflect their faces . . .). "After dark, with its baby-killing ghosts and doomed pickup truck drivers, Willowbee seemed unbearably creepy," Josie reflects.[46] The reader, too, is unsettled as "autochthonous, localized, and historically situated knowledge systems" displace modern, universalizing, Western ones. In such a so-called America, it's best to expect the unexpected, to distrust the signage, and to listen instead to one's Apache great-grandmother.

Kai Minosh Pyle draws the dystopian scenario of "How to Survive the Apocalypse for Native Girls" from the sf megatext, saving him much of the effort of justifying and explaining broad changes to society:

> Shanay's grandma is one of the best doctors, because she was trained both Anishinaabe-way and in one of the old universities before the borders broke down. She likes to joke that it's a good thing the apocalypse happened, because that way she didn't have to pay off her student loans, which were apparently a thing that, like money, used to be a big problem for people.[47]

Retrospection (money "*used to be* a big problem") makes the collapse of colonial governments and capitalism and the resumption of full tribal autonomy into a *fait accompli*, no longer a distant ideal but a self-evident response to practical problems. Kinship, *inawemaagan*, offers a readymade local communism so that, as Mark Fisher writes, even the dystopian scenario affords a small space for utopian reimagining.[48]

This defense of utopian imagination is opened to critical inspection, however, as the protagonist, Nigig ("Otter"), struggles to distinguish between emancipatory and oppressive inheritances, to know when what the Council says reflects internalized colonialism—in this case, antiblackness and the colonization of genders and sexualities—more than any authentic tradition. Here, questions of sovereignty are complicated by legacies of self-hatred: despite the authority of Kinship, as a two-spirit girl, Nigig and her kin are despised by some of their kin. In a tense scene, she confronts a hostile Council member, a woman from the Eagle Clan, about her rejection of Nigig's two-spirit friend Migizi (a name which, ironically, means "bald eagle" [Livesay and Nichols]):

> "You can't exile someone just because you don't like them," I said hotly. "Kinship—"
> "Kinship is exactly the reason why that freak had to be gotten rid of," she spat ... "They're dead, child. No one survives long outside the protective network of the Nation."[49]

Nigig's narrative, conveyed in sixteen "instructions" for how to survive the apocalypse, culminates in recognition of others taking part in the internal struggle ("Maybe this, too, is Kinship") and a final instruction: "I know now that the only way to survive the apocalypse is to make your own world."[50] All of Pyle's rhetorical craft has been in support of this utopian call to self-recreation, to the assertion of another kind of (Indigiqueer) sovereignty.

Sovereignties

The concept of sovereignty, of course, is open to multiple competing interpretations among Native activists and artists. It is interesting to note, in this inevitably too-brief survey, that Indigenous Futurist writings participating in rhetorics of incredulity, as I have termed them, seem the least disposed to raise questions about the nature of the Indigenous sovereignty that is being sought and that those manifesting rhetorics of believing seem to do so more often. Is this simply an artifact of selection, since the most reflexive and complex stories were drawn from anthologies of Indigiqueer writing? How do these rhetorics respond to different historically and/or culturally specific needs? These are questions for further scholarship. What seems more certain is that the enterprises of imaginatively and politically "mak[ing] your own world" are mutually implicated.

PART III

Toward a Multiplicative Realism

8

The Rhetoric of Fictive Science; Or, The Academy of Outrageous Books

> Recall that Poe calls diddling a science—a rational practice, achieved through testing and through doubt . . . In a way he is mocking pseudoscience by suggesting a science of falseness. But Poe may as well have meant science fiction, a genre he is often credited with starting . . .
> —Kevin Young, *Bunk: The Rise of Hoaxes, Humbug, Plagiarists, Phonies, Post-Facts, and Fake News*

If the Indigenous sovereignties that modernity had thought to destroy have proven indestructible, what becomes of the sciences that were its proudest achievement? Postmodern relativism was a relatively highbrow concern in the twentieth century; now, in the so-called "post-truth era," it seems to be a matter of terrible public urgency. It is difficult to even discern a single entity called "the public" anymore; we inhabit such different universes of discourse that it is as if we inhabited different universes (without even a pluriverse to connect us). Fantasies of secession or space colonization aside, these disparate worlds still concern one and the same Earth. The sciences that describe that reality and the trouble it is in are themselves in trouble. Thus, the Marches for Science and renewed calls for a war on all manner of outrages against it—even if those outrages issue from just across campus, in the Humanities building.

There is so much to be outraged by in the present circumstances (2017-the foreseeable future?) that we have to worry as to whether outrage itself is a precious and nonrenewable commodity. As such, the competition held by *Togakkai*, the Academy of Outrageous Books, might seem especially profligate: prizes are awarded yearly to the "best" *tondemo-bon*, books that are "amusing from a perspective that differs from what the author intends"—a perspective not unlike that which, back in the 1980s, the Church of the Subgenius brought to bear on conspiracy theory, pseudoscience, New Age hucksterism, and other contemporary varieties of "diddling." Togakkai and Subgenii alike read these productions (as if following Kevin Young's suggestion) as a kind of science fiction. Surely the pleasure of laughing at the rubes who read such "outrageous" productions in a nonscience-fictional manner must be fading when the gaudy absurdities of QAnon, promoted by media organizations reaching half the country, has become a significant factor in a US presidential election (November 2020). Under such surreal conditions, the arch attitude modeled by the Academy of Outrageous Books loses all efficacy: it becomes impossible to ironically quote "outrageous" discourses without simply *reproducing* them, becoming another vector of their viral spread. Conversely, the ironic enjoyment of the "Bad Writing Contest" run by the journal *Philosophy and Literature* in the 1990s—notable "winners," exemplars of outrageously dense jargon, included Judith Butler, Frederic Jameson, and Homi K. Bhaba—must surely have turned to ash, now that the neoliberal demolition of universities is well and truly underway. The only amusement that is to be gained, surely from a perspective different than what these authors intended, must belong to the conspiracy theorists who find in such writing further proof that universities are the purveyors of a nefarious "Cultural Marxism" aimed at destroying America.

If only! If only they could see, as we do from our positions inside the academy, the ascendancy of the STEM disciplines and the decline of such supposedly wooly-headed endeavors as History, Philosophy, and English. It is as if we were back in the heady days of Empiricism with David Hume, ready to launch a new Inquisition:

> When we run over libraries, persuaded of these principles, what havoc must we make? If we take in our hand any volume; of divinity or school metaphysics, for instance; let us ask, *Does it contain any abstract reasoning concerning quantity or number?* No. *Does it contain any experimental reasoning concerning matter of fact and existence?* No. Commit it then to the flames: For it can contain nothing but sophistry and illusion.[1]

That fire alarm has been ringing now for some thirty years. In my academic corner of the Great Sorting Machine, under the pressure of forty years' austerity now compounded by the Emergency, departments and programs are being asked to sort themselves into the essential and the extraneous, useful and useless, profit centers and cost centers: the streamlined future of the university and its dusty past. Those of us in the dustier precincts, the ones labeled HUMANITIES, are told that we need to argue for our continued relevance, to armor ourselves against the arrival of the Empiricist vandals, to clad our work in code, in numbers.

We have, in fact, been doing this coding work for quite a while now—only a great deal of contemporary theory is running on *n*th-generation patches to old philosophical code from the French 1950s-1970s, and buried in *that* software is some much older science. Of this, Marx's nineteenth-century economics and Freud's turn-of-the-century neurology are far from the dustiest relics. Take, for example, the currency of Gilles Deleuze's repurposing of Baruch Spinoza's remark about "the body":

> Spinoza offers philosophers a new model: the body. He proposes to establish the body as a model: "We do not know what the body can do . . ." This declaration of ignorance is a provocation. We speak of consciousness and its decrees, of the will and its effects, of the thousand ways of moving the body, of dominating the body and the passions—but *we do not even know what a body can do*. Lacking this knowledge, we engage in idle talk.[2]

Or rather, in Samuel Shirley's translation of Spinoza:

> However, nobody *as yet* has determined the limits of the body's capabilities: that is, nobody *as yet* has learned from experience what the body can and cannot do, without being determined by mind, solely from the laws of its nature in so far as it is considered as corporeal.[3]

Since 1677, we have learned a lot about the limits of the human body, have we not? Hasn't physiology advanced even a bit in the last three centuries? Or is Spinoza not referring to the state of scientific knowledge about human bodies at all? In other words, when we read the statement "We do not know what the body can do," are we reading a paraphrase of an outdated report on the current state of empirical knowledge ("nobody *as yet* has learned . . ."), or something belonging to another register of discourse altogether, a philosophical register, in which statements are not necessarily subject to empirical proof or disproof (as ethical discourse is generally taken to be, and as literary discourse is taken to be)?

A skeptic, a New Atheist, or simply a defender of science might regard this last suggestion as a weaselly tactic, a shifting of the goal posts: when religious beliefs run up against the hard facts of scientific knowledge, the crafty response is to reinterpret those beliefs in ways that are compatible with this knowledge and then claim that one's reinterpretation is what the content of those beliefs actually was all along. Indeed, religious discourses are well-equipped for this kind of defensive maneuver: as Murray Bookchin grumbles, "the Eastern sages talk in vague phrases that often have multiple meanings, as befits most religious teachers, who normally hedge their statements lest a prophecy fail to materialize in reality."[4] Yet a division between language-games is the peacekeeping strategy of no less a defender of science than Stephen Jay Gould (1997):

> The text of [the papal encyclical] *Humani Generis* focuses on the magisterium (or teaching authority) of the Church—a word derived not from any concept of majesty or awe but from the different notion of teaching, for magister is Latin for "teacher." We may, I think, adopt this word and concept to express the central point of this essay and the principled resolution of supposed "conflict" or "warfare" between science and religion. No such conflict should exist because each subject has a legitimate magisterium, or domain of teaching authority—and these magisteria do not overlap (the principle that I would like to designate as NOMA, or "nonoverlapping magisteria"). The net of science covers the empirical universe: what is it made of (fact) and why does it work this way (theory). The net of religion extends over questions of moral meaning and value. These two magisteria do not overlap, nor do they encompass all inquiry (consider, for starters, the magisterium of art and the meaning of beauty). To cite the arch clichés, we get the age of rocks, and religion retains the rock of ages; we study how the heavens go, and they determine how to go to heaven.[5]

This was a peace plan offered in the wake of the Sokal Affair (1996), the event that marked a new phase of the Science Wars. Ironically, "Science Wars" had been the title of the special issue of *Social Text* in which Alan Sokal published his scholarly hoax article, "Transgressing the Boundaries: Towards a Transformative Hermeneutics of Quantum Gravity." The "Science Wars" issue had been intended as a "concerted, published response to [Paul R. Gross and Norman Levitt's] *Higher Superstition*," which was an attack on "the academic left and its quarrels with science" (1994); subsequently, however, the issue became famous (or infamous) for the editors' having taken Sokal's bait and inadvertently published an article full of pseudoscientific fakery.[6] The point Sokal wanted to make was that scholars in the humanities, especially prone to poetically misreading the findings of contemporary physics, had been "transgressing the boundaries" between Gould's magisteria, illegitimately appropriating concepts and language from the sciences. Gould, a one-man United Nations, attempts to return the combatants to their territories and redraw the lines of demarcation. We may note that his peace plan looks awfully similar to Kant's Modern Constitution: on one side is nature and objectivity, on the other, social relations and subjectivity. Such a peace is perhaps only a little better than a continuation of war by other means.

The Science Wars are old, and I want to register as a conscientious objector. Here is a short list of my objections:

a) With apologies to Gould, as any history of science shows, for every Galileo persecuted by the Church, there is a Newton obsessed with angelology, and for every Darwin speculating on moral philosophy, there is a Hobbes speculating on natural history, so that the bounds between magisteria have always shifted and overlapped; even an n-state solution such as NOMA is unlikely to be workable. We should be ready to ask whether three and a half centuries have seen some empirical determination of the limits of what a body can do, as we should be ready to inquire into the relation between the Reverend Thomas Malthus' statistical charts and his doctrine of original sin.

b) Cross-magisterial critique is possible and desirable if only we take William James' "pragmatist corridor" between them. "Against rationalism as a pretension and a method," with its attempt to corral all experience into a single scheme, pragmatism . . . lies in the midst of our theories like a corridor in a hotel. Innumerable chambers open out of it. In one, you may find a man writing an atheistic volume; in the next, someone on his knees praying for faith and strength; in a third, a chemist investigating a body's properties. In a fourth, a system of idealistic metaphysics is being excogitated; in a fifth, the impossibility of metaphysics is being shown. But they all own the corridor, and all must pass through it if they want a practicable way of getting into or out of their respective rooms.[7] The problem becomes not one of partition, then, but of translation.

c) We are right to be wary of the mystification of scientific methods and findings; indeed, scientists share this worry since the "junk science" of the climate denialists and antivaxxers takes advantage of this mystified status (blackboxing) accorded to all things scientific in the popular imaginary. If Hobbes at points disavowed any natural-historical pretensions for his "state of nature" theory, it has nonetheless become sedimented in our culture to the point that many *believe* that it once existed and are *afraid* that it might return—dispositions inculcated by the official "contemporary myth" (Barthes) taught in the US schools: the pseudo-field test of *Leviathan* afforded by William Golding's SF novel, *Lord of the Flies*. If all fiction with a realistic pretense operates something like one of Zola's fictive experiments, offering its plotting as an "*argument*" that "if you place this person A in this situation B, the result will be event C," i.e., that "this is how the world works," this particular bit of simulated social science has had especially pernicious effects, such that tweets about the supposed anarchy in the streets of Portland can simply characterize it as "Lord of the Flies happening in an American city."[8]

d) Mythopoiesis is an indefinite process opening onto the pragmatist corridor, a verbal "magic" that is subject to "the 'collective revelation' of testing and discussion," not a black box closed for all time.[9] Although we certainly ought to be wary of some mythologies—for instance, Hobbes' and Golding's scientific myths—mythmaking has kept Indigenous peoples alive, and as the examples of Alfonso Herrera and Leonid Andrenko show, even "junk science" may, if handled thoughtfully and carefully, have something like the value of other sorts of junk, the kind found in your attic or basement, the kind of junk one might sort through and come up with something useful. If we are to clean house, we ought to do so slowly and carefully. For instance, one glittering item we might rescue from the cluttered attic might be the speculations of the Russian "biocosmists" of the 1920s. A. M. Gittlitz, writing for the *New Inquiry*, wants to consign them to the dustbin of history. Not only are their technological ambitions indistinguishable from the utopian aspirations of the Bolsheviks then; ironically, their dreams of space colonization and human life extension are also catnip to the neoliberal techno-utopians of Silicon Valley today. Guilt by association! An argument too quickly concluded! This is a *wasteful* mode of critique, a kind of Humean arson in the name of purification.[10]

I am tired of warding off arsonists. Do we need fewer empty words in the humanities and more data, more numbers? Very well, then. Below are five equations. Of course, as soon as we start to write equations like this, we are accused of speaking in code or of abusing someone else's. It is true that out of all of the above examples, the only two that are *mathematically* valid, to the best of my knowledge, are the first two—perhaps predictably so since one was written

I. $C' = (c + v) + s$

II. $F = c\dfrac{p}{a}$

III. $+13\sqrt{+} = \sqrt{\%}$

IV. $\dfrac{S}{s} = s = \sqrt{-1}$

V. $\displaystyle\int_{\alpha}^{\Omega} \operatorname{Terra}[x]_n = \iiiint \ldots \iint \operatorname{Terra}(x_1, x_2, x_3, x_4, \ldots, x_n, t)\, dx_1\, dx_2\, dx_3\, dx_4 \ldots dx_n dt = \operatorname{Terrapolis}$

by an economist, Karl Marx (1818–1883), and the other by a civil engineer and astronomer, Fernando Tárrida del Mármol (1861–1915). The third, which is, in fact, not an equation nor even a mathematical expression, despite the appearance of the "equals" sign, appears in *The Grammar of the AO Logic Language* (*Grammatika logicheskogo yazyka AO*, 1924) by the linguist Wolf Gordin (1885–1974). The fourth, which has been fiercely derided as "fashionable nonsense," illogic and arbitrariness disguised as math, is the work of a psychoanalyst, Jacques Lacan (1901–1981).[11] The last appears in a recent book by feminist scholar Donna Haraway (1944–), who calls it a "a story, a speculative fabulation, and a string figure for multispecies worlding."[12] Each, in their own way, makes an appeal to the authority of "universal science," of "measurement and order," and therefore represents an exercise in rhetoric.[13] Let us see how each functions within and as an argument.

Marx's Simple Formula

In *Capital, vol. 2*, Marx summarizes the process by which capital is reproduced and accumulated by the formula "M—C ... P ... C'—M,'" whereby a sum of money (M) is expended on commodities (C), including raw materials and labor, which are combined in a production process (P), the product of which (C,' or a new commodity) is sold for a greater sum of money (M'), beginning the cycle anew.[14] In the first volume, Marx breaks down the transition between C and C':

> The capital C is made up of two components, one, the sum of money c laid out upon the means of production, and the other, the sum of money v expended upon the labour power; c represents the portion that has become constant capital, and v the portion that has become variable capital. At first then, C = (c + v)... When the process of production is finished, we get a commodity whose value = (c + v) + s, where s is the surplus value ... [Thus,] the formula C = (c + v)... was transformed into C'=(c+v)+s, C becoming C'.[15]

This equation, then, is not the kind in which the product of mathematical operations yields some kind of surprise or new knowledge (as in the Drake Equation, once one plugs in specific estimates for each of its factors); it is, rather, a "tautology," as Marx quickly adds, leading us back to the already known.[16] By definition, the process of transforming a sum of capital (C) into a larger sum of capital (C') must be the same as adding something (a surplus, s) to what was already present (c + v, the capital's constant and variable components). Why restate the expression C' as the more complex expression c + v + s?

One way of thinking about it is that this is Marx writing didactically, as a philosopher-pedagogue, taking a concept and analyzing it into its components. *Capital* certainly found use as a teaching text in socialist night schools. However, it is also a persuasive text, and not only in the places where its "rhetoricity" is most visible, e.g., in his use of fantastic figures of speech (the famous examples of the wooden table coming to life or the image of capital as vampiric) or his dramatization of the reading process (as when he theatrically invites the reader to peek behind the appearances of free agreement to witness the tyranny of capital over labor).[17] On the whole, *Capital* seems dry and technical when juxtaposed with the polemical fireworks displays of *The Eighteenth Brumaire* or *The Manifesto of the Communist Party*. *Capital*'s rhetoricity is primarily a matter of concealing its own rhetoricity. It is designed to persuade primarily by what Foucault calls *mathesis*, by using formalization to reduce the complex to the simple[18]—and it does so by representing capitalism itself as a kind of abstract machine designed to process the complexities of material life (the infinite plurality of things, in all their sensuous particularity, and of their uses, as well as the messy multiplicity of human social relationships) into simplicity: *all* the qualitative features of matter are reduced (ultimately) to "the universal equivalent" of money, *all* the social classes reduced (ultimately) to two.[19] It is in this sense that Evgeniy Kuchinov suggests treating *Capital* as "a science fiction work"—we might add, a *subtractive world*.[20]

Tárrida's Simple Formula

Tárrida del Mármol's *Problemas trascendentales: Estudios de sociología y ciencia moderna* (1908) collects some twenty-six essays divided into sections on "Popular Science," "Scientific Sociology," and "Cosmology." In the first and second sections, we find speculations on the electrochemical mechanisms behind the homing instincts of animals, on "the molecular sensibility of minerals" ("many naturalists," he notes, "and especially the Mexican Herrera, have shown how indecisive are the boundaries that separate what is inorganic from what is living"), on Kropotkin's studies of mutual aid, and on the possibilities for achieving a "harmony of the passions."[21] The third section presents works closest to those he published in scientific journals such as *Knowledge & Scientific News* in the first decade of the twentieth century, replete with equations describing orbital mechanics, e.g.:

$$t = \frac{\sqrt{c}}{\pi} \times \frac{T}{\cos.a} \times \sqrt{\frac{M+m}{m+m'}} \times \sqrt{\frac{d}{D}}$$

This was Tárrida's mathematical model explaining the phenomenon now known as tidal locking, which results in satellites like the Moon showing only one face to their planets.[22] Here is the engineer's explanation of his "very simple mathematical formula for general felicity" (from his essay, "El Cero de la Autoridad [The Zero of Authority]"):

> Let p stand for the progress made at a given moment; a, the quantity of authority existing. Since felicity is in direct proportion to p and in inverse ratio of a, the formula for Felicity (F) at that moment will be: $F = c.\frac{p}{a}$, where c is a coefficient function of a certain number of finite quantities.[23]

In other words, this is a hypothesis about human happiness or an expression of the proposition that human "felicity" is increased, all things being equal (the c might as well stand for the Latin weasel-phrase, *ceteris paribus*) to the extent that "progress" exceeds "authority." The trouble here is less that it is difficult to quantify "progress," "authority," or "felicity," as political scientists are in the habit of doing pretty constantly. Rather, the troublesome c ("a coefficient function of a certain number of finite quantities") throws the relationship of all the other factors in the equation into some doubt.

What does this "formula" make clear, and what does it make murky? What could c account for? (I am reminded of Proudhon's tongue-in-cheek suggestion that exploitation is a kind of curiously persistent accounting error.) The single sign hides a multiplicity of forces or conditions. We might reflect, as Tárrida could not, that what passes for "progress" in contemporary China—the alleviation of poverty, the rise in standards of living, and access to consumer goods as the token of "felicity"—is taken to offset the growth in "authority," as central Party power is amplified by sophisticated surveillance technologies. Perhaps a, too, confers a false simplicity on the multiplicity of reasons why we comply with what others desire of us, for better and for worse: the experiential authority of skilled laborers, the moral authority of elders, and the scholarly authority of experts are condensed into this single letter, along with the authority of bosses, priests, and rulers.

But the slight air of flippancy with which Tárrida proposes his "very simple mathematical formula" should make us wonder if he isn't making a kind of algebraic joke, gently subverting the rhetoric of mathematical certainty. "The only exact sciences," he writes, "are mathematical sciences or those to which the application of mathematical principles lends accuracy"; elsewhere, he suggests, *mathesis* can obfuscate, appearing "to confer . . . the character of eternal truths" on what is really ephemeral and hypothetical.[24] However, Tárrida derives another algebraic consequence from this expression: "if a were to become zero (0)," he deduces,

> the formula for general happiness, at that time, would be
>
> $$F = c. \frac{p}{0} = \infty$$
>
> that is, a quantity greater than any quantity, however large, or, in other words, that would no longer be susceptible to increase.
>
> The object of life would have ceased to exist.[25]

This implication, he reports, displeased other attendees at the London anarchist conference where he presented his work.[26] After all, what kind of anarchist could propose that even an infinitesimally tiny amount of authority was required for life to have an objective? Yet Fernando Tárrida del Mármol, B. Sc., C.E., former professor of mathematics at the Academia Politécnica de Barcelona and former political detainee in Barcelona's most notorious prison, contributor of a regular column on scientific matters to the anarchist journal *La Revista Blanca*, called himself an "anarchist without adjectives."[27] His eyes were on the stars, but his feet were very much on the ground.

Gordin's Encrypted Message

The third equation is not an equation at all, in spite of the equals sign, nor even a mathematical expression; rather, it is a word, pronounced *ebdie-iz*, in Wolf Gordin's constructed language, AO, which he first elaborated in the early 1920s in Moscow (which he renamed "AO-grad"). This immense project, which formed the centerpiece of a 1927 exhibition in Moscow (in the breathless prose of its impresarios, the Association of Inventist Inventors, "the world's first exhibition of models and mechanism of interplanetary vehicles constructed by inventors of different countries"), deserves to be linked not only to the social history of Zamenhof's Esperanto, but to the literary history of J. R. R. Tolkien's Quenya (Elvish), H. P. Lovecraft's Aklo, George Orwell's Newspeak, Marc Okrand's Klingon, and the Dothraki of George R. R. Martin and David J. Peterson. This is not least because it is adumbrated, at least in its important principles, by the Gordin brothers' second utopian novel (or "utopiia-poema"), *Anarkhiia v Mechte: Strana Anarkhiia* (*Anarchy in a Dream: The Land of Anarchy*, 1919).

Gordin's *Grammatika logicheskogo yazyka AO* informs us that *iz* ($\sqrt{\%}$) represents an intensification of *ebdie* ($+13\sqrt{+}$), and the equals sign, Gordin writes, is a "German hyphen."[28] Somewhere, there exists or existed a dictionary of AO[29] which explains what *ebdie* means (or meant); perhaps, like the author of the dictionary itself, it was expunged from the space of the Union of Soviet Socialist Republics ("four words, four lies," as Cornelius Castoriadis put it). Like Tárrida, Wolf Gordin and his brother Abba (1887–1964) were anarchists, although they preferred to add many adjectives, such as "inventist," "sociotechnical," "inter-individualist," "panmethodological," and "universalist," seemingly in hopes that these would persuade the OGPU to regard them as harmless. Mathematization can be regarded as another strategy of protective coloration; the Gordin Brothers evolved into cryptobiotic forms of intellectual life.

Thus, Wolf, the author of this language, encoded himself into it: he signed his work "*Beobi*," or rather, "$1+01\sqrt{}$," meaning "Society-Me" or simply "Human." Indeed, although these strings of signs may seem rather coldly inhuman, Gordin/Beobi calls AO "the Language of Humanity."[30] At the same time that it withdraws meaning from the eyes of the State, then, mathematization also makes a "universalist" gesture: just as Carl Sagan suggested that mathematics could form the first foundations of communication with other intelligent life forms in the universe (*Communication* 346), Beobi and his followers believed that AO, with its numerical alphabet, could provide the basis for the first truly international—indeed, "interplanetary"—civilization.

The use of five mathematical operators (×, +, √, −, %) and six numerals (0–5) as the alphabet of "the language of a new civilization" was itself a rhetorical choice, distancing AO not only from all inherited languages but also from its "international" competitors, such as Esperanto, Ido, Volapük, or the Isotype picture-language of Otto Neurath and Gerd Arntz (Beobi, *Grammatika* 16). These rival international languages are typically constructed by "purifying" existing, historically evolved languages, combining their useful elements, dispensing with their excrescences, and thereby *rationalizing* them, like Baron Haussmann imposing geometrical order on the unruly medieval streets of Paris. AO was constructed instead from the ground up, *ab novo*, starting from first principles, like Spinoza's *Ethics*, with its quasi-mathematical deductions (proceeding "in detailed geometrical fashion" from axioms to proofs):

A.

AXIOMICS

(according to the old system, ALPHABET).

1) Vowel axiomics: ×, +, o, √, −.
Aa, Ee, Oo, Ii, Uu-Yuyu.
2) Consonant axiomics: 1, 2, 3, 4, 5,%.
Bb, Cc, Dd, Ff, Ll, Zz.

. . . .

The first alphabet, the alphabet of vowels, represents grammatical forms, and the alphabet of consonants—from a minimum of numerical relations in conjunction with the first—forms the outline on which the entire language is embroidered.

The ABC in the AO language represents, therefore, not only its phonetic basis (as in all modern languages) but also its logical basis (as opposed to pre-existing languages).

The ABC of AO is, therefore, a kind of axiomatics of the language, of its combinations, and of the operations of which the whole language consists, precisely constructed on a mathematical-logical basis. . . .

× (*a*), as a sign of multiplication, symbolizes invention due to the fact that invention alone multiplies benefits, while all other activities only repeat, imitate and exploit the first.

That + (*e*) denotes "yes" and—(*u*) negation does not require any special clarification since the + sign in mathematics denotes positive numbers and magnitudes, and the—sign (minus) negative numbers and magnitudes.

> Mathematical 0 (zero), formed by the same picture, by a fortunate chance for us, as the Latin O serves as an important and formidable symbol of the worthlessness, the nullity of the non-inventive "natural," i.e., focusing on the "nature" of civilization, the objects of which are, after they have been invented, pure zeros for man and his life, for civilization. Every object is the result of an invention, but it is no longer an invention itself but the zero limit of an invention.
>
> The symbolization of a property, an adjective, by the algebraic square root sign √ (*i*) is motivated by the fact that any invention of a new property is, as it were, an extraction, an invention of the "essence" of a thing.[31]

Ironically, in spite of the designer's intentions—like the ideogrammatic written languages of the East, but unlike any other Western language, every fundamental element of AO is meaningful—it is impossible to guess the meaning of *ebdie-iz* with any certainty. It may have something to do with the idea of "communication-invention or union," as it appears under that heading, but *ebd* (+13) is the AO word for "meta-ethics" (from +1, sign of the future, and 3, sign of the instrument). The notation +13√+=√% is now an unreadable fragment, an encrypted message from the past to a future that hasn't yet and may never arrive. Like the "letter from the stars" in Stanislaw Lem's *His Master's Voice*, it may remain an enigma, forever beyond the reach of Turing's machines.

Lacan's Black Boxes

Commentators on Lacan's work, according to Christian Lundberg, often distinguish between an early "rhetorical" period as distinct from a late Lacan who formalizes everything into quasi-mathematical expressions, or "mathemes."[32] Both premises of this periodization are open to doubt, as the early Lacan is also drawn to mathematical analogies, and all such analogies are argumentative, rhetorical. The equation reproduced here comes from his 1960 talk, "The Subversion of the Subject and the Dialectic of Desire in the Freudian Unconscious":

> I will begin with what the abbreviation S($\cancel{A}$) articulates, being first of all a signifier. My definition of the signifier (there is no other) is as follows: a signifier is what represents the subject to another signifier.
>
> This latter signifier is therefore the signifier to which all the other signifiers represent the subject—which means that if this signifier is missing, all the other signifiers represent nothing. For something is only represented to.

> Now insofar as the battery of signifiers is, it is complete, and this signifier can only be a line that is drawn from its circle without being able to be counted in it. This can be symbolized by the inherence of a (-1) in the set of signifiers.
>
> It is, as such, unpronounceable, but its operation is not, for the latter is what occurs whenever a proper name is pronounced. Its statement is equal to its signification.
>
> Hence, by calculating this signification according to the algebra I use, namely:
>
> $$\frac{S\ (\text{signifier})}{s\ (\text{signified})} = s(\text{statement})$$
>
> with $S = (-1)$ we find: $s = \sqrt{-1}$
>
> This is what the subject is missing in thinking he is exhaustively accounted for by his cogito—he is missing what is unthinkable about him.[33]

Bruce Fink complains that by quoting this passage (in a poor translation) without any explication, Sokal and Bricmont have, in effect, blackboxed it: "This passage, without any further explanation, is obviously incomprehensible."[34] Let's open the black box.

This algebraic expression serves as a demonstration in the course of a larger story Lacan is telling about the emergence of a subject, his famous "Graph of Desire," a diagrammatic structure a little akin to the "autocatalytic tree" that William Gibson and Bruce Sterling imagine in *The Difference Engine* (1990), their image for the first-coming-to-consciousness of an intelligent computer, but even more so like the first-coming-to-consciousness story that Greg Bear tells in *Queen of Angels* (1990), in which a robot probe sent to explore the Centauri system becomes conscious only when it first recognizes that the living structures it has found are *not* evidence of any sentient life, registering this with the terrible recognition: "I am *lonely*."[35] This perception of "a lack in the Other," S(A̸), is where Lacan begins.[36]

In Lacanian theory, A, the "Big Other [Autre]," is not a particular person (an "other" or *autre*, "a") but a "battery of signifiers," the sociosemantic network of the Symbolic. When you're aware that Big Other is Watching You, the sense of social surveillance has nothing to do with any specific pair of eyes but with the vast, anonymous machinery of social relations imagined as a kind of pseudo-person who is taking stock of what you do and say. It is something like Heidegger's concept of the "They," *das Mann*, which haunts the term papers of less experienced student writers: "They say that . . ." "Society thinks that . . ." This Big Other is evoked more subtly and effectively by more experienced writers in such innocuous-seeming phrases as "Of course..." "As we know..." As the Symbolic, the Big Other *is* language, and anything that can be can be expressed in language, *of course*. In that sense, language is thought to be "complete": all the signifiers you will ever need are in the battery, ready to be used. But whenever we give a name to something new, we reveal that something was, in fact, missing from the battery, that the supposedly full and seamless Symbolic network was *minus one signifier* (S =-1): hence "a lack in the Other."[37]

Lacan is careful to distinguish his "algebra" from the kind we're used to: ordinarily, his mathemes can't be operated on in any mathematical sense. In expressing the idea of the signified as $\sqrt{-1}$ (the "imaginary number" of mathematics, i), he renounces "any claim to its being able to be used automatically in subsequent operations."[38] At the same time, however, he breaks his own rule by simplifying the "equation" using the mathematical operation of multiplication, expressing $\frac{-1}{s} = s$ as $-1 = s^2$ to find $\sqrt{-1} = s$.[39] Back-translated into the language of Saussurean linguistics that Lacan begins with (signifiers, signifieds, symbolic networks, etc.), we get something like this: because the battery of signifiers is necessarily incomplete (since something or someone new can always be named), the signified of a name is an impossibility, like the imaginary number i, which functions mathematically even though it is fundamentally senseless. We could read Lacan's formula, then, as another way of expressing the idea of the "cognitive unconscious": there is an unthought substrate underlying thought, a kind of babble underlying the Babel of language as such.

Is there really anything earthshaking, however, in the revelation that no language, at any given point in its development, is capable of expressing everything that may exist or that processes of naming can bring new things into linguistic expression? Apart from any lingering Cartesians, does anyone disbelieve in a cognitive unconscious? Surely, it is more realistic, on materialist grounds alone, to suppose that the complexities of thought and consciousness emerge from the simplicity of blind and mindless processes. It is not that Lacan has said anything ludicrous or even especially provocative here; the mathematical trappings are, if anything, a mystification, a black box. They are calculated to persuade us of the authority of the Barred Author, a(A̶).[40]

Haraway's Utopian Calculus

Whereas Lacan pretends to set strict interpretative bounds to his "algebra" (no mathematical derivations allowed!) only to violate them when it suits his argument (let's make mathematical derivations!), Donna Haraway is clear from the beginning that her "multiple integral equation for Terrapolis" is, if not quite a parody of math, "a story, a speculative fabulation."[41] The premise is reminiscent of Stephen Duncombe's "metric"—

$$S_x = \frac{1}{x}\sum_{i=1}^{x}\frac{\Delta ai}{\Delta di} + S_y$$

—for the "æffectiveness" of activist art (its simultaneously *affective* and *efficacious* character):

> Am I serious? Yes . . . and no. The math works, but what it can work on is very little. At times, with relatively straightforward objectives that can be easily measured, such as increasing attendance at an organizing meeting, we might be able to use a formula like this. But such easily quantifiable objectives are few and far between, and this isn't really my point. I offer the formulas above more as a metaphor than mathematics, a heuristic tool to get us thinking about the æffect of artistic activism and how to account for it.[42]

Like Haraway, Duncombe offers a mathematical "expression" in a double sense; what it expresses is partly its own limitation, its inability to adequately capture the true qualitative complexity of an aesthetic and social phenomenon.[43] Haraway provides a "mathematical joke-exposition" of her equation:

> x_1 = stuff/physis, x_2 = capacity, x_3 = sociality, x_4 = materiality, x_n = dimensions-yet-to-come
> α (alpha) = EcologicalEvolutionaryDevelopmental Biology's multispecies epigenesis
> Ω (omega) = recuperating terra's pluriverse
> t = worlding time, not container time, entangled times of past/present/yet to come[44]

Let's assume that the "integrals" here stand not for mathematical expressions of volume but for the concept of *integration* (linked, perhaps, to her "string figure" imagery of "knots . . . ramify[ing] and doubl[ing] back in many attachment sites").[45] Terrapolis, as a utopian alternative to the disintegration enforced by the division of life into nature and society, animal and human, *terra* and *polis* (or *zoe* and *bios*), would then represent a mode of "recuperati[on]" for damaged life, a reintegration of what has been sundered.[46] If this is accurate, we may find Haraway's mathematical joking to be overly clunky (altogether too heavy with trendy theoretical references) or overly lyrical (too allusive and playful to be very useful for the materialist analysis of the socioecological crisis), but it is more complex than the formulas of Marx and Tárrida, more honest and modest about its epistemological claims than Lacan's or Wolf Gordin's.

Haraway's mathematical imagination certainly runs farther than the imaginary of her checkbook, which has sent small donations to a string of liberal and centrist candidates and causes in recent months (November 2020). (Abject fear has prompted me to do similarly.) Perhaps Liza Featherstone is right to suggest that "there may be less than meets the eye" when it comes time for Haraway's theoretical abstractions to do work in the world.[47] Just the opposite when it comes to the simplicity of our simplest equations, where there is much more than meets the eye: in their lives, Marx and Tárrida both personally risked something, staked something on the firm conviction that $C' = (c+v)+s$ or that $F=c\{p \text{ over } a\}$. Can Terrapolis' pluriverse be rescued from the limitations of its creator?

Fictive Sciences

How is it that Victorian sciences—the science of history, the science of the psyche—can lose their strictly scientific status in the eyes of the international scientific community while retaining a certain quantity of truth, even acquiring truths that were not available to their creators? Science fiction may have something to teach us here, as it is customarily in the business of inventing not only fictive technics, such as the "cyberspace matrix" of William Gibson's Sprawl trilogy, but also entire fictive sciences, such as the "psychohistory" of Isaac Asimov's Foundation trilogy. Just as "cyberspace" came into being as a sign without a referent, acquiring its referent belatedly, so fictive sciences—such as the theoretical exobiology of Leonid Andrenko—may someday be verified, just as nonfictive sciences—such as the plasmogeny of Alfonso L. Herrera—may lose their status. Even faster than the traffic between fictive and real technics or sciences is that between fictive and real religions, of course: to take but two examples, the Church of All Worlds from Robert Heinlein's *Stranger In a Strange Land* became an actual, living branch of paganism, and L. Ron Hubbard's Dianetics, promoted by SF editor/impresario John W. Campbell, Jr., has flowered into the Church of Scientology, counting tens of thousands of devotees. While Scientology is primarily successful as a money-making operation, it is by no means clear that fantastic religions are never of use to their adherents: surely I was not the only child to whisper, at night, Patrick's Rune from Madeleine L'Engle's *A Swiftly Tilting Planet* or the Litany Against Fear from Frank Herbert's *Dune*. May not fictive sciences be judged by the same pragmatist standard—by the "felicity" of their performative effects rather than by their accuracy of reference to some objective state of affairs?

Why not, in short, a meta-fetishistic relationship to Theory? If the formula for fetishism is Octave Mannoni's "je sais bien, et quand même" ("I know very well . . . and yet . . ."), why not lucidly admit:

- I know very well that there is no objectively existing "mirror stage" that can be verified by empirical investigation across cultures,[48] and yet I find that there is much to be learned from investigating the connections between images, self-identity, mimesis, the gaze, and identification with others.
- I know very well that *Capital* represents, at best, an idealized model of a capitalist economy that sins drastically by omission, and yet I find it (non-exclusively) useful as a demonstration of the capitalist system's fundamental incoherence.[49]
- I know very well that Saussure's semiology omits reference so that semiosis appears as a strictly self-referential process without beginning or end, casually "eras[ing] the world,"[50] and yet it can be important to attend to the ways in which signs operate *as if* independent of embodied realities.

All of these theories are incomplete in ways that limit their claims to truth, so that adopting any of them as a perspective means adopting, in Žižek's terms, a fetishistic stance.[51] "The magical decree," as Kenneth Burke points out, "is implicit in all language"—i.e., language itself is fetishistic—in so far as "the mere act of naming an object or situation decrees that it is to be singled out as such-and-such rather than as something-other."[52] Meta-fetishism, in this sense, emphatically does *not* entail an uncritical stance, but it does entail a certain "principle of charity" toward theory, based on the principle that charity begins at home: we are always operating under the condition of incomplete knowledge, operating *as if* certain assumptions were true (or rather, true enough for practical purposes).

Wolf Gordin anticipates us here. In 1921, writing under his AO name, "Beobi," he pens a tract on epistemology, *Gnoseologiya*, heavily influenced by the pragmatism of William James. He asks:

> If truth is effectively utilitarian and temporary, then why call it true, i.e., why add the concept of timelessness and supersufficiency to it? If truths serve only

as "working hypotheses," then they are no longer true, but plans, drawings, practical activities. Science, then, can no longer be science. In any case, there is no place for theoretical science anymore. If it is all about practice and utilization, then theoretical science must give way to applied science, as applied science, in its turn, must be transformed into pure technics.[53]

We needn't go so far as to say, with Wolf and his brother Abba, that all science is mere "sociomagic" in order to appreciate the invitation to leave behind something that Murray Bookchin would call "scientism"—an ideology that claims a "neutral" subject position, detached from any particular "social context," eschewing "philosophical generalizations" in favor of what is quantifiable and measurable, but also claiming a universality that overrides the claims of any local knowledge or subjective desire.[54] This empiricist (and imperialistic) ideology has scarcely weakened in the century since the Gordins wrote of it; indeed, in the last generation, it has blossomed into a militant (and militarist) "New Atheism." The Science Wars of the 1990s, it seems, were only a proving ground for the rhetorical weapons used on the cultural front of the War on Terror in the 2000s. It is in search of an exit strategy from this war that we now turn.

9

How the Gordin Brothers Escaped Western Gravity

> A man breaking his journey between one place and another at a third place of no name, character, population or significance, sees a unicorn cross his path and disappear. That in itself is startling, but there but there are precedents for mystical encounters of various kinds, or to be less extreme, a choice of persuasions to put it down to fancy; until—"My God," says a second man, "I must be dreaming, I thought I saw a unicorn." At which point a dimension is added that makes the experience as alarming as it will ever be. A third witness, you understand, adds no further dimension but only spreads it thinner, and a fourth thinner still, and the more witnesses there are the thinner it gets and the more reasonable it becomes until it is as thin as reality, the name we give to the common experience.... "Look, look!" recites the crowd. "A horse with an arrow in its forehead! It must have been mistaken for a deer."
> —Tom Stoppard, *Rosencrantz & Guildenstern Are Dead*

How do we reach any consensus on what is real? In his bid to end the Science Wars with a sort of peace treaty (*War of the Worlds*, 2002), Bruno Latour gives a brief account of modernity itself as a kind of military process, a campaign aimed at "unifying and pacifying" an unruly planet.[1] In constructing a "disenchanted," unitary representation of nature, science had set limits to all political disagreement:

> The world had been unified, and there remained only the task of convincing a few last recalcitrant people who resisted modernization—and if this failed, well, the leftovers could always be stored among those "values" to be respected, such as

cultural diversity, tradition, inner religious feelings, madness, etc. In other words, the leftovers could be gathered together in a museum or a reserve or a hospital . . .

[However,] if nature had the immediate advantage of imparting unification, it also had the serious drawback, in the eyes of its very promoters, of being fundamentally devoid of meaning . . . The modernists themselves were fully aware of this, and even acknowledged it with a sort of sadomasochistic joy. "The great scientific discoveries," they were glad to say with a shudder, "are incessantly wrenching us from our little village and hurling us into the frightening, infinite spaces of an icy cosmos whose center we no longer occupy." Ultimately, though, this was not a matter of choice: modernization compelled one to mourn the passing of all one's colorful pretensions, one's motley cosmologies, of all the many ways of life with their rich rituals. "Let us wipe away our tears," the modernists liked to declare, "let us become adults at last; humanity is leaving behind its myth-imbued childhood and is stepping into the harsh reality of Science, Technology and the Market. It's a pity but that's the way it is: you can either choose to cling to your diverse cultures, and conflicts will not cease, or, alternatively, you can accept unity and the sharing of a common world, and then, naturally (in every sense of the word), this world will be devoid of meaning. Too bad, love it or leave it."[2]

Many decided to be hurled and learn to love it, embracing secular nationalist modernization programs that Khaldoun Samman compares to the titular device of H. G. Wells' *The Time Machine*, relegating the particularities of their cultural and religious heritages, their stories, costumes, and dances, to the realm of values-to-be-preserved.[3] Others, unable to construct their own time machines quickly enough or unwilling to bear its considerable costs, were genocidally murdered, herded into reservations, and turned into exhibits like the dioramas of physical anthropology in the Museum of Natural History.[4] Nature itself would be segregated out into preserves, carefully denuded of natives, and subsequently imagined as "pristine."[5]

Yet any well-made time machine can also be operated in reverse. The fascination of modern art with the "primitive" is well known; less well known is its affinities with anticolonial and anarchist movements.[6] Modern thought, no less than modern culture, draws on ethnographic sources to imagine another world: the worldly, assimilated Franz Kafka encountering the "backwards" Jewish immigrants from the East (*Ostjuden*), the nobleman and scientist Peter Kropotkin encountering Tungus and Mongol "savages," discover in them visions of another world, a possible way out of the Great Sorting Machine of modernity.[7] It is thus that Daniel Colson calls our attention to "the affinity of anarchism with a much broader human thought and experience, which goes beyond the historical and geographical limits of European modernity."[8] That is what I want to explore via an obscure utopian science fiction novel by two obscure writers, who signed a lot of their major works collectively as the "Brat'ya Gordinii," the Gordin Brothers: Abba and Wolf.

The Song of the Stone: Antifetishism or Metafetishism?

A bit over a hundred years ago, in postrevolutionary Moscow and Saint Petersburg, a couple of middle-aged Jews from the Belarussian sticks imagined rock music and put it into their utopia: that is to say, *literally*, the music made by rocks. In "the land of Anarchy," thanks to the invention of "a special 'ether,' which conveys with the quickness of thought the slightest vibrations, movements, swells of sounds," everyone can hear the rocks singing.[9] "Let those who have ears hear," declares their spokesperson, in a subversive echo of the Gospels:

"Let those who have ears hear the song of being, the boiling song of life, the thundering song of mirth. *We* all hear it . . .

"Take, for example, a rock," said the man from the land of Anarchy, "who usually sings of peace. Her whole song is the apotheosis of immobility. [. . .]

"She sings about this:

I am all at peace;
Immobility—is the spirit of life.
And the word "cease!"
Among the word-flock, is a shepherd's pipe.

"We can express it better without rhyme, without a rhythmic line; otherwise it's hard to understand.

"She praises immovability:

And forever I lie,
In rest is the meaning, the purpose of being.
The contemplation of permanence catches the fish.
In the unchanging, in stillness, in standing still with my whole self.
In motion, death, decay, dust;
In movement, the satanic, the springtime, the tides . . .
In upheavals, change, destruction, horror, fear . . .
And creativity is sleeping, always quietly, sweetly childish at the bottom of the
   dream.

> *Always in place*
> *This is my creed.*
> *There is no leaven in the dough,*
> *In valley and on mountain I stand.*
> *The one who knows how to resist is a hero.*[10]

"You people talk like our poets wrote!" protests one of the viewpoint characters, a newcomer in the land of Anarchy, but the "man from the land of Anarchy" insists: "No. These are not images, not figures of speech. That is what we hear . . . for us, everything literally sings."[11]

I am reminded here of Seo-Young Chu's "science fictional theory of mimesis," which speaks of lyrical language as the "torque" applied to a resistant reality in order to bring it into representation.[12] This neatly captures both the science fictional and the lyrical force of the Gordin Brothers' writing, prose that is designed to "speak from beyond ordinary time" to propel us out of our own subjectivities and into something other.[13] It is utopian language, futurist language, reminiscent of the radical poetry of a Pablo Neruda addressing the continental plate as if it might answer, in its "lost tongue," with its "mouth/of mute stone."[14] But science fiction literalizes the metaphorical: where Neruda's lyrical declaration is mortgaged to metaphor, the Gordins' protagonist insists that "This is not the contrivance of a poet . . . It is a fact."[15]

For the Gordins, indeed, everything *is* singing—living, purposive, expressive. This has less in common with Neruda's modernism than with the Indigenous worldview of the Ojibwe or Onödowá'ga:' peoples, who have long listened to the rocks.[16] In this respect, *Anarchy In a Dream: The Land of Anarchy* pushes against the very boundaries of science fiction itself, as Stanislaw Lem, together with the Marxist critic Darko Suvin, defines the genre *against* "myth," "the folk (fairy) tale," and "fantasy": whereas "the world of a myth or fairy tale is ontologically either *inimical or friendly* toward its inhabitants, never *neutral*," science fiction is cognitively grounded in "the real world, which may be here defined as consisting of a variety of objects and processes that lack intention, that have no meaning, no message, that wish us neither well nor ill, that are just there."[17] If science fiction is limited to words about a wordless world, cold and senseless, then perhaps, as Evgeniy Kuchinov suggests, the Gordins' *Anarchy In a Dream* merits the label of "extro-science fiction."[18]

Abba and Wolf Gordin championed technics but opposed science. If this is hard to understand from a standpoint that sees technology as applied science as well as from a standpoint that sees science and technology as inseparable, the key difference between them, for the Gordin Brothers, lies in the scientific concept of *laws*.[19] Submission to "laws of nature," they argued, parallels and reinforces submission to social laws, epitomized by the unquestionable Law of God the Father and King, *Avinu Malkenu*. They declared a revolt against both "religion and science," "the kingdom of heaven and the natural kingdom":

> angels, spirits, devils, molecules, atoms, ether, the Laws of God/Heaven and the Laws of Nature, forces, the effect of one body on another—all this is fictional, created in the image of society (sociomorphically).[20]

"Sociomorphism" is the constellating tendency: the projection of the social order onto the order of the cosmos itself. The scientific imagination, no less than the religious, calls upon the available materials of the local social imaginary to describe what it reads in the heavens and the earth:

> God and Nature are made in the image of man (anthropomorphically). The Eskimo [*sic*] creates them by hunting for a polar bear (the world came from a polar bear), the Jews create them according to their crafts (god as a joiner, a carpenter); Newton, Kant, and Laplace construct a Nature after the pattern of European mechanics, Darwin and Spencer—after the pattern of English horse-breeding...[21]

And it is no wonder, as the two grew up together in the company of the premodern State: "God is the image of the absolute monarch of Asia; the laws of heaven, the laws of the stars, the astrology of Assyria and Babylonia."[22] Where the old anarchist slogan had declared, "*Ni dieu, ni maître!* [No god, no master!]" the Gordins adopted as the device for their extended "pan-anarchism": "*Nyet ni boga, ni prirody!*" ("No gods, no nature!")[23]

Their anarchist extro-science fiction attacks the very premise of what Bruno Latour calls "the Modern Constitution," with its foundational "separation of powers": the splitting of the world into "two entirely distinct ontological zones: that of human beings on the one hand; that of nonhumans on the other."[24] This is a profound anarchism: not only do the Gordins protagonize their utopia in an anarchist way—the five newcomers to the Land of Anarchy, refugees from our world, represent not only "Worker(s)" and "Individual(s)," but also "Youth," "Oppressed Nation(alities)," and "Wom[e]n," which the Gordin thought of "the 'union of the five oppressed'"[25]—but they also protagonize the nonhuman world, refusing what Daniel Colson identifies as the first of "three illusions" to which "anarchism is opposed": "the illusion according to which the human being and humanity are (by their essence) radically separated from nature, from the things and beings that surround them, which it must appropriate and instrumentalize."[26] This profound anarchism, in Latour's phrase, had never been modern. As Graham Harvey comments, "Stone personhood presents a particular challenge to the modern Western worldview which presumes that rocks are the primary, archetypal form of inanimate matter."[27] The non-ordinary time from which the Gordins speak, in other words, may be not only a utopian elsewhere but a uchronian elsewhen, a *never-modernity* that doesn't belong to the West and its logic of progress at all.

The Gordins' singing rock brings to my mind an image from Murray Bookchin's *Ecology of Freedom*. As a necessary prelude to imagining a social future that isn't merely the "the extension of the present . . . enlarged rather than challenged,"[28] Bookchin proposes looking to the past, to "the outlook of so-called primitive or preliterate communities," which he names "organic societies because of their intense solidarity, both internally and in relation to the natural world."[29] That he *only* engages them within the framework of "the past," their voices mediated by the accounts of white anthropologists, and never as people living and struggling in the present, demonstrates the enduring hold of settler racism over his social imagination. Nonetheless, his appreciation of the radical difference represented by their nonmodernity, their refusal to break the world apart, is profound. Finding "utterly alien" the technocratic imaginary of modernity, with its treatment of the nonhuman world as mere material-to-be-appropriated or inert "substance," people like the Anvilik Inuit presupposed instead a "coequality" of subject and object that amounted to a refusal of that distinction: when carving ivory, an Inuit carver would "[turn] it gently this way and that way, whispering to it, 'Who are you? Who hides in you?'. . . Instead of compelling the fragment of ivory to become a man, a child, a wolf, a seal, a baby walrus, or some other preconceived object, he tried subconsciously to discover the structural characteristics and patterns inherent in the material itself."[30] This deeply social conception of labor, technics, and nature—a relationship that Marx and Freud would call "fetishistic"—characterizes the Gordin brothers' "land of Anarchy," too.

This is odd, in part because the Gordins seem otherwise to be on a campaign against fetishism. *The Land of Anarchy* is in many ways a novelization of their nonfictional *Manifest Pananarkhistov* or *Pananarchist Manifesto*, published the year before, in which the brothers call for "liberation from prejudice, from primitive fetishistic fear."[31] This sounds about as modern as it is possible to get—in fact, it's a three-decades-early echo of Adorno and Horkheimer's famous diagnosis of the modernity represented by the smoking ruins of the Nazi death camps, the modernity which "aimed at liberating men from fear and establishing their sovereignty" but which spelled "disaster triumphant."[32] The Gordin Brothers' "pantechnical" paradise is perhaps the epitome of Adorno and Horkheimer's "instrumental reason," the rationality that grasps everything as an instrument, a means to an end: "Here, in the kingdom of technology, of labor, there is almost no nature."[33] Super-advanced technics was to overcome both God and Nature, those Spinozan synonyms.

Here is where things get complicated. The "fetishism" that Abba and Wolf Gordin were mainly concerned to oppose was not the Inuit carver's apprehension of the ivory as a living, communicative agent, but—quite to the contrary—modern science, which they saw as the "superstitio[us]" worship of "nature, molecules, atoms, ethers, matter, energy, causality and evolution [etc.]," and along with it, Marx's "scientific socialism," which they dubbed "social fetishism," "superstitious socialism," "sociomagic," or simply "the magic of parliamentarism-Sovietism . . . The magic of laws, decrees . . . the magic of parliamentary speeches": "Shame and disgrace to the shamers of Utopia, Engels and Marx, the sorcerers of the scientific-socialist black book."[34] In particular, they are opposed to reductivism and to the concept of natural "laws," which they see as a "sociomorph[ism]," a projection of the modern State onto the structure of the universe: "the laws of nature are the laws of the State," and where an Inuit might construct the spirit of nature in the image of a polar bear, "Newton, Kant, and Laplace construct a Nature after the pattern of European mechanics, Darwin and Spencer—after the pattern of English horse-breeding (natural and sexual selection in the image and likeness of the artificial selection practiced by English stud farms)."[35] This is not far from Bookchin's critiques of our tendency to read the *findings* of modern science through the lenses of capitalism, war, and governmentality, projecting these into a "stingy" Nature structured by a "war of all against all" into "'alpha' male" hierarchies, etc.[36] In short, it is "the foolishness of the *European* savage" with which the Gordin Brothers are occupied.[37]

Technics, for the Gordin Brothers, were an *alternative* to science, to scientific laws, and to the administered world of modernity—a pragmatist process of getting along in the world, for which "the important thing," as the Man from the Land of Anarchy remarks, "is that everything works":

> We have nothing but technology; this is our religion and our science. We threw away all "laws," all doctrines, all hypotheses and basic principles. We set out to try, to play, to create, to desire. And we educate our youth: to do, to do, to do. "Knowhow" [*Umet'*]—that is our symbol of faith.[38]

In contrast to the "high modernism" of "supreme self-confidence about continued linear progress . . . [via] scientific understanding of natural laws," as James Scott defines it, the Gordin Brothers found their utopia on an advanced version of what Scott calls "*mētis*," "the knowledge that can come only from practical experience."[39] Pantechnics "is sensual, it is muscular"; it is knowledge-in-physical-relationship rather than knowledge as the accumulation of abstract laws.[40] Informal and local in character, *Umet'* or *mētis*—a "pragmatic and highly concrete knowledge"—refuses to separate theory from practice (the second "illusion" to which anarchism is opposed, according to Colson).[41]

This kind of "pragmatism" also typifies, for instance, the Ojibwe way of being in the world, for which the people are "to live by their wits, and not be tied down by any one set way of doing things": "In this regard, as the story of Wenabozho and the skunk demonstrates, even a fart can be useful."[42] And the resistance to abstraction is another point of commonality between Indigenous ways of thinking and the Gordin Brothers' utopianism: the agency of rocks, birds, and cars is not to be treated in a de-specified way, absent considerations of place, person, and situation.[43] Graham Harvey recounts how an anthropologist once asked an Ojibwe elder near the Beren's River in Manitoba, "Are *all* the stones we see about us here alive?": "[The elder] reflected a long while and then replied, "No! But some are."[44] Just as the Land of Anarchy's five suns "each had its own living solar self," while the Gordins use universalizing language to describe nonhuman agency—"*everything* literally sings"—the sensuous particularity of the rock's poetic voice suggests that this rock has a living stony self, that we are listening to *this* rock. And to confirm it, the man from the Land of Anarchy tells our protagonists: "Know that all objects are individual, because each object has its own self . . . As far as each rock is a rock, its song is 'rocky,' but as far as it is individual, so is its song peculiar, personal, original, individual."[45]

This strange act of audition is, however, not treated as magic, not even in the casual style of magical realism, even if it is reminiscent of the kinds of stories Gabriel García Márquez might have heard from his grandmother. Instead, the Gordins deploy the classic explanatory rhetoric (techno-mimicry) of science fiction: "for hearing we invented a special 'ether,' which conveys with the quickness of thought the slightest vibrations, movements, swell of sounds."[46] This rhetoric serves further to rationalize the even greater strangeness of the future anarchists' ability to talk *back* to the world:

> "Look, I'll put this in terms of your primitive mechanics: every word forms waves—every word is special waves. Different words vibrate, furrow air in different

ways and cause waves in different ways, differing from one another in size, width and originality. Why do you admit that these waves should disappear without a trace, without taking any action! Why don't you admit that these waves can be disposed of, used, systematized... and harnessed, like forces, like engines!"[47]

And so, like "some kind of fakir, a magician, a conjuror"—in the eyes of our astonished protagonists—the man from the Land of Anarchy explains that "We are technicians":

"We act, we don't philosophize. We just say to a tree, for example, a centuries-old oak tree, 'Go and cover a man sleeping on a hill with your shadow,' and it will go and do what is asked of it. Of course, all this must be said in his language, in the language of trees and in his special oak dialect."[48]

This "pantechnics" is quite clearly, so far beyond our understanding that it has triggered Clarke's Third Law: "Any sufficiently advanced technology is indistinguishable from magic." The narrator reflects that "Here the word 'why?' had no place," for "it was afraid of the rays of miraculous technology."[49] The Land of Anarchy is "mysticism *made real*, a fable *brought to life.*"[50]

Why this strange proximity between futuristic vision and the survivance of tradition?[51] How has the Western quest for mastery over nature reached its end in the abolition of the difference between nature and artifice, in the (happy) undoing of the "disenchantment of the world," the restoration of lost "magic"?[52] How, in this world crammed with fantastic machines (among them, a flying mountain—"Why doesn't it fall?"—and a giant "floating smile" that you can sail in), has the Great Sorting Machine ground to a halt?[53] How is it that these two Westerners should attempt to escape the gravitational pull of the West? And *do they really*?

Well, yes and no. In one sense, for them, the Modern Constitution has been abrogated, and with it, all the laws, natural and social. On the other hand, they are the epitome of Western High Modernism: they want to make the promise of humanism good by humanizing the universe to overcome nature by merging with it technologically! The protagonists, representing the rejects of the Great Sorting Machine, are nonetheless marked by the abstract machine of allegory, which also does the work of "purification": Woman *only* represents women, and therefore is not oppressed by racism, ageism, and so on; Youth *only* represents the young (and therefore is by default masculine, like all the protagonists other than Woman); poor Oppressed Nation (named in the plural!) has to bear an outsized burden of representation, standing in for "any nation, as long as it's oppressed. I am an Armenian, a Hindu, a Latvian, a Lithuanian, a Belarusian, a Jew, a *Tzigane* and a black Negro"[54] . . . etc. The Gordins' concept of the multiplicity of struggles is promising but inadequate; it approaches but does not reach the concept of intersectionality (did they not imagine that one could belong to multiple categories at once?). And I am not confident, for all their lyrical power and utopian ambition, that they are capable of imagining a fully decolonized world, much less a way to get there. "We're thinking of staying in this country forever, for life," our protagonists say; "Is it permitted to come here as a stranger, a settler?"—and receive the answer: "We have no strangers, we have no settlers. Every five who come to us are very welcome among us."[55] This cuts to the very heart of the *mésentente* between white anarchist anticolonialism and Indigenous decolonialism: the too-often inability of white anarchists to understand Indigenous demands for sovereignty, to understand even the very concepts of *indigeneity* and *settlement*. Both of the Gordin Brothers were close to Labor Zionism in their youth, and as adults, in their own flight from oppressions, they emigrated to lands stolen from others, dying as colonizers.

But that list of Oppressed Nations is still interesting in its hybridity. (Like the protagonist of Philip K. Dick's *A Scanner Darkly*, face blurred behind a "scramble suit" signal, Oppressed Nation "stood there as if one, as if many. You couldn't see them, you couldn't make him out, you couldn't distinguish him").[56] We might notice that the Gordin Brothers themselves fall into at least two of its categories: they are Belarussians, and they are Jews. Moreover, raised in a town full of ecstatic Hasids as well as militant Jewish workers, they are the renegade sons of a famous rabbi, the "Gaon" Yehuda Leib Gordin, teachers of Hebrew, and readers not only of Tolstoy and Bakunin but—as they coyly allude to in the novel—of Kabbalists like Chaim Vital and Yitzhak Luria; "Kabbalah . . . [is] an anticipation of pantechnics," remarks the man from the Land of Anarchy.[57] Leonid Heller flatly states, "The Gordins *are* Kabbalists."[58] And perhaps this has everything to do with the *affinity* between their singing rocks and the speaking stones of the Ojibwe and the Onödowá'ga, their ability to "*find* [themselves] in the other and to find the other in [themselves] as already there."[59]

To be clear: affinity is not identity. "Jews, natives, evermore the simulations of the other, have endured the miscarried cause of modernity," writes Gerald Vizenor, but adds: "but not for the same reasons."[60] The pogroms that swept the Gordins' world, horrific as they were, were not identical to the genocidal violence endured by Indigenous peoples in the Russian Empire as in the Americas.[61] There is a long and ugly history of Jewish writers appropriating Indigenous identities, from Jackie Marks' long and lucrative impersonation of a Native American ("Jamake Highwater") to Franz Kafka's "wistful" fantasy of becoming a "Red Indian" as well as the current Zionist campaign to rebrand Jewishness as an oppressed "indigenous" identity.[62] The Gordins do not participate in this particular colonial masquerade, and I do not want to place them into it. As Daniel Colson explains, what "anarchism calls analogy and affinity . . . arise . . . from *repetition*" but also, crucially, "from *discontinuity* and *difference*."[63] It is in this sense that affinity undercuts the operations of the Great Sorting Machine so that the radically nonidentical can recognize one another without falling into the trap of Hegel's violent colonial process of recognition, the Master-Slave dialectic. Affinities, analogies, are what make the Union of the Five Oppressed immediately intelligible, but they can also be quite strange and even troubling, as when Gerald Vizenor considers that while "clearly, aliens are not *indians* by conception or description," nonetheless "the experiences of abductees and captives are similar."[64] As fugitives from the world of Western gravity, in the words of Evgeniy Kuchinov, the Gordins' novel is perhaps a machine for "becoming an alien on Earth."[65] Perhaps, if only for a fleeting moment, this writing-machine opened up a wormhole between many worlds?

Pluriversality, or the Architecture of Cosmism

Maybe what *The Land of Anarchy* does, in the end, is simply to raise the question of how anarchists might bridge cosmos with cosmos, as well as the question of how universalizing anarchist imaginaries are tied to particularities of culture; maybe it challenges us to rethink our cosmopolitics, to imagine yet again a world within which many worlds fit. This is, in fact, what the Gordin Brothers tried to imagine under the name of "cosmism."

Nearly thirty years before Stalin would denounce Jews as "rootless cosmopolitans," ironically, the Gordins deconstructed the concept of "cosmopolitanism." In *Sotsiomagiya i Sotsiotekhnika* (*Sociomagic or Sociotechnics*, 1918), they suggest "unmask[ing] this deceptive magic word":

> Cosmopolitanism . . . means not one nation, but all nations, not just a country, but all countries, in a word,—cosmos. What will happen to the cosmos? Will this cosmos be free? The other half responds to this: politism. Politism means there will be a State. Thus, the whole cosmos will constitute a single condition of slavery . . . this is an almost literal translation of the word "cosmopolitanism." There will be a worldwide state, worldwide violence, worldwide dictatorship,

worldwide famine, worldwide decrees of punishment, and worldwide prisons. But, if the State is violence, if the fatherland is—a theft, then what is a global State, if not global violence, and what is a global fatherland, if not global theft?[66]

The Land of Anarchy, in the Gordins' novel by that secondary title, was nourished by another source: a "sea [that] was called cosmism."[67] "If anyone were to harm one of the many, many nations that we have living side by side as neighbors with others," explains the Man from the Land of Anarchy, "the storm would fume, and the sea of Cosmism would rise up from its marble shores."[68] In less mythopoeic terms, their *Pananarchist Manifesto* (1918) defined the goal of "cosmism" as "the liberation of ethnic groups," and hence "the total annihilation of the modern territorial or imperialist order"[69]—and as a means to this, in the words of their earlier manifesto, *Sefer el ha-golah: A megile tsu di yidn in goles* (*A Book for the Jews in Exile*, 1908), "the solidarity of interest of the oppressed," "unity of the oppressed nations" without the imposition of that unity from above.[70] The unity of this cosmos minus all politism—Deleuze and Guattari's "stranger unity that applies only to the multiple," perhaps?—is, significantly, to be supported by "a general, international language."[71]

Here, we see the impetus for Wolf Gordin's AO, a universal language systematically stripped of national particularity—the most "high modern" of all languages, perhaps, albeit in the service of a complete disassembly of the modern global order. As Michael G. Smith points out, it was "a remarkable achievement," not only "in terms of its novelty and simplicity" but in its will to abolish some seemingly fundamental linguistic institutions: "True to the anarchist ethic . . . [AO] altogether dispensed with gender (signifying male oppression), as well as possessive cases and possessive pronouns and the genitive case (signifying property relations)."[72] In AO, thus, there really would be no way to say "my boots" or "my house," much as in the fictive "Pravic" language invented by the anarchist revolutionaries in Ursula K. Le Guin's *The Dispossessed* (1973). But this is nearly an afterthought. Wolf's "violent diatribe against the genitive" in his *Grammar of AO* (1924) is aimed not only against the grubbier forms of belonging—property, family—but against the very notion of "origins":

> The genitive, which plays such a colossal role in all the languages of the old civilization, is totally absent in AO. AO considers the genitive to be a survival of geneticism (originism), fetishism and mythologism.
>
> AO, as the language of the new civilization, rejects any concept of metaphysical origin (origin of the world, of the body, of the mind, and of society), as the

fundamental prejudice of every religion and every science, a socio-morphism, akin to landed property, an aristocracy of blood, lineage, class, status.

The New Civilization does not have its eyes stuck on the back of its head; it does not ask "from where," it does not have the reactionary Esperanto word "*de*" [from, of, belonging to]; it asks where to, "toward what," what to apply this to, looking to the future.[73]

The assumption built into Gordin's language is thus not only sociopolitical—the assumption of "the New Civilization of all-invention, which always has to do with the results of labor, work, invention"—but at the same time, properly *ontological*: it assumes that what is most fundamentally real is not "the noun, found readymade," and the inert object which it would denominate, but "the verb," the ongoing process, action, creation.[74] Its aim is to complete, for the first time, the Enlightenment project of demystification (Max Weber's *Entzauberung*, de-magicking), the elimination of "sociomagic" or "fetishism," to accept the wager implicit in Nietzsche's remark that "we have not got rid of God because we still have faith in grammar":

> The languages of ancient civilizations serve as archives of all the remains of cultures; they are museums of superstition, fetishes for those who speak them. The language of the new civilization must defetishize everything. The nutritive source of all fetishism (in a broad sense), which coincides with animism, is the "inhuman" language (i.e., the ancient language, unworthy of the new Humanity), and the essential source of pollution is the "old" verb.[75]

The "new civilization" is thus to be founded on a new relationship between words, their speakers, and the world they inhabit. "The overarching goal of AO," notes Smith, "was to achieve the greatest meaningful and logical 'correspondence' between words and concepts and things as possible, essentially creating a classification scheme, a catalog of the human experience."[76]

Wolf Gordin's goal, in the construction of AO, was nothing less than the establishment, for the first time, of a true harmony between mind, word, and world. In this respect, AO falls more directly in the tradition of "philosophical languages" such as John Wilkins's (1668), with its forty categories designed to represent every fundamental human concept. We might remember Wilkins's *Essay towards a Real Character and a Philosophical Language* from the satires to which it has been subjected—notably, from Jonathan Swift's unflattering portrait of the Laputan savants' "school of languages" in *Gulliver's Travels* (1727), but even more famously, from Jorge Luis Borges's sendup of Wilkins in his essay, "The Analytical Language of John Wilkins" (1941). Having exposed certain inevitable "ambiguities, redundancies, and deficiencies" in Wilkins' scheme, Borges claims to be reminded of "a certain Chinese encyclopedia entitled *Celestial Emporium of Benevolent Knowledge*":

> On those remote pages it is written that animals are divided into (a) those that belong to the Emperor, (b) embalmed ones, (c) those that are trained, (d) suckling pigs, (e) mermaids, (f) fabulous ones, (g) stray dogs, (h) those that are included in this classification, (i) those that tremble as if they were mad, (j) innumerable ones, (k) those drawn with a very fine camel's hair brush, (l) others, (m) those that have just broken a flower vase, (n) those that resemble flies from a distance.[77]

This parody gave Michel Foucault a laugh—"the laughter that shattered . . . all the familiar landmarks of my thought"[78]—and provided the perfect anecdote for his famous preface to *The Order of Things*, his obituary for the human sciences and the modern *epistème* from which they arise. The obvious illogic of Borges' fictive taxonomy, for Foucault, only renders visible and questionable the logic of every real taxonomy, every "universal science of order," from paleontology to political economy.[79]

Isn't AO's cosmism, then, just another exhibit in the museum of modern ambitions—an *n*th-generation retread of Galileo's proposition that "this grand book, the universe . . . is written in the language of mathematics"?[80] Galileo, too, wished to eliminate "the animal" to "[take] away ears, tongues, and noses" until "there would remain only figures, numbers and movements."[81] Might it not merely present an especially zealous version of the modern project of "purification," embodying "the moderns' motto 'No past'"?[82] And what could smack more of classical humanism—a thought founded in the figure of "Man," that "invention of recent date . . . be[ing] erased, like a face drawn in sand at the edge of the sea"—than Gordin's grandiose program for a "language of mankind," or even, in the title of Wolf Gordin's *Plan Chelochestvo* (1921), a "Plan for Humanity"?[83]

Maybe so. But there is another aspect of AO that is very much rooted in the "past" in tradition: in the original sense of the Hebrew word, a *kabbalistic* aspect. The conversion of letters into numbers is, of course, part of the nonmodern practice of *gematria*, Hebrew numerology—a source of great theological creativity. And if the aspirations of AO are of a piece with the sonic and semiotic technology of the Land of Anarchy, the technics that allows one to bid an oak tree to cast its shade over a man sleeping on a hill, then its nonmodernity is even more pronounced: it participates in an *epistème* prior to that in which "things and words were to be separated from one another," one in which a "profound kinship of language with the world" yet prevails, so that word and thing, signification and force, are not distinct.[84] What if, instead of placing AO on the heap of exhausted modernist concepts, we read it as an exercise in *gematria* after the manner of Isaac Luria (who, like the Man from the Land of Anarchy, was in the habit of conversing with trees)?

If the numerological magic of our age has made possible increasingly rapid and supple machine translation, we may not feel the need for an international language in quite the same way as the Gordin Brothers did a century ago, but have we solved the riddle of cosmism, learned exactly how to think "humanism after the death of Man"?[85] Rather than reading the rationalized architecture of AO as the linguistic counterpart to the Bauhaus architecture of uniform glass-and-steel towers that still dominates the "cosmopolitan" metropolis, I would like to place it in conversation with the humbler, more traditional, but also more inventive architecture codified by Christopher Alexander, Sara Ishikawa, Murray Silverstein, Max Jacobson, Ingrid Fiksdahl-King, and Shlomo Angel as *A Pattern Language* (1977). Here, too, is an attempt to find certain human universals—"problem[s] which [occur] over and over again in our environment" paired with time-tested "solution[s]" to those problems—and to articulate these "patterns" within an overarching grammar of embodied practice. One such problem, for instance, is the optimum ratio of room area to ceiling height; Alexander et al. synthesize their "solution" from the classical architecture of Palladio, traditional Japanese building practices, and the anthropologist Edward T. Hall's research on spatio-social relationships, advising a variety of ceiling heights for different psychological and social needs (intimacy/formality, solitude/community).[86] Another is the problem of "political alienation," the mystification of power by its removal from public view and popular control; the "solution" that Alexander et al. put forth—one that has been rediscovered repeatedly across social and historical differences—is to devolve governance to a human scale, decomposing cities (based on early estimates of the size and shape of social networks!) into self-managing communities of five to ten thousand.[87] The grammar mediating between these two "patterns"—one apparently a mere matter of aesthetics, the other a political matter—is clarified by Jacques Rancière's insight that politics *is* aesthetics, a matter of how experiences are distributed.[88] *A Pattern Language*—note the indefinite article: it is "one possible pattern language" out of many—is an AO-like project in its aspiration to a certain universality, but tempered by a respect for tradition and vernacular wisdom.[89] If this is a "cosmist" architecture, then it resembles James' pragmatist corridor.

What would happen if we read the Gordin Brothers' cosmist project not only as defetishizing but also—particularly as seen through the frame of their visionary fiction—as *refetishizing*? The defetishizing impulse is certainly on full display, as the Land of Anarchy is a space in which old mystifications have lost their force: "work" and "science," for instance, appear as empty, abandoned verbal fetishes. "We have no workers," the Man from the Land of Anarchy informs his guests. "Work for us is fun, amusement, a nice pastime."[90] When he is asked about the scientific basis of the technology that has helped abolish work, he can only laugh: "You're funny! You see, we don't recognize science. We consider explanation to be a superstition; the important thing is that everything works. I can give you a look at this kind of 'machinery,' if you like, but it would be of little use. There's not much to it."[91] The tour of utopia is thus a dispelling of the old forms of social mystification undergirding both capitalism and Marxism, the kind of thing they attack as "sociomagic." At the same time, as the Gordin Brothers repeatedly tell us, their utopia effectively *is* a magical realm, having fallen under Clarke's Third Law:

"Here, the myths of antiquity walk in broad daylight. Here dreams, visions out of legend, are embodied in acts, in facts, in bodies, in reality," Oppressed Nation said.

"Here, one gets lost among the fabulous technical possibilities, and the possible and the impossible merge and blend together forever," said Worker.

"Here, Reality straddles Dream, spurs him, drives him at full speed, and he, dashing and frisky, swallows the earth and the heavens, and from under his hooves fall sparks, flashing and winking out, to reignite a particle of existence, a particle of present life. Dream rushes with all his strength in a wild, violent, stormy trot, but Reality catches up with him. She's on his back. He stops and glances around.

"Here, Dream has gone wild, bucked, reared up, but Reality is firmly, solidly saddled on him. She spurs him on to even greater efforts.

"Dream has climbed the mountain of imagination, leaping over the abyss of illusion—and Reality is right there. On his back, she holds fast to his mane, which is flying in all directions.

"Dream is unbridled, indomitable. He is a stranger to all restraints. He hates the roads and paths.

"Here he rushes, a whirlwind that has broken free from the cage of time, that has snapped the chains of causality and laws, across the fields, lands tilled by the plow of the Miraculous and sown with seeds of mysticism and inspiration, flying like an arrow of eternal discontent, a strong, creative longing in the morning and a clear, sunny sorrow at noon, flying, rushing; his quick, slender legs take flight from the land of utility and profit, surging to the edge of reveries, fables, mysteries, marvels, but Reality, a tried and tested rider, stays firm in the saddle."[92]

This is not only a paean to advanced technics; it is a lyrical critique of "causality and laws," of "utility and profit," a visionary image of a society that has left these behind. The ultimate "miracle" of the Land of Anarchy is not merely technical, as Oppressed Nation points out, but "sociotechnical," a matter of ingenious "social construction."[93] The cosmism that liberates Oppressed Nation is complemented by the communism that liberates Worker, the "gynism" that liberates Woman, the "pedism" that liberates Youth, and the anarchy that liberates Individual. These are "social miracles . . . social wonders which, embodied in institutions, walk in broad daylight, the real, living children of actuality."[94] Here, the protestations of the Gordin Brothers in *Sotsiomagiya i Sotsiotekhnika*—"the path to nonscience doesn't mean a return to the prescientific, to barbarism, but . . . a step forward to the post-scientific, to Pantechnics"—seem to fall flat: in pragmatist terms, in terms of lived experience (or in terms of "practical, sensual-muscular" realities, in their phrase), don't the prescientific and the post-scientific resemble one another rather closely, as the rhetoric of *Anarkhiia v Mechte* announces at every turn?[95] Perhaps we could draw a corollary to Clarke's Third Law: just as "any sufficiently advanced technology is indistinguishable from magic," so *any sufficiently advanced sociotechnics is indistinguishable from sociomagic.*

Let us then intervene again in the Gordin Brothers' discourse to insist, with Kenneth Burke, that what is wanted is not the removal of sociomagic from discourse, a purified technics of the word, but a magic perpetually fashioned and refashioned by "the 'collective revelation' of testing and discussion."[96] Such a sociomagic would *be* a sociotechnics, an effective practice of renegotiating embodied realities, of redistributing the sensible and reshaping the real.

Sociomagical Counterpower

There is, of course, a dark side to this utopia. The harmony and tranquility which preside over the Land of Anarchy are subtended by a threat, an *or else* . . . that is explicit, not subtextual. As the Man from the Land of Anarchy explains, the system of fivefold symbolism that frames their world—the Five Oppressed (Worker, Oppressed Nation, Woman, Individual, and Youth) are complemented by five liberatory principles (Anarchy, Communism, Gynism, Cosmism, Pedism), each of which has its own allegorical statue seated atop its own mountain (Equality, Brotherhood, Love, Freedom, Creativity), towering over five seas (Communism, Cosmism, Gynism, Anarchism, Amorphism), illuminated by five suns (Laborer, Aterritorial, Beauty, Singularity, Youth)—serves as a check on any imbalance, any violation of the order of things.[97] "In our country," he tells the newcomers,

> if you offend a child, then the fifth sun [the sun of Youth] will go out, all the other four will follow, and all the seas will leave their shores, and all the statues will leave the mountains and leave, and there will be an "end" to the land, an earthquake such as the world has not yet seen, such as man's imagination hasn't yet imagined, and all the foundations will collapse, and all will be destroyed, and there will remain just one raging, wild element: water and darkness.

Similar cataclysms would result from violations of the other principles, we are told. However, this is not intended to alarm:

> "Do not be afraid. This has never happened. It will never happen. This country stands, or rather hangs on five harmonies. Every violation of harmony threatens death.
> "But harmony and order shall never be broken and chaos shall never come."[98]

We may note that this scenario is almost the direct inverse of Ursula K. Le Guin's "Omelas," the flawed utopia sustained by a single child's suffering! Yet this somewhat gruesome reminder of mortality in utopia (*et in Arcadia ego*) might sound to us like a discordant note, undermining the otherwise perfect (not to say rigid) harmony in advance. We could deconstruct the utopian image quite easily; isn't the preeminence of the spoken word, the phonocentrism, a dead giveaway?

However, we might look at this from another perspective. In the Land of Anarchy, everything is artifice, the mountains, seas, and suns no less so than the statues; the terror of the cataclysms is thus owing to something that the people themselves have created. This paradox—collective human action producing something that in some respects constrains collective human action—is perhaps just the paradoxical knot of the social itself. The fivefold *or else* looming over the Land of Anarchy might be exactly what we should expect to find in an authentically classless, stateless society: a sociomagic/sociotechnics by which the society holds itself to its own contract by means of its own imaginary.

David Graeber points out that such imaginaries, in fact, tend to appear in association with real, living classless and stateless societies—that is to say, egalitarian indigenous societies such as the Piaroa, Tiv, or Malagasy, where "extreme forms of symbolic violence" perversely serve as the guarantors of material, social peace:

> Of course, all societies are to some degree at war with themselves. There are always clashes between interests, factions, classes and the like; also, social systems are always based on the pursuit of different forms of value which pull people in different directions. In egalitarian societies, which tend to place an enormous emphasis on creating and maintaining communal consensus, this often appears to spark a kind of equally elaborate reaction formation, a spectral nightworld inhabited by monsters, witches or other creatures of horror. And it's the most peaceful societies which are also the most haunted, in their imaginative constructions of the cosmos, by constant specters of perennial war. The invisible worlds surrounding them are literally battlegrounds. It's as if the endless labor of achieving consensus masks a constant inner violence—or, it might perhaps be better to say, is in fact the process by which that inner violence is measured and contained—and it is precisely this, and the resulting tangle of moral contradiction, which is the prime font of social creativity. It's not these conflicting principles and contradictory impulses themselves which are the ultimate political reality, then; it's the regulatory process which mediates them.[99]

Once again, the hypermodern and its "primitive" Other turn out to be strangely adjacent to one another. It's as if the ultimate development of the machine, *the machinification of nature itself*, were also a return to Rousseau's "state of nature," the recapitulation of Marx and Engels' "primitive communism."

The Pantechnical Garden, or the Systematic Derangement of the Senses

In Jeannette Ng's *Under the Pendulum Sun*, nineteenth-century English missionaries to the land of the Fae, Arcadia, find that not only is everything there *unnatural*, it is actually *artificial*: dancers at a ball are elaborate clockwork automata, the air is mined from distant frozen seas and mountains and hauled to market, snow is shaken out of an old woman's featherbed, frost has to be painted on windows and autumnal colors painted on leaves, and so on.[100] Of course, in all of its strangeness, Arcadia also resembles England: our English protagonist, Cathy Helstone, recalls her surprise at finding that the "natural" landscape of her beloved Yorkshire moors is artificially cultivated by burning.[101]

Things are similar in the Gordin Brothers' fairyland, where, as the Man from the Land of Anarchy tells us,

> "Everything that you see with your eyes, everything that you hear with your ears, everything that you perceive, everything is made, created. Here, in the kingdom of technology, of labor, there is almost no nature.
> "Everything here is artificial . . ."
> "Really? And the sky? And the earth? And the trees?"
> "Everything! Everything! Our world is our deed, our work."[102]

In this respect, too, "the ancients . . . are much closer to us than your scientific understanding of the world," for in their cosmogonies, "they taught that the world, the universe, everything that exists is the result of a creative act."[103] The Land of Anarchy is thus "pantechnical" (and "post-scientific") in more than one sense.[104] We are told repeatedly that "all this is quite natural" and, at the same time, that "everything is . . . artificial, creative."[105] The Gordins' utopia-poem thus realizes at once Deleuze and Guattari's declarations that everything is machines and that everything is nature.[106] It stands on the other side of the Modern Constitution.

The "technicization [*teknizatsya*]" of everything follows a logic not unlike that once described by Marshall McLuhan, who regarded technologies as "extension[s] of human sense or function."[107] And so it is that the Five Oppressed follow the Man from the Land of Anarchy through a series of pantechnical "departments" dedicated to the five senses.[108] It is thus that they learn that the anarchists' advanced technics has allowed them to discern the speech of rocks and trees. More unexpectedly, however, they find that at the entrance to each of these departments, that sensory modality *disappears*. Upon entering the Visual Department, they find that "appearances" have abruptly ceased that they are surrounded by a disturbing blankness:

> "You see, in order to overcome visionality, we had to overcome sight itself," the man from the land of Anarchy said convincingly.
> "I refuse to understand!" I exclaimed.
> "It's very simple! We proceed from the fact that first we destroy and then create, for the spirit of creation is in destruction."[109]

The Gordin Brothers imagine a reconfigured sensorium on the basis of a cancellation of the senses, visibility on the basis of invisibility: "In the Visual

Department, there's nothing to see!" "Our first step along the path of overcoming visionality," explains the Man from the Land of Anarchy, "proceed[ed] from the point of its negation; we made the objects that surround us unseen, invisible. We destroyed visibility."[110] It is this that invites comparisons with the magical practices of the Jewish mystics.[111] The suppression of sound is experienced first as an absolutely terrifying condition:

> I began to fear this silence, deep as the meaning of life.
> Everything had gone numb, like a seal of death....
> "What is happening?" I wanted to break out of this circle of the impossible, of the incomprehensible, that enclosed me.
> "We must break through it!" my whole being shouted in me.
> "I'm close to insanity," I thought.
> "I'm losing consciousness," the thought pierced me.
> "What is this?" Everything was crystalized in that single question.[112]

The rhetoric of silence, unconsciousness, death, incomprehensibility, insanity, and impossibility—accompanied by affects of anxiety and mounting panic—dramatizes what Evgeniy Kuchinov calls *fictio audaciae*, the daring of utopian fiction.[113] What the Five Oppressed are passing through, in this moment, Kuchinov calls, following Quentin Meillassoux, a "non-Kantian world of the third type," a world in which the abrogation of scientific law threatens not only scientific knowledge but consciousness itself.[114] In this way, the Gordin Brothers defy the blackmail of the Enlightenment, in which a submission to "law"—both the laws subtending the phenomenal world and the laws of the rational despot—is the price of the free use of reason; their aim is to dispel anxiety, to assure us that even the most thoroughgoing anarchy is survivable, livable. Thus, the scene of panic is juxtaposed with a description of the Man from the Land of Anarchy as an embodiment of joyous affect:

> As before, the expression of the man from the land of Anarchy was completely clear, suffused with the gaiety of play.
> The seal of joy was visible on his forehead.
> Little doves of merriment nested in his eyes, cooed in his pupils, fluttered, dived, flew in and flew out, inscribing circles in the air, magic circles of insouciance, insouciance in his bright smile.
> And he somehow danced with his hands. Rather, not with his hands, but with the fingers of the hands that he held stretched out, a little tensely, in front of him.

> It seemed to me that he was playing some unknown song on an aerial harp, the strings of which form light breezes, trembling, sounding soundlessly in this melodious, feeble silence.
>
> I watched his fingers.
>
> They capered, dancing the drunken dance of a drunkard's roundelay.
>
> They twirled like springtime.
>
> They chased each other, caught up with one another, embraced, kissed, touched and parted.
>
> The mystery of the fingers.
>
> The game of life and death and love was in the fingers.[115]

The Man from the Land of Anarchy is, of course, trying to communicate with them by sign language—"we all know how to speak it," he explains, once they have moved out of the silent zone, in case they find themselves in "an inaudible environment . . . underwater . . . or when there is too much noise."[116] Here, we find a strange entanglement with eugenic logics. It seems unlikely that "the game of life and death and love" in the Land of Anarchy still includes involuntary deafness, for almost nothing about existence remains involuntary: among these perfected bodies, illness is no more, and death itself is a matter of volition.[117] Nonetheless, the utopians of the pantechnical future cannot understand why anyone would rather do without another means of expression, a beautiful invention.

The lyrical description of the dance of the fingers heralds another discovery: the pantechnical abolition of nature (which also naturalizes everything) seems to open up the possibility of new kinds of desire.[118] Descriptions of the Man from the Land of Anarchy from the standpoint of the Individual are suffused with homoeroticism:

> He was divinely beautiful. He was peculiarly beautiful. He was a fairy tale of magical beauty, told by the evening penumbra between two fields, green and agitated by the kiss of spring and wind.
>
> He had a special harmony. The silent consistency of lines.
>
> He had even more beauty than strength, even more strength than beauty.
>
> In him, force, might, courage merged, sang their song to the last note, with beauty.
>
> He breathed beauty; the breath of his beauty was palpable. Tangible in one's bare hands.
>
> And his power was subtle, a distant whisper, echoing gently, quietly, in the high mountains of the all-triumphant, of omnipotence.

And his power was quiet, like a ray of the moon reflected in the midnight waves of the spring-loving, peaceful lake in the heart of Youth.

And the strength of his power was like a certainty, like an inevitability that flies down the road to the land of doom, to the edge of fate.

And his power was deep, like an intimation and a premonition, born from the bottom of the abyss, a trembling anticipation of the unknown, the untested.

And his power was sublime, like the flight of the eagle of salvation, soaring above the clouds of suffering, soaring through the skies of compassion, bathed in full light.

He stood ahead of us. He was all one single "forward!"

So much courage, so much determination.

And so much tenderness, softness.

The embodiment of the doctrine that it is the gods who give birth to the gods.

The embodiment of a heart that has surpassed itself in warmth and coldness, in gentleness and cruelty.

The willful and the passive splash in the sun of his harmony, in the swell of his features, in the curves of his whole body.

"Whence comes such beauty, and whence such directness of spirit?" I could not help asking myself.

Such is their life, such are they.

Life without worries, the brow of being without a single wrinkle, without a single crease. Working without work, without bondage, was what had given birth to this wholeness, this perfection of inner and outer, this consonance of body and soul, this attunement of flesh and spirit.

And his beauty was quiet, like the whisper of the beloved on the breast of the beloved, a leaf on the branch of a blossoming apple tree in the hour of morning, in the moment of awakening.

And his beauty was as subtle as the trace of a creeping thought on the dawn-crowned mountains of ancient religious wisdom.

And his beauty was mighty, like the tempest of insurgent time upon the boundless sea of chronicles, which sinks the ships of creeds, smashes them, and buries their splinters in the maelstrom and foam of the everchanging waves.

And his beauty was as deep as the clear but distant waters of the well of revelation, vouchsafed to travelers in the forests of vocation, with which they water their lathered and tired horses . . .

And his beauty was lofty, like the resurrection of the sound of the horn of victory over death, over nature, over impossibility.

"Let us sit here, beside this spring!" said the man from the land of Anarchy, and the sounds of his voice traveled, sending ripples across the quiet surface, the Ode I had composed in my mind to his beauty and harmony.[119]

Unbridled Dream, it seems, has leaped over the Law of the Father, its canons of the acceptable and the abominable, the permitted and the forbidden. It has done so in language that nonetheless echoes Biblical cadences, uniting stereotypically masculine traits ("force, might, courage") with stereotypically feminine traits (his body is aestheticized, a spectacle of "beauty," "harmony," "consistency of lines," and characterized by "passiv[ity]" and "quiet" as well as action and splendor). The body imagined here is a natural body (compared to green fields, birds in flight, "a leaf on the branch of a blossoming apple tree in the hour of morning"), a technical body (representing "victory ... over nature"), an allegorical body (compared to "the tempest of insurgent time upon the boundless sea of chronicles, which sinks the ships of creeds") and at the same time, an embraceable, carnal body (we are led to imagine the aftermath of a scene of love, "the whisper of the beloved on the breast of the beloved"). In a world of impossibilities made *lyrically actual*, impossible desires and loves find a place.

Perversion and Levitation

Antigravity is perhaps the original sin of science fiction. Well-known is the dispute between Jules Verne, author of *De la Terre à la Lune* (1865), and H. G. Wells, author of *The First Men in the Moon* (1901). Where Verne had extrapolated carefully—reasoning that a powerful enough cannon would have sufficient force to overcome the Earth's gravity entirely, propelling a projectile to the Moon and that a large enough projectile could contain human travelers—Wells had declared the problem of gravity a nonissue by fiat, basing his lunar voyage on the discovery of a gravity-blocking metal, Cavorite. This drew an impassioned critique from Verne:

> I make use of physics. He invents. I go to the moon in a cannonball discharged from a cannon. He goes to Mars [sic] in an airship, which he constructs of a metal which does away with the law of gravitation. *Ça c'est très joli* ... but show me this metal. Let him produce it.[120]

In his preface to a collection of his "fantastic stories" (1934), Wells belatedly replied: "By the end of last century it had become difficult to squeeze even a momentary belief out of magic any longer ... I simply brought the fetish stuff up to date."[121] For Wells, extrapolation and other such techniques are indeed only so many rhetorical devices, "an ingenious use of scientific patter" to "trick [the reader] into an unwary concession to some plausible assumption."[122]

"The Laws of God/Heaven and the Laws of Nature," the Gordin Brothers declare in their "Pananarchist Manifesto," are "fictional, created in the image of society (sociomorphically)."[123] In their fiction, the laws of gravity, like the Law of the Father, are no more. The Pantechnical Garden in which all the wonders of the Land of Anarchy are displayed is tens of thousands of *cubic* miles in size—"Items are not only placed side by side, but also one above the other in the air."[124] It is thus best explored by antigravity shoes:

> "What is this material? Silk or what?"
> "You can call it silk, if you like."
> We put on our "shoes" . . . [and] tried to step up into the air. Our legs obeyed, our bodies obeyed; we began to rise . . . It was easier to walk through the air than on the ground.
> "Still, this is very strange," I said . . .
> "I don't understand! Why don't I fall down?" said Worker . . .
> "There is nothing incomprehensible in the land of Anarchy. Understand, finally, that the world is not to be understood at all. You are in new shoes, but still encumbered with old views and concepts," the man from the land of Anarchy grinned.[125]

This non-gravitic world is, by Žižek's standards, a scene of sheer "perversion": the Gordin Brothers have imagined a world "unencumbered by the inertia of the Real," loosed from "the Real of human finitude," i.e., no longer structured by the fear of death or sexual difference.[126] "What the pervert enacts," Žižek writes, "is a universe in which, as in cartoons, a human being can survive any catastrophe; in which adult sexuality is reduced to a childish game; in which one is not forced to die or to choose one of the two sexes."[127] He identifies this perverse enactment with the suspension of time and gravity in *The Matrix*, e.g., in the famous scene of Neo and Agent Smith battling while hovering in midair, in "bullet time": there, reality is treated as a mere "virtual domain" in which you can change the "rules" at whim, which feels like absolute freedom, but the price of this illusory freedom is "an utter instrumentalized passivity," allowing oneself to be used like an object. In short: surrender to patriarchal gender norms, or surrender to capitalism![128]

Among the inescapable "rules" that cannot and must not be tampered with, for Žižek's fellow Lacanians, are the laws of sexuation. Another kind of barring of the subject: since "there is no sexual relationship," there is no place in Lacanian theory for a Tiresias who has known what it is like to be of the other gender. Such are the limitations of a theory that cannot countenance the lived reality of directors Lana and Lilly Wachowski, these impossible people who have passed the Lacanian "bar" in becoming women. So much the worse for theory.

In the Dangerous School, or the Pedagogy of Failure

In 1908, in their native city of Smorgon—a Jewish enclave caught in the dubious interzone between Lithuania, Poland, Belarus, and the Russian Empire, long since destroyed by pogroms, wars, and the Shoah—the Gordin Brothers founded "Ivriya," a *heder metukan* or "reformed school" for the study of Hebrew as a secular language. While still under the influence of labor Zionism (represented then by the Tzeirei Zion movement), they inclined increasingly toward anarchism. The town's pious grumbled that it was a "*heder mesukan*," a "dangerous school" for fomenting radicalism among the young, which was not wrong.[129] As prodigal sons of the famous rabbi Yehuda Leib Gordin, they rejected not only the mystical piety of the local Hasidim but also the rationalized Haskalic religion of their father, instead drawing inspiration from the works of Proudhon, Stirner, and especially "that Bakunin of pedagogy, L. N. Tolstoy."[130]

In particular, their educational experiment was inspired by the example of Tolstoy's school at Yasnaya Polyana, at which wrote Tolstoy, "only by my moral influence did I compel the children to answer as I wished them to do."[131] Was it possible? They found that a crowd of children that has been encouraged to make a sound (to demonstrate the Hebrew word for "sound") will not be so quick to stop that a Socratic or "catechizing" approach no longer worked—and yet, as they recount, "We were surprised by our students."[132] The very failures of instruction in this new environment were instructive. So Tolstoy himself had found:

> I had the intention in this first lesson of explaining wherein Russia differs from other countries, her borders, the characteristic feature of its government; to tell who was the reigning monarch at this time, and how and when the Emperor mounted the throne.
>
> TEACHER. Where do we live? in what land?
> A PUPIL. At Yasnaya Polyana.

SECOND PUPIL. In the country.

TEACHER. No; in what land are both Yasnaya Polyana and the Government of Tula?

PUPIL. The Government of Tula is seventeen versts from us . . .

TEACHER. No; Tula is a government capital, but a government is another thing. Now what land is it?

PUPIL (WHO HAD BEEN IN THE GEOGRAPHY CLASS). The land is round like a ball . . .

TEACHER. Where does Russia end, and where do the other countries begin?

PUPIL. Where you find the Germans.

TEACHER. Now, then, if you should find Gustaf Ivanovitch and Karl Feodorovitch in Tula, would you say that this was the land of the Germans, and therefore it must be another country?

PUPIL. No; it's where you find a whole lot of Germans.

TEACHER. Not necessarily; for in Russia there is a land where there are a whole lot of Germans. Johann Fomitch here comes from there, and yet this land is Russia. How is that?

Silence.

TEACHER. It is because they obey the same laws as the Russians.

PUPIL. How do they have the same law? The Germans do not attend our church, and they eat meat in Lent!

TEACHER. Not the same law, perhaps, but they obey the same Tsar.

PUPIL (*THE SKEPTIC SEMKA*). Strange! Why do they have a different law and yet obey our Tsar?

The teacher feels the necessity of explaining what a law is, and he asks what it means to obey a law, to be under one law . . .

The lesson lasts two hours. The teacher is . . . forced to the conclusion that these methods are unsatisfactory, and that all that he has done is perfect rubbish . . . *Raseya*, "Russia," *Russkoi*, "Russian," remained the same unconscious symbols of *mine*, *ours*, something vague and indeterminate. *Zakon*, "law," remains to them an incomprehensible word . . .[133]

Count Tolstoy finds that without the violence of authority, he cannot manage to teach the peasant children what nations are, what a government is, what law is. The "catechism" of questions and answers leads not to understanding but to greater mystery: even though the children readily parrot the ambient ideology

(Germans are outsiders, God loves Russians only), they cannot be made to understand the basis for national differences (the invisible inward essence of Russianness) or imperial unity (the invisible overarching framework of the law). What had appeared to him as basic knowledge, as facts and reason, is revealed as a collection of "unconscious symbols . . . something vague and indeterminate," empty abstraction, a nonsense. The children have defeated him.

By the evidence of their prewar tracts, *Sistema Material'noy i Otnositel'noy Yestestvennosti* (*The Systems of Material and Relative Naturalism*, 1909), *Seferot Ha-Iledot* (*Children's Books*, 1912), *Undzere Khiburim* (*Our Essays*, 1912), and *Ha-Sderot Ha-Iledim* (*The Order of Children*, 1913), the Gordin brothers might have read Tolstoy's account as an experimental proof of the futility, the bankruptcy, of schooling based on the inculcation of abstractions—what Paulo Freire famously dubbed the "banking method" of teaching: "Four times four is sixteen; the capital of Pará is Belém."[134] If the rod is needed to teach such things, they can't have much meaning beyond their disciplinary function. "The absence of punishment in the school," as they would write in the revolutionary years to come, "is the condition for the successful development of pedagogical technology."[135]

In 1918, Abba Gordin (signing his work simply "Br. Gordin") published a children's novel, *Pochemu? ili Kak muzhik popal v stranu Anarkhiya* ("*Why?*": *Or, How a Peasant Reached the Land of Anarchy*), a fable about the young peasant Pochemu (whose name means "Why"), who undergoes a long series of misadventures in which he continually plays the part of the Bad Pupil, the hapless dope who can never understand the lessons his society has to teach him. Thus, when his father takes him to church—"a big house, with domes and crosses"—he asks: "Papa, who lives in this hut?" "God lives in this hut," comes the reply. But this simple explanation does not satisfy Pochemu's insatiable curiosity: "And why do we live in a small house, while God lives in a big house? Why isn't he better off sitting in heaven?"[136] Caught up in a political demonstration, hauled before a judge, Pochemu is asked his name: "Why do you need to know? After all, it's me who's in prison, not my name." After the laughter in the courtroom dies down, the judge replies:

> "We must know in order to judge you."
> "Why are you judging me, if I'm not judging you?" asked Pochemu.
> General laughter.
> "The king has ordered you to be judged."
> "And who told the king to order that?" asks Pochemu.

"No one tells the king, he does not obey anyone."

"Why are you obeying him?"

"I am not a king," said the judge.

"And if you don't obey, will you become a king?" Everyone falls silent, perplexed.

"Silence," said Pochemu with enthusiasm, "I knew it! He who does not obey anyone is his own king."[137]

Cast out by his own society, Pochemu wanders, ending up by finding the Land of Anarchy, where the tables are turned, and he becomes Tolstoy, the Teacher. Here, every norm that goes without saying at home is received as an incomprehensible mystery. In a dialogue with a scholar from the Land of Anarchy, property predictably becomes the subject of considerable comic misunderstanding, as Pochemu is baffled in his efforts to explain what the word "theft" means:

"Well, for example, these are my boots; let's say somebody takes them."

"How are they yours? They are boots and nothing more. And why should he? Let him go to the warehouse and take it."

"Well, let's say this is my house . . ."

"This is 'your house'? I understand 'your son,' 'your daughter,' 'your father,' 'your mother'—but how will this person take them for himself? And the house—how is it yours? Did you give birth to it or what?"

The old scholar writes hastily in the book: "There is a belief among savage peoples that they give birth to their homes, and that they have them, each has his own house, like we have our own mother."[138]

Despite the humor of the imagery conjured up by passages like this, even if *Pochemu?* can be considered "the first written literary utopia in Russia after the revolution," it would be perhaps understandable that it should be "totally forgotten today," completely absent even from scholarly studies of Russian SFF and utopian literature. Nothing here is truly new.[139] Yet according to Boris Yelensky's Russian-Revolution memoirs, this was "the one pamphlet that Abba Gordin ever wrote in his life that had real potential as propaganda." Left out to read for Alexei, a young worker from the countryside, it elicited urgent questions: "What is anarchism? Where is the 'Kingdom of Anarchism'?" The young man then sought copies of *Pochemu?* to circulate in his village (134). The pedagogy of counterpedagogy—naïveté as method, logical demonstrations that fail dramatically, lessons that end in bafflement—is surprisingly effective.

A Pragmatist Counterpedagogy, or Magic without Spells

If Tolstoy's experiments in Christian schooling at Yasnaya Polyana showed the Gordin brothers what could be done in practical terms, it was perhaps Stirner's radically antireligious philosophy that helped them to make sense of it. "In childhood," Stirner observed, "liberation takes the direction of trying to get to the bottom of things, to get at what is 'behind things'... therefore we like to smash things, like to rummage through hidden corners, pry after what is covered up or out of the way, and try what we can do with everything."[140] All that resists, all that is hidden, "*natural powers*" no less than "the mysteriously dreaded might of the rod, the father's stern look, etc.," threatens to overwhelm the child, and is, therefore, an enemy to be fought against: "in childhood one had to overcome the resistance of the *laws of the world*."[141] The Gordin Brothers, too, believed that from the child's perspective, nature itself, with its enigmatic and incontestable "laws," is an enemy. "In their early pedagogical writings," as Evgeniy Kuchinov explains,

> they make a curious observation: there is nothing natural in pedagogy. There is no natural desire of the child to know the laws of nature; anyone who begins to learn, acts primarily as a burglar [*vzlomshchik*], a lawbreaker [*narushitel' zakonov.*]. Every child begins with the desire to rewrite the world again. This desire, of course, is suppressed by the fundamental NO: it is impossible to begin with the invention of the new; you must first submit yourself to the existing laws of nature.[142]

As anti-naturalist counterpedagogues, the Gordin Brothers were convinced that Tolstoy had to be improved on and that new techniques were needed. Thus, the Gordins decided to teach Hebrew first through "commands":

> For example "*shev* [שב]" used to be translated to the child. "*Shev*" means "sit down." Today, we are wiser, and we do not translate it to the student's language. We do not exchange a contract for a contract, a word for a word, but rather explain it within the context of the concept, that is, we tie the sum sounds of "*shev*" to the concept of sitting... the teacher shows the student the form of the word "*shev*"—and sits down; "*amod* [עמוד]" [stand]—and stands.[143]

This might seem paradoxical: isn't antiauthoritarian pedagogy specifically a rejection of a society structured around the giving and receiving of commands? Indeed, some have concluded grimly that "emitting, receiving, and transmitting order-words" is the primary function of language as such: "Language is not made to be believed but to be obeyed, and to compel obedience."[144] Yet such dedicated antiauthoritarians as Paul Goodman and David Graeber have pointed out that the very avoidance of commands in Western middle-class speech serves, not necessarily to put power relations on a footing of equality, but primarily to *conceal* them; we add "please" and "thank you," language once reserved for feudal vassals to their lords, to make it seem as if a request rather than a command is being voiced, even when compliance is absolutely expected and required.[145] Conversely, imperatives do not always imply an imperial speaker: "An 'imperative,'" as John L. Austin remarks, "may be an order, a permission, a demand, a request, an entreaty, a suggestion, a recommendation, a warning ('go and you will see'), or may express a condition or concession or a definition ('Let it . . . '), &c."[146] They are the primordial ancestors of abstractions, lived antecedents to the cognitive: the primitive precursor of the ineffable concept, *nothingness*, and its more everyday instantiation as *not*, is the cry, "Don't!"—close cousin to the word of political rebellion: "No!"[147] "Imperatives," Goodman argues, "never were sentences talking about the world, but were direct actions," a fundamental category for anarchism.[148] Thus, in *Anarkhiia v Mechte*, the inhabitants of the Land of Anarchy are experts in the technics of performative language but use "no command[s]"; "we do not have 'incantations [*zaklinaniy*]' or 'spells [*zagovorov*],' only a 'request [*pros'ba*]' and an 'entreaty [*moleniye*].'"[149] Likewise, the AO language conspicuously and intentionally lacks a command form.

In keeping with their "sensual-muscular" pragmatism, the Gordins intended to enlist the body in learning:

> Until Pestalozzi came, the bent ear (mechanical method) prevailed. Pestalozzi came and promoted the idea of investigation (the observation theory). Fröbel came and put a brake on the absolutism of studying the constitution of the limbs (work and play). Now, republicanism of the limbs (work and play) came to the fore. The child is not full of eyes like the angel of death. The child has eyes that see, ears that hear, hands to work, feet to run, and a palate to taste, etc. Natural learning must be from the perspective of "all my limbs speak."[150]

The body, of course, is always enlisted in learning, "train[ed] . . . to carry out tasks, to perform ceremonies, to emit signs," including the signs of gender: traditional schools from the nineteenth century to now, teach children first "to sit" facing the authoritative speaker at the front of the room, and then, if they are girls, "to sit *properly*."¹⁵¹ But Ivriya's pedagogical technology more closely anticipated the so-called "Do As You Please" schools of Homer Lane (the Little Commonwealth, 1913–1918) and A. S. Neill (Summerhill, 1921–), the ones that C. S. Lewis so fiercely caricatures in *The Silver Chair* (1953) as "Experiment House": "the people who ran it . . . had the idea that boys and girls should be allowed to do what they liked. And unfortunately what ten or fifteen of the biggest boys and girls liked best was bullying the others."¹⁵² So, in Ivriya, according to one witness,

> the children were free to do everything that was right in their eyes. They sang and danced at school, painted and sculpted, and engaged in all kinds of work and practical thinking. They didn't sit at all on their benches, they ran around all day long, and didn't read a book, even for a short while.¹⁵³

The work of learning was transformed into attractive play—indeed, into *plays*: "The kindergarten will in the future become the theatrical kindergarten."¹⁵⁴ The Gordins created just such a playhouse, an organization of verbal-kinesthetic *performances*. Ivriya's children were thus constituted not as spectators but as "actors": both players and agents.¹⁵⁵

Another aspect of performance was crucial to the Gordins' thought and practice, then and later. In learning the Hebrew language by enactment, rather than as a "constative" system for the description of a static reality, the children of Ivriya apprehend it as first and foremost "performative."¹⁵⁶ This was linked to the specifically *secular* nature of their *heder metukan*: to teach Hebrew as a language of everyday life rather than a frozen order of words for ritual recitation, they had to abandon the method of beginning with "catechis[tic]" memorization of the names for things ("parroting and aping!") in favor of one in which children "are invited to create new creations."¹⁵⁷ And it is in the nature of performatives that they do not *represent* states of affairs but *create* them. Here, the brothers' "new pedagogy" opens up onto a new ontology, a pragmatist ontology in which "the value of any opinion or idea is not to be found in its abstract truth, but rather in the potential or actual ability to change the material of its environment":

> All the systems, all the theories that are created by humanity to explain and understand the world are based on the search for the "nonexistent [*ha-ayin*]"—on the desire to invent or create something that does not exist in reality.¹⁵⁸

No God, No Nature: or, The Anarchist Space Program

NI BOGA, NI PRIRODY! So proclaimed the showcase in the storefront at 68 Tverskaya Street in Moscow on the evening of April 24, 1927, under the mysterious compound word *VSEIZOBRETAL'NYA (ALLINVENTIVE)*. "Lit more dazzlingly than the rest," according to one visitor's account, and drawing a crowd, the glass case opened onto "a fantastic landscape . . . against a black and blue background, a generously starry sky," traversed by a streaking rocket and angular slogans: "We cosmopolitans will invent ways to the worlds!" "We will invent everything through the AO language!"—and crowned by "a bewildering inscription: 'The First World Exhibition of Interplanetary Vehicles and Mechanisms.'"[159] Milling about rooms full of exotic diagrams and models, visitors may have missed the fact that they were on the premises of an anarchist organization, shortly to vanish along with the ever-dwindling tolerance of a State apparatus that was by no means fooled by shows of outward ideological conformity, such as the staging of an exhibition of Soviet ingenuity on the ten-year anniversary of the February Revolution.[160] The displays are enough to make you forget where you are: "Having taken only a couple of steps," wrote one witness, unwittingly anticipating Neil Armstrong, "it was as if I crossed the threshold from one era to another—the cosmic era . . ."[161] Here are elaborately labeled cross-sections of space stations, rockets with wings like dragonflies, atomic propulsion systems, a Martian diorama . . . "And then," as another Soviet will recount fifty years later,

> the display of "The International Language AO." A stunning theme! "The language of logical concepts," with an alphabet of eleven letters depicted by algebraic signs! The minimum number of letters makes the AO language the lightest language in the world, which is why it will have to become a human language.

Or rather, as the Soviets will have to say, "let us say in advance, their expectations were not met, and they could not be justified."[162] Nonetheless, there it is: simultaneously the most abstract exhibit and one of the most striking in its manifesto-like appearance. "Students who study and speak the language 'AO,'" it proclaimed,

> . . . are *cosmopolitans* (citizens of the universe) who have expressed the desire to embark on interplanetary travels, just as citizens of a country took a risk in riding, for the first time ever, a steamship, train, hot air balloon, or airplane.

1) Between the concept and the thing must be COHERENCE.
2) Between names (words, syllables, sounds) and concepts, there must be a CORRESPONDENCE in accordance with the "AO" phonological system.
3) It follows that the names (words, syllables, sounds) must be in accordance with things. More generally: Language must be in correspondence (relationship) with the world.

Conclusion: the language "AO" therefore derives its justification and existence from the absence in modern languages of any relationship between words and things-concepts and, even more, between sounds and things-concepts.[163]

And the First World Exhibition of Interplanetary Vehicles and Mechanisms derives its justification and existence from this one exhibit.

The impresarios of this spectacle, the Interplanetary Detachment (*Mezhplanetnogo Otryada*) of the Association of Inventives-Inventists (*Assotsiatsii izobretateley-inventistov*, abbreviated as AIIZ), were in fact associates of Wolf Gordin, arrested and deported nearly two years earlier, whom they hailed, in absentia, as "The Inventor."[164] These men and women were "*AOistov*" ("AOists"), practitioners and devotees of the "language of mankind," which they now presented as the first truly "cosmic" language, suitable for communication with other forms of intelligent life in the universe.[165]

The countercultural atmosphere of this event and its organizers cannot be understated, and it might not be entirely unfair to compare it to that of a *Star Trek* convention. The Inventor's books, L.-Ron-Hubbard-esque tracts on "Gnoseology" and "Inventor-Nutrition," displayed alongside works on the theory of space travel, were signed by a mysterious string of mathematical symbols: "$1+01\sqrt{}$." These were to be pronounced "*Beo-bi*," meaning, in AO, something like "Man of Society," from the words *beo* ("society, humanity") and *bi* ("me," the suffix for personal names)[166] "You are ultraidealistic," declared a poem written for the occasion by Alexander Sergeevich Suvorov, who had chosen the AO sobriquet "$\sqrt{2 \times 1}\sqrt{}$" or "*Ibtsabi*" ("Invent Yourself"): "You must recreate everything,/With your inventiveness/You will conquer the Universe."[167] The AIIZ's enthusiasm for neologism, as Asif Siddiqi notes, led them to coin terms for things that did not yet exist, such as astronautics: "*zvezdoplavaniia*," literally "astrodynamics," a derivation from "*vozdukhoplavaniia*" or "aerodynamics." We might think of this as the 1920s equivalent of William Gibson accidentally coining the term "cyberspace" in a story for a science fiction magazine (1982); indeed, Siddiqi also notes that our word "astronautics" derives from "*astronautique*," which would be coined by the Belgian science fiction writer J.-H. Rosny aîné in December 1927, thousands of miles away in Paris, where the members of another newly-formed society for the promotion of space voyages were trying to come up with a *name* for the thing for which they wished to award a prize, reaching for the eighteenth-century word *aéronaute*, a portmanteau suggesting that the Montgolfier brothers in their hot-air balloon (1784) were to be imagined as sailors of the sky.[168] The French engineer after whom the astronautics prize was subsequently named, Robert Esnault-Pelterie, had answered the AIIZ's request: there in the exhibition hall on Tverskaya Street stood a display with his original liquid-fueled rocket designs.

The traffic between technoscience, fantasy, and political ideology is very fast and thick on Tverskaya Street (which the AOists have renamed "AOulitsa" or "AO-street," just as they have renamed Moscow "AOgrad"). Nor is it all a matter of dreamy enthusiasm, the effervescence of Soviet culture, in the words of China Miéville, still enjoying the last days of "a protracted *sumerki* [summer], a long spell of 'liberty's dim light,'" before the slide into "despotic degradation."[169] For one thing, even if it had not (and never would) have as many speakers as Esperanto, AO had developed an actual constituency, becoming one of the languages spoken in the Tolstoyan commune "Zhiz'n i Trud" ("Life and Work"); there, if it had not quite "gripped the masses," AOist theory seems to have gripped a handful, in the Marxian phrase, "becom[ing] a material force," part of the lived everyday practice of a radical community.[170]

Civility and War

The Soviet State, having been until only recently just a theory, was very much in danger of losing its grip on the masses in those years of Civil War (1917–1922), New Economic Policy (1921–1928), and international isolation after the collapse of revolutions in Europe (1924–1928). Anarchist revolts in Ukraine (1917–1921), at the Kronstadt naval base (1921), and in the heart of Moscow itself (1917–1918) had indeed passed from "the weapon of criticism" to "criticism by weapons," and even after the pacification of the remnants and before the onset of the Stalinist Thermidor (1924), the space for social and ideological experimentation was beginning to close. In order to escape what Deleuze and Guattari might call the "paranoiac machine" of the State, a repressive engine in which "bureaucratic or fascist pieces are still or already caught up in revolutionary agitation,"[171] the Gordin brothers, having been enthusiastic participants in the first phases of the revolutionary process, proliferate strategies:

- Taking a series of increasingly conciliatory *stances toward the State*:
    - *ecstatic embrace* of the revolutions of February and October 1917: "Comrade anarchists! The great historic hour has struck" (Nov. 11, 1917);[172]
    - *intransigence*: "All representation is a brake on Freedom. Participation in elections is tantamount to treason" (Nov. 14, 1917);[173] "The mask must finally be ripped off [from] the Marxist charlatans" (1918);[174]
    - *acceptance*: "Socialism must beat the bourgeois world in order to be beaten by Anarchism in its turn" (1918);[175] "We, anarchists-universalists, take an active part in Soviet construction, considering it a socialist initiative that lies on the path to anarchism-universalism" (August 16, 1920);[176]
    - *taking a distance* from the State's enemies: "We completely reject the disorganizing methods of struggle, rebellion, terror and partial expropriation" (August 16, 1920);[177]
    - *declaring loyalty*: "Anarchist revolution will be possible only after victory for socialism in a number of countries and states. All dreams and thoughts about the establishment and implementation of anarchism, bypassing the stage of socialism, going straight from the capitalist mode of government and production to the anarchist, we reject as utopian" (Dec. 1920)[178]
    - *disavowing practice* in favor of "pure" theory: "He [Lev Chernyi] is a theorist; how do his philosophical categories threaten you [the State] on a practical, political plane?" (1921)[179]
- Writing for, editing, and/or publishing a series of increasingly ephemeral and small-circulation *periodicals*:
    - The Petrograd Beznachalie: Organ Soyuza pyati ugnetennykh, anarkhistov-kommunistov (Without Authority: Organ of the Union of the Five Oppressed, Anarchists-Communists) of Petrograd (Oct. 25, 1917, one issue only)
    - The Anarkhiia of Moscow (September 13, 1917–July 1, 1918)
    - The Burevestnik (Stormy Petrel) of Petrograd (Nov. 11, 1917–May 21, 1918)
    - Six issues of Universal: Politika, Filosofiya, Ekonomika, Iskusstvo, organ of the Moscow Section of Anarchists-Universalists (Feb.–Dec. 1921)
    - Four issues of Cherez sotsializm k anarkho-universalizmu (Through Socialism to Anarcho-universalism) of Moscow (April–July 1921).
- Joining or forming a series of increasingly exclusive *organizations*:
    - the Petrograd Anarchist Federation (1905–1924)

- the Moscow Federation of Anarchist Groups (March 13, 1917–April 1918)
- the First Central Sociotechnicum (1918)
- the Petty Secretariat of the Northern Regional Union of Anarchists (1918)
- the Association of Pan-Anarchists or Union of the Five Oppressed (1918)
- the Kursk Federation of Anarchist Groups (November 1918)
- the Moscow Anarchists' Union (1919–1920)
- the All-Russian Section of Anarchists-Universalists (Summer 1920–1921)
- the Moscow Section of Anarchists-Universalists (December 1920)
- the Organization of Anarchist-Universalists (Interindividualists) (1921–1925)
- the Association of Inventives-Inventists (1925–1927)…

• Coining a series of increasingly allusive *neologistic labels* for their ideology: "pan-anarchism," "anarchist-universalism," "interindividualism," "sociotechnics," "pantechnics," "panmethodology," "allinventiveness"…

In short, having lost the revolutionary footing of 1917 and instructed by the raids, arrests, and forced closures of April-May 1918 on the limits of Bolshevik tolerance for dissent, the Gordin Brothers had retreated, in practice, into the imagined shelter of a Kantian social contract that had already been nullified. The new monarchy would not allow them to argue about whatever they wished in exchange for obedience.

A smaller utopian dream was hatched in the midst of all this ferment: "Build the Sociotechnicum!" the Brothers exhorted. Both school and self-managed community, the Sociotechnicum would regard itself as sovereign, outside of the State's authority, "ectoarchic," in something like the manner of autonomy actually won half a century later by the universities of Greece. In this sense, it might foreshadow la paperson's notion of the "*fourth world university* … a placeholder for the places of epistemology that are autonomous from the university,*"* corresponding to "the 'fourth,' autonomous form of civil society."[180] The First Central Sociotechnicum, dedicated to the proposition that "social apparatuses can be invented, perfected and cultivated artificially,"[181] was to be a kind of alternative to universities and to University Discourse, an advanced kind of *cheder mesukan* for the practical, muscular-sensual investigation of new forms of social life free from coercion. Both in this school and in the worlds to be hatched from it, they proposed, "the empty place of punishment … will have to be filled with social technology."[182] But history, in the charge of its designated representative, the Soviet State, had other plans.

The "Allinventives" on Tverskaya Street knew the score. AIIZ organizer Ivan Stepanovich Belyaev, who had chosen as his AO-name $5+3051\sqrt{}$ or *Leedolbi* ("Seeing the Fulfillment of His Hopes"), had already done a stint as a political prisoner. That summer, tensions between Wolf Gordin, the Inventor, and the Soviet government reached a peak. In July, frustrated by the repeated denial of government permission to travel to Mexico to spread the word of AO, *Beobi*-Gordin went on a hunger strike and was joined (against his wishes) by *Ibtsabi*-Suvorov. Four days into the strike, a certain "anti-Soviet" poster placed in the window of 68 Tverskaya prompted the authorities to crack down. Having heard from a comrade that the police were arresting everybody at 68 Tverskaya, the Inventor went down there, as he told his interrogators, "because I wanted to finally resolve the issue."[183] Meanwhile, AOist comrade Efim Moiseevich Serzhanov ($1\sqrt{}\times+51\sqrt{}$ or Biaelby, "Inventor of Life") had been hauled before the OGPU for participating in the hunger strike. Both *Beobi*-Gordin and *Ibtsabi*-Suvorov were subjected to a "forensic psychiatric evaluation": the inspecting physician decided that they

> do not show signs of a clearly expressed mental illness, but are psychopathic personalities such as schizoids with a kind of formal thinking, with a penchant for resonant-philosophical constructions, with a special kind of isolation and aloofness from concrete reality. These psychopathic features explain both the fanatical fascination with the newly invented language and the negativistic way of protesting them against the authorities, expressed in a hunger strike. These characteristics do not make them irresponsible and not liable. Neither need be hospitalized.[184]

This is not far from the nonprofessional opinion of former anarchist turned Bolshevik Victor Serge, for whom Wolf Gordin ("the 'Beobi Man'") is the hapless victim of an "*idée fixe*." Serge can read in the "inventist" tracts little more than "a delirious fantasy, a perpetual dream rising to the heights of lyricism . . . energy, violence, and vehemence, all of it expressed in a language sprinkled with scientific-seeming barbaric neologisms."[185] The very proliferation of texts intended to ground AO in the authority of knowledge (the "gnoseological") instead testifies to its counterfeit (merely "scientific-*seeming*") status, its fantastic character (departing from "concrete reality"), its disreputability ("violence," "fanaticism"); they demand not to be *read* as rational signs, referring to a public order of things, but read *through* as delusional symptoms of a private disorder. Marx and Engel's "scientific socialism" having taken on the material force of the State, all of its "utopian" rivals are now definitively outmoded, and this becomes the task of the psychiatric State apparatus: the *Zhiz'n i Trud* commune will be banished to Siberia, and its land and buildings will become the Kaschenko state hospital, a reformatory for "excitable epileptics, inaccessible schizophrenics, irritable idiots, and unstable psychopaths" in which "the main method of treatment is work."[186]

Rehabilitation, or a Forgetting

Abba and Wolf Gordin escaped the Great Sorting Machine of the OGPU, but where did they escape *to*? At the end of January 1926, Wolf Gordin sailed into the port of Victoria, British Columbia, then made his way to Chicago, where his sister Masha received him; eventually, he seems to have settled in Los Angeles, where he gave lectures on "cosmopolitism and humanitism," wrote essays on relativism for the *Journal of Philosophy*, and contributed articles in French to the anarchist journal *L'En-Dehors*. In 1942, he appeared in the journal of Linus Pauling, who received a call from Pilgrim State Hospital, where "a certain Vladimir Gordin, who uses the name of Bibi Beobi, was admitted to the hospital due to 'acts of strange behavior,'" claiming that he worked with Pauling at Cal Tech (Pauling only recalls having met him to discuss "his views on biology"); perhaps Beobi never really got away from the psychiatric State at all. A personal ad in the *Los Angeles Times* in May 1961—shortly after the spaceflights of Gagarin and Shepard, and a few days before Kennedy's announcement of the Moon program—announces a pamphlet titled *The Bluff and Barbarism of Science With Its Swinish Nature* by Bibi Beobi (price: fifteen cents), which apparently argued that "neither the Planets and the Moon nor the Sun nor the Stars do exist as heavenly bodies." After that, he disappears from the records, surfacing once more in a final publication, *The Plight of the Woman and the Rape of Truth: The Inventist-Temporarist-Everybodist-Nikedist-Feminist Manifesto* (1973), also under the name "Bibi Beobi"; he is buried, sans obituary, under this final alias. Just weeks after his older brother, Abba, arrived in San Francisco in 1926, having made a harrowing getaway via Manchuria and Shanghai; settling in New York, then eventually emigrating to Israel in 1958, he died in the country founded by the allies of his Zionist terrorist nephew, David Raziel, leaving behind a vast inheritance of manuscripts in English, Yiddish, and Hebrew (never again Russian). Perhaps Western gravity claimed them in the end, geopolitically speaking.

The thought of the Gordin Brothers was untimely, arriving both too soon and too late. And so it is that they have been excised from historical memory, read either as a distraction from the central narrative or as a transcendent exception to it. It has been easy to forget Abba and Wolf Gordin; as in the case of Alfonso L. Herrera, we can forget them in the very act of remembering them if we remember them as nothing more than the inconsequential eccentrics noted briefly by Victor Serge or Paul Avrich, "thinning" their reality until they appear as nothing more than the grotesque but merely aberrant figure of the horse with the arrow in its forehead. We could forget them in yet another way if we rehabilitate them, after the manner of Allan Antliff, by reading them as the discoverers of poststructuralism *avant la lettre* so that our encounter with them appears as a mysterious and transcendent experience, like the glimpse of a unicorn in the woods.[187] They were, in one sense, ahead of their time in their grasp of social construction, their questioning of metaphysics, and their refusal of reductionist analyses of social identities and social struggles. Yet they were also, as we have seen, importantly *behind* their time and *against* it, proposing a future that in some respects resembled the very past that "high modernisms" sought to abolish. It is when they are read not as antifetishists and futurists but as *metafetishists*, as *nonmodern* thinkers, that the uniqueness of their contributions starts to emerge.

PART IV
Fantastic Politics in an Imperiled Pluriverse

10

Unicorn Rhetoric

> Jo marveled at the relatively strange aerodynamics at play in the unicorns' flight. Which is what you tend to marvel at when your gay dads spend most of their breakfast talking about drag, when drag is a force that acts in the opposite direction of a moving force—and also the subject of their favorite TV show.
> —Mariko Tamaki, *Lumberjanes: Unicorn Power!*

The works of the New Black Fantastic and the Gordin Brothers demonstrate some of the ways in which science fiction, fantasy, and horror can intervene in the public, political realm. What I want to suggest now is that the public, political realm is *itself* becoming increasingly science fictional, fantastic, and horrific in character; their figures and tropes pervade the political world.

What does that look like in practice? Let us consider as an example the rhetoric of UNICORNS.[1] If rhetoric is, on at least one rough definition, the political art, the art of finding or creating allies, we might take the UNICORN as an example of superb artlessness: after all, as Graham Harman notes, even invisible neutrons "have more and better animate and inanimate *allies* testifying to their existence than do unicorns."[2] But if rhetoric, more broadly construed, is the art of doing things with words and images, we can see that in so saying, Harman breathes rhetorical life into UNICORNS by giving them something to *do* in the world, transforming them into an example of something seemingly entirely fantastic that nonetheless could potentially lay claim to a kind of reality. Consequently, eco-socialist Andreas Malm cites this passage from Harman's book to furnish an example of the kind of "epistemological nihilism" to which the so-called ontological turn leads (and therefore its political irresponsibility).[3] While we might sympathize with Malm's call for "climate realism," we can observe that he inadvertently

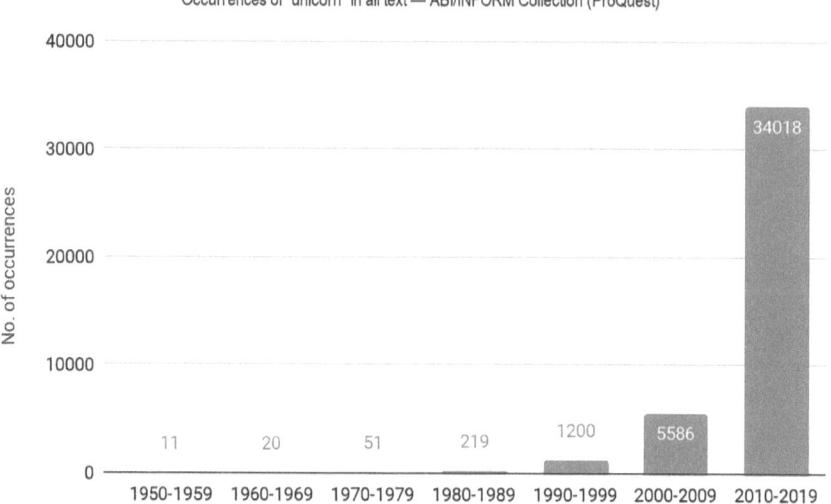

Figure 4.1.1. A proliferation of "unicorns."

demonstrates Harman's point: if "whatever *acts* in some way is real," then it is just *insofar as they have effects*—e.g., "an effect on the structure of fairy tales and the stuffed animal collections of children"—that "unicorns cannot be excluded from the picture."[4] In Malm's writing, the effect of citing "unicorns," is to ridicule, to expose, to censure; they are the smoking gun of fashionable nonsense, relativism, the death of reason. The UNICORN *does* so many things: it incriminates; it unmasks; it delegitimizes. It is rhetorically powerful, and as such, *it is politically alive.*

UNICORNS, of course, belong to the deep springs of mythology, a realm of supposed stasis: archetypes are just what we take to be "timeless" images, symbols with seemingly permanent meanings. We can look them up in books: they symbolize moral and sexual PURITY, etc. Nonetheless, in the last generation or so, they have proliferated, both as images and in verbal expressions, achieving a new kind of media ubiquity. As a rough and ready gauge of this change, we can see that the term has quickly spiked in newspapers (fig. 4.1.1)—an indication that discussions of and in the contemporary world are, for some reason, becoming UNICORN-saturated. This is notable even in comparison with other pop-culture fantastika: at the moment when fantasy writers Holly Black and Justine Larbalestier published the collection *Zombies vs. Unicorns* (Summer 2010), zombies were just coming into ascendancy, to enjoy unparalleled popularity through the Obama administration, but in 2017, they were surpassed by UNICORNS (fig. 4.1.2).

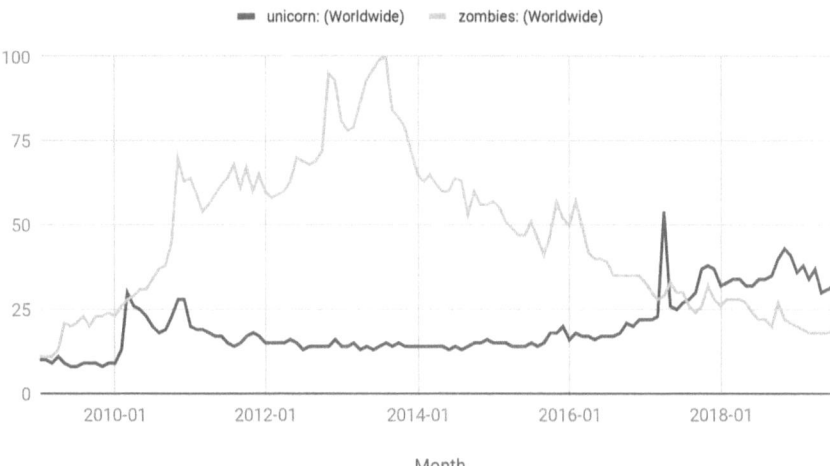

Figure 4.1.2. Relative public interest in unicorns (blue) and zombies (red) over time, 2009–2019, as measured by number of Google searches for "unicorn" and "zombies." Data source: Google Trends (https://www.google.com/trends).

All of this may make it *more* difficult to say what they mean, what kind of semiotic work they are doing, and for what and whom. It's worth a closer look. Here follows a brief Unicorn Cultural History of the last decade.

Perhaps the verbal usage that is easiest to identify uses UNICORN as shorthand for RARITY, a compressed analogy. For instance, *Urban Dictionary* records a use of "unicorn" as slang for the ineffably perfect "dreamgirl," the subject of an endless (and possibly DELUSIONAL) quest.[5] RARITY makes UNICORNS a figure for DESIRE in more than one sense: since 2013, startups valued at a billion dollars or more have been called "unicorns," so investors speculating in exotic tech stocks, hoping that previously unknown companies will turn out to be the next big thing are also called "unicorn hunters" or "unicorn chasers," reinforcing the sense of UNICORN as a synonym for both RARITY and MAGIC in association with DESIRE—meanings readymade for a product such as Paris Hilton's "Unicorn Mist" brand rose water spray (2017, $29/bottle).

Rose water minus the UNICORN branding can be had for less than a tenth of that price, of course; what Hilton is marketing is perhaps better described as an emotion, an affect. Therapist Kathleen Smith's *The Fangirl Life: A Guide to Feeling All the Feels and Learning how to Deal* (2016) offers girls and young women a professional perspective on the behavior of their own "fangirl brain[s]":

Close your eyes and think about a fangirl moment where you were puking rainbows: a first kiss between characters, your first real-life glance of a celebrity, or a spoiler that blew your mind. Got it? Now latch on to that image like you're trying to use it to fly to Neverland, but keep your feet on the ground. Can you feel the buzz of that unicorn vibe? That's dopamine at work. Dopamine is a feelgood chemical, and your brain wants as much of it as it can get.[6]

Here, UNICORN becomes an adjective meaning something like ECSTATIC. It is a feeling, as the word's etymology records, that removes one from oneself (*ek-stasis*: "outside-standing," i.e., an out-of-body experience, being "beside oneself" with excitement), lifts one off the earth.[7] It produces a giddiness, a swooning, that guardians of propriety have traditionally seen as dangerous: where might such flying girls and women go? At the same time, a "unicorn vibe" might also place the ECSTATIC subject in the thrall of others: as Smith warns, "The problem is that when you skydive into feels too much, you forget what it feels like to have some good solid ground beneath you. The desire to replicate the jump over and over can lead to obsession or addiction. A giant splat."[8] Not coincidentally, UNICORN branding often appears on Ecstasy tablets (price hovering around $20/dose).

We can find the same UNICORN imagery at work, not in the DESIRING pursuit of love or money or in the affect of ECSTASY but in the assuaging of *disgust*. Confusingly, on April Fool's Day, 2009, the niche catalog *ThinkGeek* offered a "Unicorn Chaser," a MAGICAL drink:

> We've all been there. You are innocently Twit-blogging on the Interscape, logging a few hours on Facebook, or checking your email and you click on a link without thinking. Suddenly, you are confronted with an image or video so horribly nauseating it makes your eyes bleed. Whether it be pictures of someone's overstretched nether regions or a video of two young ladies sharing substances they oughtn't—your mind begs for cleansing (or a swift death)! ...
>
> Introducing, the Unicorn Chaser—a drink shot specially formulated to cleanse your mind and soul. Featuring a perfect blend of vitamins, herbs and minerals (each selected for its body purification, mood elevation, stomach calming, and other beneficial qualities), the Unicorn Chaser is a life saver. Chug it within one minute of viewing the offending internet image (really, as fast as possible) and in mere seconds you will begin to feel better.[9]

ThinkGeek's "Unicorn Chaser," in turn, was a tribute to *BoingBoing* blogger Xeni Jardin's coinage of the term "unicorn chaser" (2003) for a *visual* antidote to especially disgusting or creepy images encountered on the internet: posts of safely soothing material, suitable for CHILDREN. Here, the old connotations of PURITY and MAGIC evoke a concept of restored INNOCENCE.

INNOCENCE and the CHILDLIKE are now preeminent connotations of UNICORNS, especially in their countless appearances in the realms of commerce and entertainment. UNICORNS function especially well as marketing devices: UNICORN-branded toys, games, clothing, and other paraphernalia for kids and teens flew off the shelves in 2015, "the Year of the Unicorn,"[10] and by 2017, trendspotters were announcing

> the very peak of the unicorn trend—the tippy top of a vibrant, magical rainbow where Katy Perry songs are on repeat and the clouds are made of dreams and cotton candy. On Instagram, feeds are flooded with users showing off everything from their unicorn body glitter (also unfortunately nicknamed "unicorn snot") to their unicorn bagels, pool floats, macarons, sequined jeans, grilled cheeses, high-top sneakers, iPhone cases, nail decals, and sushi rolls.[11]

While 2017 did indeed mark a moment of Peak UNICORN, disenchantment has so far refused to set in: by the start of 2019, a commentator laments that "the website Etsy lists 226,000 unicorn items for sale. There is no escaping them."[12] (Update, Apr. 30, 2023: 385,856 items.) Even otherwise subpar commodities are transmuted by the touch of the MAGIC horn: in 2017, the gargantuan Starbucks chain, merely "the latest brand to capitalize on the social media-famous unicorn food trend," rolled out a special five-day-only drink, the purple-pink "Unicorn Frappuccino," drawing a flurry of sales, despite the fact that its flavor is described as "like watered-down cotton candy," or more lyrically, "like sour birthday cake and shame." A food columnist shrewdly infers that the taste isn't the point: "This drink exists only to be Instagrammed, hashtag unicorn emoji, hashtag magical."[13] UNICORNS seem uniquely suited to an economy of rapidly circulating images.

Cannier marketers recognize UNICORNS more specifically as signifiers of 1980s-1990s nostalgia thanks to their association with the animated series *My Little Pony: Friendship Is Magic* (2010–2019), a 1980s reboot—not coincidentally, "Rarity" is also the name of one of the UNICORN characters—and to their association with the glittery kitsch aesthetic of Lisa Frank Incorporated. In *Class* (1983), Paul Fussell identifies the entire New Age kitsch universe of "plush unicorns, brass unicorns, 'Porcelain Revolving Musical Unicorns,'" etc. as the aesthetics of the American "prole catalog"; he does not note that the racial marking of this catalog is as white as the mythical beast itself.[14] In the current context, UNICORNS are also signs of a nostalgia for CHILDHOOD in general, albeit almost inevitably in the particular form of cisgender white girlhood: "This is a harkening back to mom's childhood," in the words of one corporate officer, "when everything was all rainbows and unicorns."[15] Such associations are clearly in play in Brie Larson's 2017 directorial debut, *Unicorn Store*, starring Larson as "an emotionally stunted failed artist who is fairly certain she's about to receive the titular magical creature": "*Unicorn Store* weighs the value of childlike innocence against the needs to face the realities of the adult world."[16] What can this ADULT world be, other than the world defined by an endless quest for commodities—and that most unattainable of UNICORNS: stable, long-term employment at a living wage? In the words of a viral Tumblr post by another young "creative" entering the ADULT world: "I AM LOOKING FOR A UNICORN (AND A STEADY JOB)."[17]

The implicit tension between ADULT and CHILD is already palpable in the phrase "unicorn store," which condenses all the contradictions entailed in the use of a mythical creature with an "exclusive, magical aura" for mass marketing.[18] There is something disturbing in the attribution of "linguistic sex appeal" to the word UNICORN when it is paired with the observation that this DESIRABLE quality is anchored in CHILDHOOD—as a *Boston Globe* trend piece suggests, quoting lexicographer Peter Sokolowski: "Anything that lodges itself firmly in children's minds is great fodder for metaphors in which some quality of the object is transferred."[19] In 2014–2015, UNICORNS' associations with CHILDLIKE INNOCENCE, another form of fantastic PURITY, were both reinforced and subverted by the appearances of Miley Cyrus, Ariana Grande, Katy Perry, and Taylor Swift in UNICORN onesies, performances "both polymorphically sexual and innocent," as Florian Cramer observes, directed to a hegemonic cishet male gaze.[20] For Arianna Davis, such spectacles "raise the question of whether or not unicorns are just infantilizing us all—transforming adults into women who titter in baby voices and take business meetings over multicolored coffee drinks topped with dollops of sprinkled fluff."[21] At the same time, the associations of UNICORN with FEMININITY and DESIRE make it available to (and are reciprocally made available by) the use of UNICORN imagery in queer camp aesthetics: thus, in 2014, Trans Students Educational Resources (TSER) debuts the "Gender Unicorn," a playfully-designed tool conceptualizing gender along five distinct axes ("Gender Identity," "Gender Expression," "Sex Assigned at Birth," "Physically Attracted to," "Emotionally Attracted to").[22] Thus also, a 2018 anthology of essays by writers with LGBTQ+ parents is titled *Raised by Unicorns*—"obviously a take on the old adage 'raised by wolves,'" notes the editor, as well as an allusion to the "unique[ness]" of LGBTQ+ individuals.[23] This converges, once again, with the positive sense of RARITY, to which a number of uses of UNICORN by people of color have referred, often in combination with MAGIC: in the title of her 2017 memoir, standup comedian Tiffany Haddish nominates herself *The Last Black Unicorn*—not a tribute to Audre Lorde's *The Black Unicorn*, apparently, but to her own "magical pussy"—and on his 2011 track "Radicals," rapper Tyler the Creator declares: "I'm a fucking unicorn, and fuck anybody who say I'm not."[24] Where UNICORN kitsch serves to place the bearer within an aura of belonging, signaling membership in a specifically white, cisfeminine collectivity, the same imagery can also signal sexual/racial DIFFERENCE and OUTSIDER identity.[25]

RARITY, in a negative sense, forms the nexus of another cluster of associations with UNICORN, and as such, it becomes part of rhetorics of reclamation or disavowal. An African-American entrepreneur profiled in *Forbes* feels as if he is, or is seen as, "a black unicorn, which is unfortunate."[26] In 2010, similarly, four women form a multimedia production collective under the name "Team Unicorn" to mock the notion that "like unicorns, geek girls aren't supposed to exist."[27] Here, RARITY is intensified to the point of another shift in meaning: UNICORN becomes synonymous with the MERELY IMAGINARY, the simply NONEXISTENT. Experienced as an identity, being a UNICORN is an impossible and intolerable condition: a native Hawaiian ethnobotanist tweets, on behalf of other Kanaka Maoli scientists protesting the Mauna Kea telescope project: "We are not unicorns, we are human."[28] In 2010, confusingly, *Urban Dictionary* logged a definition for "Unicorn Hunter" in the social context of polyamory communities: "Derogatory. An individual or couple seeking a unicorn or HBB [Hot Bi Babe]."[29]

A similar pattern turns the positive valuation of RARITY into the negative attributions of MERELY IMAGINARY and DELUSIONAL in a plethora of cases—one of the most common verbal and iconic usages. Richard Dawkins' New Atheist manifesto, *The God Delusion* (2006, reprinted 2016), takes "the invisible, intangible, inaudible unicorn" as a paradigmatic case of the resistance of beliefs in MERELY IMAGINARY objects to RATIONAL disproof, a rhetorical function doubtless aided by the "kitsch" character of UNICORNS: unlike God, the UNICORN does not solicit belief.[30] "Continuing their proud nation's great tradition of being simultaneously wackier and sadder than all other nations," writes blogger Ian Chant in 2012, "North Korean archaeologists have announced the discovery of what they claim is the lair of a unicorn."[31] This later turns out to be a distortion: government archaeologists had discovered references to a different legendary beast from Korean mythology, the Kirin.[32] No matter: the self-evidently ludicrous, backward, and DELUDED character of North Korea does not need to be established for a US readership. In 2015, the *New York Times*' Business Desk declares that the stock valuations of "unicorn" tech start-ups like Theranos are "justified" but notes with concern that "'unicorn' also suggests both the ambition and the absurdity inherent in the tech industry today—the idea that connecting the right bits and bytes might result in magical, mythical beasts . . . One reason Silicon Valley is simultaneously mocked and idolized is that people in the industry spend most of their days looking for corporate impossibilities."[33] Three months later, Theranos, having been hailed the previous year for its nine billion-dollar stock valuation, begins losing value as its technology claims come under scrutiny, becoming the first "unicorpse"— a cautionary tale, as more than one analyst remarked, of what Charles Mackay called *Extraordinary Popular Delusions and the Madness of Crowds*.[34]

Here, we find that UNICORNS can also come to function as signs not only of the merely DELUSIONAL but of the actually FRAUDULENT, a commonplace of political discourse. Discussions of US foreign policy in the face of the Syrian civil war, for instance, are peppered with UNICORN imagery. Echoing the language of Middle East "experts," in the course of the GOP presidential debates of 2015, Senator Ted Cruz accused interventionists seeking "moderate rebels" to support in Syria of hunting for "a purple unicorn," and was echoed by Fox News analysts in his turn, while center-right pundits like Jeffrey Goldberg suggested that overthrowing Bashir al-Assad was as REALISTIC as "want[ing] all CHILDREN to have access to unicorn rides."[35] Liberal discourse, perpetually affirming its commitment to neoliberal standards of RATIONALITY and PRAGMATISM, also makes use of this trope, as when a policy analyst is quoted to the effect that Trump's economic advisers are "looking for a magic unicorn."[36] More substantively, UNICORN can be used to indict the UNREALITY of an entire system: the "unicorn economy," legal scholars Amy Deen Westbrook and David Westbrook observe, is designed to make FRAUD ever more economically PRAGMATIC, particularly as the reliance on private funding from a handful of ultrawealthy investors allows "unicorns" to skirt the regulatory regime governing publicly-traded corporations.[37] Merging the economic and the political, a recent op-ed for *The New York Times* announces that we are living in "the Age of the Unicorn"—i.e., an era of "massive fraud," for which the inflated valuation of Theranos serves as a case in point—and dubs Trump "the unicorn-in-chief."[38]

Neighboring these tropes of hyperbole and ridicule, we find added negative associations with the FRIVOLOUS. A reviewer for the rightwing punditry site *Breitbart*, incensed at the description of novelist Benjamin Kunkel as a

"Marxist intellectual," writes in 2014 that "there is no such thing . . . Marxism is navel gazing while masturbating while riding a unicorn."³⁹ Here, the evocation of narcissism and masturbation links an ostensibly perverse, unnatural sexuality to the motif of DELUSION: the Marxist appears as a regressive Don Quixote, unable to orient to reality or an external sexual object. Similarly, in the long runup to the 2016 Democratic National Convention, innumerable memes identify Bernie Sanders' proposals to make healthcare, daycare, and/or college tuition free as a UNICORN ride, and in 2019, a chorus of conservatives label the "Green New Deal" proposed by Representative Alexandria Ocasio-Cortez and Senator Ed Markey a "promise of free unicorn rides for all" designed for "a magical world . . . powered by unicorn farts," a plan to "fly around on unicorns instead of airplanes" or replacing fossil fuels with "fairy dust we get from Tinkerbell."⁴⁰ The motif of DELUSION suggests a counterconcept of *realism* or PRACTICALITY, albeit, in this case, a thoroughly DELUDED counterconcept. Such usages don't always come from the same quarters, politically speaking: thus, BBC coverage of a 2018 climate science meeting in Korea quotes "a seasoned climate expert" as cautioning against reliance on unproven carbon-capture technologies, calling them "carbon unicorns," echoing Australian renewable-energy entrepreneur Simon Holmes à Court's dismissal of carbon-capture: "Well, if carbon capture worked in the power sector, I'd be one of the first to invest. If it was the unicorn that it promises to be—a technology that can capture carbon and safely put it underground for a reasonable amount of money and do so competitively—then I'd be first in line for that."⁴¹ Of course, the counterconcept of RATIONALITY/PRACTICALITY this UNICORN projects is that of a cost-benefits balance sheet, an extension of the economic logic currently pumping carbon into the atmosphere as if there were no consequences, no tomorrow.

But this is not the end of the story. There are numerous, often tacit and iconic uses of UNICORN as FRIVOLOUS in an ironically affirmative, positive sense. In the Fall of 2011, as if in a direct riposte to the rhetoric of "free unicorn rides" that has served to enforce the neoliberal consensus, UNICORN rides were actually available at Occupy Wall Street (OWS) (fig. 4.1.3). Just after the Occupy movement passed the seven-month mark (April 18, 2012), OWS-affiliated website *Occupy May Day* called for a general strike via a visual allegory: a unicorn branded with the initials of the anarcho-syndicalist Workers' Solidarity Alliance rips the head off of "WORK," personified by the infamous "pepper spray cop" of UC Davis, Lt. John Pike. When rightwing news outlet *The Blaze* seized on it as evidence of Occupy's murderous intentions, WSA responded: "It's absurd. But I'll tell you what's not absurd: Unicorns killing

Figure 4.1.3. A pink unicorn tricycle faces off against police at Occupy Wall Street. Photo by Katrina Brees.

work . . . We know now, after years of struggle, that capitalism can only be abolished—" in a knowing nod to *My Little Pony*—"through the power of Friendship."[42] Here, UNICORN FRIVOLITY is enlisted in a rhetoric drawn from the queer camp imaginary: bracketed by irony, the image suggests CHILDLIKE INNOCENCE and ultraviolence at the same time, cheerfully refusing in advance *The Blaze*'s moralistic framework. It operates in the TRANSCENDENT manner of Harold Beaver's "homosexual signs," expressing "the desire of the subject never to let itself be defined as object by others but to reach for a protective transcendence."[43] A similar radical-camp gesture is visible in the name chosen by the radical journalists' collective Unicorn Riot, formed in 2015. Here, the rejection seems to be of the RATIONALITY and PRAGMATISM comprising the ideology of "fair and balanced" mainstream journalism, beholden to the very corporate interests it should investigate: pooling equipment and skills and making decisions by consensus, Unicorn Riot's members have provided frontline coverage of Standing Rock and exposed the plans of white supremacist group Identity Evropa.[44] If there is a CHILDLIKE INNOCENCE here, it is all the better to witness the nudity of our Emperors.

It is not impossible, however, to enlist UNICORNS in the service of domination; indeed, the association with PURITY and WHITENESS already suggests one way to do just that. It is CHILDLIKE INNOCENCE and FRIVOLITY, however, that play an unexpected role in the operation of white supremacist UNICORNS. In a paradoxical relationship with the soothing aesthetic of the "unicorn chaser," the alt-right imaginary, what Andrea Nagle calls "the image- and humor-based culture of the irreverent meme factory of 4chan and later 8chan," is one of "images that 'cannot be unseen.'"⁴⁵ At the same time, though, alt-right and alt-light crusades have repeatedly targeted media products like *My Little Pony: Friendship Is Magic* for ostensibly sullying the PURITY of their INNOCENT CHILDHOOD memories with so-called "Social Justice Warrior" (SJW) concerns and perspectives, even if only to complain that the titular heroine of the rebooted *She-Ra* series (2018-) was insufficiently sexy.⁴⁶ But even this stance of indignation is secondary to the purely ludic, gratuitous cruelty of 4chan/8chan. Johan Huizinga describes the "ludic" as creating a "magic circle . . . within which special rules obtain," a kind of Other World: "Inside the circle of the game the laws and customs of ordinary life no longer count. We are different and do things differently." Importantly, he notes: "This temporary abolition of the ordinary world is fully acknowledged in child-life."⁴⁷ The college parties at which blackface is *de rigeur*, the Laugh Factory audience enjoying Daniel Tosh's rape jokes, are just such MAGIC circles; so are many of the special zones of the internet, where "people say and do things . . . that they wouldn't ordinarily say and do in the face-to-face world."⁴⁸ Irony, as a playful mode of thought and discourse, also suspends the laws, making possible both the "benign disinhibition" of self-exploration or community formation *and* the "toxic disinhibition" of anonymous malice with impunity; it allows the edgelord-ironist to "hover" over every finite commitment to meaning, as the German Romantics put it.⁴⁹ Fashwave or neo-Nazi appropriations of *My Little Pony*'s unicorn Twilight Sparkle dare us to take them literally *and* mock us for doing so (figs. 6 and 7). When someone logs on under the name "Aryanne's Jewish Sex Slave" to post a meme of Twilight Sparkle to the *DeviantArt* forum with the legend "EQUALITY IS A LIE," this is not an "oppositional reading" of *My Little Pony*, even of such a slightly-tongue-in-cheek kind as Brandon Morse attempts for *The Federalist* (pointing out that in a double episode from Season 5, Starlight Glimmer's vision of "enforced equality" is rejected—striking a blow, in Morse's eyes, against "social Marxism").⁵⁰ Kathleen Richter performs just such a reading, from roughly the inverse political position, for *Ms.*, pointing out that, at least in Season 1, "the only black ponies . . . are slave ponies to the white pony overlord."⁵¹ But all such oppositional readings are still beholden to the rhetorical RATIONALITY of public interpretation, relying on textual evidence aiming at persuasion. The fashwave shitposter publicly affects not to care about what anyone else may think or feel: cruelty is MAGIC, allowing the troll-magician to float up above the world, to fly, to take a UNICORN ride. UNICORN MAGIC is thus available to fascists as well as to anarchists, to antisemites as well as to queer activists.

What can we conclude from this overly long and radically incomplete survey beyond throwing up our hands at the semiotic mess we've found? UNICORN, it seems, can mean anything, and its opposite. Indeed, the real rhetorical power of UNICORNS might be, as Cramer argues, their power to "endure contradictions," to rise above them, as if they could use the speed of electronic circulation and replication to outrun the binary logic of all the codes, evading the law. Yet that very evasion might take on dramatically different senses, operating on behalf of the most generous, emancipatory, and genuinely utopian impulses, or as extensions of the worst forms of "cynical reason": those modes of "seduction" and "coercion," the fomenting of addictions and the systematic destruction of every alternative, that constitute "capitalist cynicism," and that which, in its "narcissistic and brutal" character, we can recognize as "what we *call fascism*."[52] In this respect, the swirl of competing meanings around the single image of the *unicorn* raises a question not frequently asked: not "how does the political enter the realm of fantasy?" but "how does the fantastic enter into the realm of politics?"

This is the question which the next chapter will address.

11

The Fantastic from Counterpublic to Public Imaginary

> Yesterday, as I have been suggesting, fantastika recognized the world. Today, it may be, the world is eating the fantastic.
> —John Clute, *Pardon This Intrusion*

On March 11, 2020, the day the World Health Organization declared the novel coronavirus outbreak a pandemic, journalists took to their laptops to advise the public that the 2011 film *Contagion*, which had leaped to the second-most rented title in Netflix's catalog, was not, in fact, a good guide to "[what] we are headed for" and admonish against acting as if this were "a fictional zombie apocalypse," even as others compared the spectacle of a dramatically stilled metropolitan downtown to "the opening scene from *The Walking Dead*."[1] The dramatic decrease in air pollution led some to think of the infamous snap of the fingers by which Thanos had executed his neo-Malthusian genocide in *Avengers: Infinity War*.[2] Business pundits opined that "we might be heading toward a 'Star Trek' future, though 'Mad Max' is entirely possible," while parenting blogs considered the possibility that "we [are] headed into a Lord of the Flies/Hunger Games situation here."[3] That evening witnessed a cascade of nearly simultaneous media events: the issuing of new travel bans, the cancellation of the NBA's entire season, and, memorably, the appearance of the former Republican Vice Presidential candidate (once famous for asserting that the Affordable Care Act mandated "death panels" for the elderly and disabled—a fantasy that, eleven years later, was emerging as the dark truth of the for-profit US healthcare system) doffing her bear costume to deliver Sir Mix-A-Lot's 1992 rap hit on *The Masked Singer*. The implausible incongruity of these events in juxtaposition with one another, the impression they gave of sheer *irreality*, resisted integration into any normative model of reality;

they called for another "kind of awareness," a perception that Istvan Csicsery-Ronay suggests "we might call *science-fictionality*, a mode of response that frames and tests experiences as if they were aspects of a work of science fiction."[4] And so it was that a single tweet supplied the moment with a perfect science-fictional caption: "Sarah Palin singing Baby Got Back as the world burns. This is the darkest Timeline."[5]

In assessing a bit of popular discourse like this in terms of *science-fictionality*, we are aided by an emergent understanding of science fiction as an increasingly "more diffuse" entity—as Broderick's "mega-text" or "intertextual encyclopaedia of tropes and enabling devices" loosed from any canonical or authorial origins and circulating freely through popular culture, so that science-fictional "metaphors and models" (like "the darkest timeline") can appear not so much as an allusion to a specific media text, made within a specific fan community (like fans of the non-SF series *Community*, as we shall see), but simply as part of popular consciousness, a way of seeing the world: the "widespread normalization of what is essentially a style of estrangement and dislocation," Csicsery-Ronay writes, "has stimulated the development of science-fictional habits of mind."[6] As science fiction loses its generic specificity, moreover, these habits of mind pertain less to any discretely bounded genre, distinguishable by immanent structural criteria (such as cognitive estrangement) and a distinct readership (a socially estranged fandom), but to a "fantastic supergenre" encompassing fantasy, horror, comics, tabletop and live-action roleplaying games (RPGs and LARPs), and videogaming, to be theorized in terms of hybridity rather than purity.[7] This hybrid "post-genre fantastic" threatens to subsume even "realism" itself: if the fantastic is not the name of a distinct object (a genre) with special properties (cognitive estrangement) but just a process of exercising figurative "torque" on reality, then even mimetic fictions can be reread as "a 'weak' or low-intensity variety" of the fantastic.[8] New realities like COVID-19 and climate change resist representation with such violence that they seem to call for "science fiction seeing."[9] Now, in the violence of our historical "world storm," we have to reassess the cultural position of the fantastic in our culture and its political functions.[10] We are forced to do this because—in the world, and not only in theory—it has vanished from its previous cultural location and reappeared everywhere.

Within a relatively short time, in other words, the fantastic seems to have ascended from subculture to mass culture.[11] *Entertainment Weekly* declared 2007 "The Year of the Geek," and to judge by the receipts from San Diego Comic-Con, every year since has been another such.[12] What was once a truly "minor" literature, the preserve of small communities of fans, has become not only "big business" but a new "cultural dominant."[13] Once, in the twentieth century, tabletop fantasy RPGs, science fiction comic books, and horror films were played, read, and watched at a certain distance from the noisy space of official political life. In the twentieth century, a fandom, as mediated through its magazines and conventions, was "restricted to some" but also "accessible for money," "concealed" but also "outside the home," and "circulated in print" but with features curiously characteristic of oral cultures, including the production of "tacit and implicit" ingroup knowledges; in Michael Warner's terms, they were the stuff of *counterpublic spheres*, set at a quantitative and qualitative remove from the public sphere in which the *polis* conducted its collective affairs.[14] What happens, then, when the fantastic moves from counterpublic to public space?

Perhaps this was not such a quantum leap. We might observe that practitioners, fans, and impresarios of the fantastic have long exaggerated their own marginality.[15] Where their Romantic forebears imagined themselves as *poètes maudits* wasting away in noble poverty, exiles from the world of conformity and commerce, the writers of superhero comics or swords-and-sorcery novels have imagined themselves as persisting in their art in the face of the "mainstream" world's ignorance, contempt, and indifference. "If poets are the unacknowledged legislators of the world," declared Bruce Sterling, riffing on Shelley's "Defence of Poetry," "science-fiction writers are its court jesters ... Very few feel obliged to take us seriously."[16] This perennial positioning of the fantastic and fan cultures in relation to a "mainstream," an imagined field of cultural legitimacy bearing the dignified titles of "literature" and "art," is in many respects question-begging, referring to institutions whose power has been seriously diminished in the age of mass reproduction—that is to say, for as long as the fantastic has existed as such.

Nevertheless, even if elite literary or artistic institutions are themselves "marginal" to the culture industry in this century, it remained more or less true, at least through the last century, that the fantastic was relegated to "the margin of the margin," in the words of Samuel R. Delany.[17] A Stanley Kubrick, John Updike, or Roy Lichtenstein who picked up the elements of science fiction, fantasy, or comics, respectively, was said to be "slumming," debasing his own art form, or else "elevating" the genre, transmuting the base matter into cultural gold. The common assumption was always that "the Genres," as Díaz dubs them, represented the particular, whereas Literature (and its lowercase shadow, the "mass literature" of romance, western, and detective novels) represented the universal (the one in its claim to eternal value, the other in its broad immediacy). For Leslie Fiedler, science fiction had originated as "the property of a minority-pop audience, which tended to read it, on the one hand, *against* classroom literature, and, on the other, *instead of* rival pop forms."[18] For quite some time, the fantastic and its fan cultures *did* occupy something like a counterpublic sphere.

It may be objected, on grounds deriving directly from the newer genre theories, that this diffusion of the fantastic has been long in the making. In the field of cinema alone, the cultural significance of the fantastic was great enough to draw the attention of elite critics (Carl Jung, Kingsley Amis, Leslie Fiedler, Susan Sontag) by the 1960s, and as far as the market was concerned, Martin Scorcese had already lost the argument against it nearly the moment he arrived in Hollywood: the wave of big-budget special-effects film franchises touched off by *Star Wars: Episode IV—A New Hope* (1977) never really ended, enduring only the lightest of lulls during the indie boom of the 1990s before returning with a vengeance. The mass marketing of science fiction novels, as Sarah Brouillette notes, was strong enough by the mid-1980s that late work by the surviving giants of the Campbell Era routinely made the bestseller lists.[19] When Catherine Belsey wrote, in her widely-adopted textbook, *Critical Practice* (1980), that "classic realism" is "still the dominant popular mode in literature" as well as in "film and television drama," this was already at the cost of effacing the difference between the realisms of George Eliot and D. H. Lawrence, on the one hand, and those of Tolkien and Wells on the other.[20] While Routledge reprinted this textbook in 2002, retaining this assertion, it had not aged well. Nearly another two decades on—even given the staying power of the detective novel and the cop show, for instance, or the rise of genres such as the pseudo-memoir and reality TV—Belsey's assertion would now be difficult to maintain. Even if published science fiction and fantasy never approached the popularity or profitability of their cinematic counterparts, the same could be said of "mainstream" literary writing, which has nonetheless merged with the field of the fantastic so frequently since the second half of the twentieth century that even the terminological scaffolds hastily erected around the zone of hybridization—e.g., "span fiction," "interstitial arts," "slipstream"—beg the same question with which Thomas Frank once challenged "alternative music": "alternative to *what*?"[21] In other words: if the fantastic was ever culturally marginal, by the end of the twentieth century, it was already thoroughly imbricated in the hegemonic culture in the shadows of which it allegedly stood.

Even if we can recognize precedents, preparations, and premonitions, the change has still caught us off guard. The present dominant status of the fantastic in mass culture has emerged swiftly enough to leave us with epistemic whiplash. If our scholarship is ready to abandon stable, crisply defined taxonomies of genre, to conceptualize genre as something rather more fluid, are we really ready for the "evaporation" of the very center/margins, inside/outside, and high/low distinctions once taken to structure a "mainstream" culture?[22] What happens to our understanding of science fiction, horror, or fantasy when the categories of audience and fandom that once served as such a handy substitute for clear and distinct immanent criteria—e.g., "science fiction is a category constructed by what readers expect"—are no longer merely "mutable and unstable" but unavailable?[23]

The scope of this mainstreaming of the formerly marginal, the transformation of the fantastic from a subcultural to a hegemonic status, is only apparent if viewed from a transmedia perspective. For instance, sales of adult science fiction and fantasy (SFF) peaked at over twenty-five million books in 2009, thereafter to fall into a slump from which they have never recovered. Nevertheless, by 2018, E-book and self-publishing platforms had more than made up for the 50 percent drop in unit sales: the number of adult SFF books bought in *all* formats from *all* sources topped fifty million.[24] These numbers, in turn, were colossally outstripped by young adult science fiction and fantasy, where sales of the *Harry Potter* books alone (1997–2007) reached half a billion units globally, a success loudly echoed by the *Twilight* series (published 2005–2008, 150 million copies sold) and the *Hunger Games* trilogy (published 2008–2010, 100 million copies sold).[25]

The comics industry, meanwhile, had experienced an early speculative bubble at the beginning of the 1990s, with a bust lasting for the rest of the decade (fig. 4.2.1). By 2015, however, the losses following the bursting of the early 1990s speculative bubble had been more than recuperated,[26] largely through sales associated with film and TV adaptations; Image Comics was resurrected almost solely on the strength of the hit TV series, *The Walking Dead* (2010-). The real profits, of course, were to be realized in moviemaking. Lionsgate and 20th Century Fox figured out how to spin box office from young adult science fiction and fantasy (*Divergent*, published 2011–2013; *The Maze Runner*, 2010–2016; etc.), Hulu and Amazon from dystopian fiction of an older vintage (*The Handmaid's Tale*, published 1985; *Man In the High Castle*, 1962), Warner Brothers and Disney from older fantasy epics (*Lord of the Rings*, published 1937–1955; *The Chronicles of Narnia*, 1950–1956), Sony, Disney (again), and (again) Warner Brothers from the plots and characters of some seventy years of comic books (*V For Vendetta*, 1982–1989; *The Avengers*, 1963; *Batman*, 1939; *Superman*, 1938 . . .). By 2013, Christine Folch could report in *The Atlantic* that "of Hollywood's top earners since 1980, a mere eight have

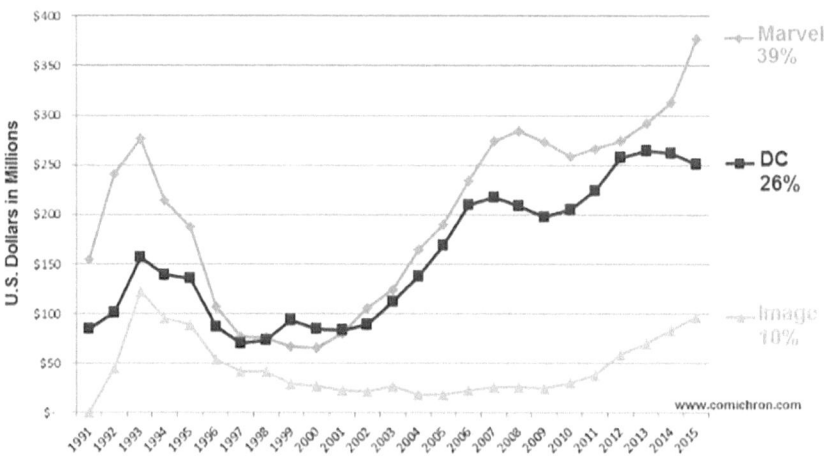

Figure 4.2.1. Comics sales by year for three major publishers, 1991–2015 (Sam Otten). The losses following the bursting of the early-1990s speculative bubble have been more than recuperated.

not featured wizardry, space or time travel, or apocalyptic destruction caused by aliens/zombies/Robert Downey Jr.'s acerbic wit" (fig. 4.2.2).[27]

This speculation in the speculative, however, was not confined to big-budget media spectacle. Entertainment fad cycles come and go, of course, as do trends in publishing, but the commercialization of geek culture has a farther and deeper reach, as it ceases to be merely the media product being sold but becomes the means of selling everything else, "from breakfast cereal to pickup trucks."[28] Media corporations have been learning from fandoms' heretofore unusual patterns of consumption how to extend their profitability in a digital media environment. "Fannish behavior has entered mainstream audiences," notes Kristina Busse, and Henry Jenkins, Sam Ford, and Joshua Green concur: "behaviors that were once considered 'cult' or marginal"—creating active audience communities that produce free publicity via social media—"are becoming how people engage with television texts."[29] Old fan institutions like the Con—most notably San Diego Comic-Con (1970–)—have become as central to the entertainment industry as CES is to the tech industry. By 2010, Patton Oswalt observed, even nonfantastic media products—prestige TV dramas like *The Wire* or competitive reality shows like *Top Chef* and *American Idol*—were being consumed in a style that had originated in fan counterpublics, in the realms of the fantastic: since, in the age of social media and online communities, what were once "hidden thought-palaces" are now "easily accessed websites, or Facebook pages with thousands of fans," there is a fandom for nearly everything, and "everyone considers themselves otaku about

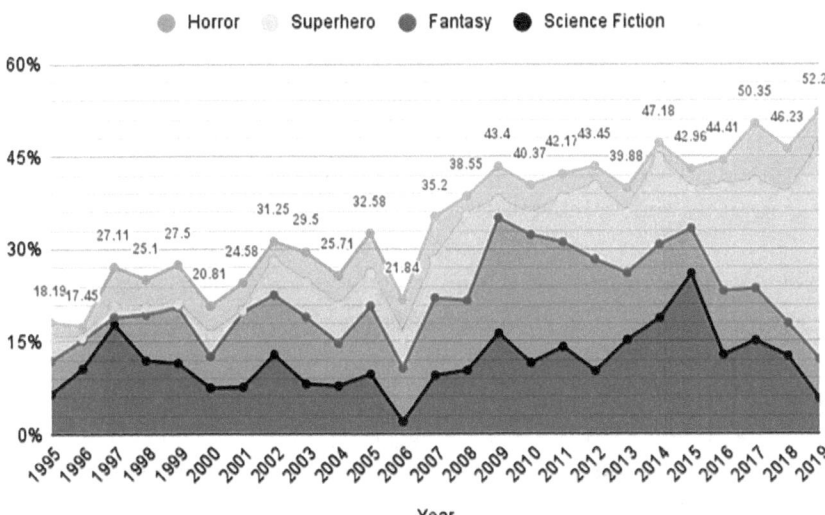

Figure 4.2.2. Percentage of Hollywood movie market share for genres of the fantastic, 1995–2019 (Nash Information Services, LLC).

something."[30] The diffusion of fannish, geekish, "otaku" *forms of life* proved a perfect survival strategy for mass-media capital in the age of social media and audience hyper-fragmentation.

As the fantastic has passed from cultural margin to cultural center, its formerly minority aesthetics becoming hegemonic, they have provided an increasing fraction of the social imaginary through which power and identity are negotiated, through which the possible is conceived and "the art of the possible" is practiced.[31] In 2007, the last year that San Diego Comic-Con (attendance surpassing the one hundred thousand mark for the third year in a row) sold admission badges onsite, the year *The Big Bang Theory* premiered on CBS, another dimension of geek mainstreaming became visible.[32] Politicians, seeking as always to tune their rhetoric to popular tastes, started *speaking geekish*. In that election cycle, *New York Times* columnist Dave Itzkoff noted, "the worlds of science fiction and presidential politics" had already become "closely aligned" as candidates and pundits vied to show their pop-culture credentials.[33] If this science-fictionalization of US politics had been anticipated by late twentieth-century discourse around the nuclear arms and space races,[34] it had become floridly, spectacularly visible by the rise of "God-Emperor Trump," in the coinage of his alt-right fandom (borrowed, in turn, from Frank Herbert's *Dune* trilogy and from the RPG *Warhammer 3000*). Congressional representatives now signal their political brand by dropping allusions to *Watchmen, Handmaid's Tale,* and *V For Vendetta* cosplay have become ubiquitous protest vehicles, police and military vehicles (as well as millions of civilian bumpers) are adorned with the Punisher's death's-head symbol, systems of privilege are explicated by widely-read blog posts referencing video games' tropes and difficulty settings, public policy discussions are shaped by *The Walking Dead*, and entire nations now conduct psychological warfare online via *Game of Thrones* memes ("Sanctions Are Coming," Trump warned Iran in November 2018; Iran replied in kind) and the carefully engineered spread of negative social media posts about the latest *Star Wars* films.[35]

One marker of the degree to which the fantastic has permeated the public imaginary—especially, though not at all exclusively, in the US—is the proliferation of references to fantastic narratives and tropes in political discourse. Take, for example, the popular catchphrase: "This is the darkest timeline." While lifted (in abbreviated form) from the dialogue of a famous episode of the TV comedy Community, it evokes scenarios like that of Amazon's series adaptation (2015–2019) of Philip K. Dick's *The Man in the High Castle* (1962), in which the US, having lost World War II, is carved up by the Japanese and German empires, and the timeline in which it won is the stuff of science fiction. Dick's themes are echoed by the first-person shooter video game *Wolfenstein II: The New Colossus* (2017), also set in an alternate-timeline 1962, in which the Nazis rule America. The *Community* episode in question, first aired in October 2011, condenses a number of disparate elements from fantastic genres—most notably, RPGs (in the episode, a dispute over whose turn it is to get pizza is settled by a dice roll) and time-travel narratives like Man In the High Castle (one of the disputants warns the die-roller that "you are now creating six different timelines," all of which are duly played out to their absurd conclusion). In widespread social media use, it has become shorthand for a sense of horrified, baffled disbelief in the face of the twin historical realities of climate change and resurgent global fascism. It apprehends this "spectacle of disintegration" (Wark) as senseless, void of meaning, unreal; reality is elsewhere, in another edition of the universe where the "good guys" won as they were meant to do. Google searches for "the darkest timeline" peaked, of course, in November 2016.

"Darkest timeline" discourse might thus seem to be the fantastic expression of those for whom history is a nightmare from which they are trying to awaken—e.g., US white liberals following the 2016 election.[36] As a stance of "rejection," this then would represent, as a means of coping with the radically unacceptable, a refusal to accept that liberal democracy had failed on its own terms once again, just as it did in the 1930s. Thus, for Adam Gopnik, author of *A Thousand Small Sanities: The Moral Adventure of Liberalism* (2019), everything from Trump's election as President to the bizarrely botched Oscar award ceremony can be explained by "a simple but arresting thesis: that we *are* living in the Matrix, and something has gone wrong with the controllers." This preservation of small sanities by means of large insanities finds its echo in the vogue for the so-called "simulation hypothesis" among tech-industry barons such as Elon Musk. The Wachowskis' cinematic premise has become a favorite recreational speculation for the "virtual class" of wealthy neoliberals, the latest phase of their Californian Ideology, which is itself compounded largely of a science fictional imaginary, as Richard Barbrook and Andy Cameron note: The *New Yorker* reports that "many people in Silicon Valley" are now "obsessed" with this notion, and that "two tech billionaires have gone so far as to secretly engage scientists to work on breaking us out of the simulation."[37]

We might also remark on the transformation of political *language*, specifically, under the influence of the fantastic. For instance, we have seen the *Mad Max* film franchise turned into a verb, "madmaxing" (used to mean, variously, "adapting to postapocalyptic conditions," "looting," "reengineering civilian vehicles for combat," "violence," or all of the above) as well as the proliferation of expressions such as "zombie banks," "zombie laws," "zombie computers," and "zombie economics" enabling the "us[e of] the living dead as a pop culture hook for promoting political and policy ideas," and so on.[38] "Fantastic language," in the words of Marleen S. Barr, is "no longer a linguistic subaltern," the ingroup discourse of fan counterpublics, but "the lingua franca of twenty-first century America's political terrain" and arguably the world's, the universal language in which political life can be apprehended, described, negotiated, and recreated, for better and for worse.[39] The universal realm that is the *polis*, the public sphere itself, is now not only *represented in* works of the fantastic but increasingly operates *through fantastic representation.*[40]

The use of imagery drawn from science fiction, fantasy, horror, superhero comics, and RPGs/LARPing to express an estrangement from reality is also strangely prevalent among the prime beneficiaries of the "new normal." Thus, the strange cultural afterlife of *The Matrix* is also unfolding in the darker precincts of the alt-right, where an induction into fascist and misogynist ideology is called "redpilling," a reference to the famous scene in *The Matrix* where the rebel Morpheus invites a new recruit to leave behind the false world for the truth: "I imagine, right now, you must be feeling a bit like Alice, tumbling down the rabbit hole ... You take the red pill and you stay in Wonderland and I show you how deep the rabbit-hole goes."[41] For some, the deceptive virtual reality is "political correctness," and the "truth" is that white men are biologically destined to rule; for others, the red pill banishes the supposed "illusion of woman's subjugation," laying bare the underlying "reality" of "men's exploitation by human women" and the cold logic of the competitive marketplace.[42] Those who have not taken the red pill and discovered "how deep the rabbit hole goes"—often by traveling down what is commonly referred to *as* a "rabbit hole," a recursive series of increasingly extreme YouTube videos—are described via the vocabulary of RPGs and computer gaming, as "NPCs," non-player characters, only apparently human and alive, hence dispensable.[43]

The point of this comparison between liberal and fascist uses of the fantastic is not to make any false equivalences: there remains no liberal counterpart to such terrifying fascist *passages à l'acte* as the murders in Christchurch, El Paso, or Poway, certainly. However, there is a perverse symmetry in the way that the fantastic expresses their mutual estrangement from the reality that contains them both. In the face of the disaster of 2016, US liberals have clung to narratives of Russian electoral hacking, attributing the present explosion of racist and misogynist violence to "bots" and "troll armies" rather than homegrown, endemic, institutionalized evil. Liberals thus shore up their belief that fascism is fundamentally unreal, a passing tangent in the long arc of a history that bends toward incrementally ever-greater justice and inclusivity, while fascists, for their part, posit the drives toward exponentially greater exclusionary cruelty and domination as the fundamental reality—using *more or less the same fantastic imaginary*.

For some prime examples of the ambiguous political uses to which that imaginary is being put, we might look to LARPing and cosplay. Events such as the March 4 Trump in Berkeley (March 27, 2017) and the Unite the Right rally in Charlottesville, Virginia (August 11–12, 2017) drew new attention to the role of fantastic and imaginative play in fascist violence, as demonstrators outfitted themselves with gaudy homemade shields, armor, and weapons reminiscent of the world of Renaissance Faires. Indeed, planning documents for Unite the Right explicitly reference the Society for Creative Anachronism (SCA) as a source of information.[44] This "medieval cosplay fantasy come to life" was further augmented by digitally altering photography of these "alt-knights" to splice them into stills from *Captain America: Civil War* and the film adaptation of Frank Miller's xenophobic historical fantasy, *300*.[45] All of these fantastic surfaces literally served to shield alt-right participants from the consequences of violence—both by allowing for a certain degree of indulgence in violent action and by ensuring that most of it fell short of the full expression that modern weaponry would have permitted—and to enable a kind of *fantasy-augmented* violence: before driving his car into a crowd of counterdemonstrators, killing Heather Heyer, driver James Alex Fields, Jr., was photographed standing in a pseudo-Greek "shield wall" formation, carrying the black insignia of the neo-Nazi group Vanguard America.[46]

Here, however, we can observe that LARPing and cosplay are also resources available not only to liberals and fascists but also to the global movements of radical contestation that have resurfaced in this century. These movements, too, have found in the fantastic a crucial "lens" for surveying the landscape of possibility.[47] For prison abolitionists like Walidah Imarisha, who are routinely told that their goals are an unachievable fantasy, "all organizing is science fiction": "Whenever we try to envision a world without war, without violence, without prisons, without capitalism, we are engaging in speculative fiction."[48] If *Dungeons & Dragons* provided a "storytelling apprenticeship" for a generation of writers, in the words of Junot Díaz, it also provided "welfare kids" with opportunities to "succeed, be powerful, triumph, fail and be in ways that would have been impossible in the larger real world"—collectively, in an open-ended collaborative process.[49] Much as Augusto Boal proposed a "Theater of the Oppressed," finding in dramaturgy a space for the "rehearsal" of transformative social action, Imarisha and brown propose the fantastic as an "exploring ground" or "laboratory" for experimenting with the possibilities of resistance and power "without real-world costs."[50] While the Nazi-slaying aesthetics of *Wolfenstein II* may have served as a cynical marketing device or a moral alibi for indulgence in aestheticized violence, its affordances also enabled players to test out their willingness to transgress the limits of liberal tolerance and civility in "[the] use of violence to combat fascism."[51] Conversely, Laurie Penny suggests that the widespread incorporation of *Harry Potter* into liberal protest aesthetics—e.g., shouting "Expelliarmus!" at the armored police sent to break up Occupy encampments—underscores their perplexity at the fact that "adults [are] so much crueller than we were promised before we became them."[52] Where both liberals and fascists use the fantastic as a means of *derealizing* reality—to evade it, in the first instance, and to impose an ideological fantasy over it in the second—antifascist radicals are finding in it, instead, a toolbox for probing possibilities and impossibilities, exploring historical conjunctures, and projecting complex social structures: in short, a means of *engaging with* reality.

Indeed, on the far left no less than on the far right, LARPing and cosplay have also served as a "rehearsal" in Boal's sense—preparation for future action—and as "recreation" in the sense of quasihistorical reenactments, imaginative time travel. Thus, in October 2011, Miriam Rosenberg Roček addressed the crowd at Occupy Wall Street in full turn-of-the-century cosplay as "Steampunk Emma Goldman," declaring that "I have travelled through time to be here with you . . . The New York Times said this Saturday that any attempt on the part of police to clear this plaza would result in the resurrection of Emma Goldman. Too late."[53] Steampunk anarchists and feminists thus engage not in a nostalgic hermeneutics of retrieval, for which the past is eternally past, the subject of perpetual "mourning" and "infinite debt," but in a willfully "anachronistic" recreation of the past, a movement for which the past can indeed "return"—not, contra Marx, as tragedy or farce, but "remade, modified . . . making us the contemporaries of and participants in a reality that never ceases to be present."[54] In so doing, it strikes at the counterfantasy of "capitalist realism" and "statist realism," the sense that "there is no alternative" to be found even in history: that the arrow of teleology runs directly to the present, without any other "timelines" pointing to more desirable might-have-beens.[55]

Steampunk protest's capacity for enriching our sense of the real by retrieving the possibilities that have been passed over within it—repressed might-have-beens—is complemented by the use of cosplay to transpose present realities with the has-never-been. For instance, in widely-publicized nonviolent protests in 2010, Palestinians in the border town of Bil'in struggling against the expropriation of large amounts of the village's agricultural land by the construction of the so-called Separation Barrier dressed (and body-painted) as the alien Na'vi from James Cameron's film *Avatar*. The creative and ludic character of the protests might have been intended in part to baffle and disrupt the ordinary action repertoire of their Israeli military opponents, but it functioned even more effectively as a means of communicating with the world, mobilizing a set of *fantastic images* in a public political speech-act, in hopes of evoking collective action repertoires embedded *in* the fantastic narrative of *Avatar*.[56] By positioning Palestinians as the besieged alien Na'vi, they positioned the armored, technologically and militarily superior forces of the State of Israel as the forces of the human colonizers, the Resources Development Administration and its brutal Marines. Since the dominant reading of the film emotionally enlists its audiences on the side of the Na'vi, the protestors could use this scenario to propose that *their* international audience should not only see the Israeli Occupation and Wall as unjust but—like Jake Barnes, the film's human hero—join the rebellion.[57] This seems to have been in part successful: the Occupation continues, and the Wall, a potent symbol and instrument of the Palestinian people's isolation, stands, but its construction was rerouted in 2011 to return six hundred dunams of the village's land.[58] Mediating not only their specific demands but their *entire situation* through a fantastic narrative made them, paradoxically, immediately legible to the world.[59]

In all of these examples, we can see that the role of the fantastic in contemporary political life can't be understood only in terms of discrete texts with discrete political meanings. Nor are *The Man in the High Castle*, *Wolfenstein II*, *The Matrix*, *Dungeons & Dragons*, *Captain America: Civil War*, and *Avatar* to be understood merely as media products waiting to be appropriated by active audiences, passively receiving their meaning from political forces; they are now *part of* the field of contending forces.[60] They number among the central constituents of the repertoire of metaphors, clichés, tropes, stock images, and myths through which political meanings *of any kind* are articulated. Scholarship in all of these genres must come to terms with their transformation from so many subcultural objects to constituent elements of a global mode of subjectivity. As the fantastic has gone from a subcultural to a mass cultural phenomenon, it has moved from a counterpublic sphere to the public sphere, becoming a structuring part of its imaginary.

When and how did this transformation take place, and what are its implications? The groundwork for this mainstreaming of geek culture, its accession to the public sphere, was doubtless laid first by the growth of computing and telecommunication, then by the Dot-Com Bubble of 1994–2000, which together placed what had heretofore been a relatively small and tertiary sector of the working class—coders and other tech workers—at the center of the new "cognitive capitalism."[61] The reorganization of corporate capital was belatedly followed by a flow of cultural capital: even if geek identities remained to some extent "stigmatized," geek genres became newly "capitalized."[62] However, it would be implausible to assume that economic centrality automatically translates to cultural centrality: the centrality of the finance sector has not produced a popular culture filled with images of heroic stockbrokers, bankers, and securities analysts, for instance, even if it is permeated with the values, concepts, and action repertoires that constitute capitalist realism. The translation of geekdom's "minor language" into something approaching majoritarian heft and scope was arguably mediated by the introduction of a vast new mediation: the internet. As most of the rest of the developed world (and, far more slowly, the developing world) flowed into this new sociotechnical environment, the newcomers found themselves within not only a technological but a *cultural* infrastructure that reflected the experiences and sensibilities of its geeky architects—overwhelmingly "white, male, tech-savvy . . . [and] passionate about geeky topics."[63] As the public sphere increasingly merged with this vast mediation, the imaginary of geekdom diffused throughout its space.

If Chu is right, however, a parallel transformation has taken place on the side of life itself. It is not only that contemporary science confronts us with objects of knowledge (such as twenty-six dimensional "strings") outside of all our everyday experience; rather, our everyday life is increasingly populated with inassimilable "hyperobjects" such as climate change or the global economy a panoply of other "cognitively estranging referents," from hybrid cultural identities to exotic forms of securitized debt instruments, and the infinity of seemingly unprecedented, disconcerting, and "disruptive" realities that constitute the "new normal" (increasingly, *no* normal).[64] All of this real strangeness lends itself to representation in terms of the unreal, the fantastic. Thus, in describing the role played in the 2008 financial crisis by the Black-Scholes formula (part of an esoteric mathematical model used in pricing certain securities, which effectively "means . . . using the future to calculate the present, rather than using the present to predict the future"), Joshua Clover resorts to metaphors drawn from the world of superhero comics ("*retcon*, an annealing of logical fissures in a given backstory after they have cracked open into system-threatening incoherence in the present and, further, threaten the ongoing or futural narrative") and science fiction ("time travel"): nothing else can capture the "necessary unthinkability" of the crisis.[65] Similarly, to capture the strange existence-under-erasure of Native American lives and lands in the US, Joseph William Singer draws on the science fictional concept of "parallel universes": the world in which "tribal sovereignty . . . has always existed" inexplicably "coexists" with the world of "the United States of America, where conquest was completed and tribal sovereignty does not exist," "occupying the same spacetime continuum and exerting substantial force on events in the real world."[66] Traumatic experiences of violence, loss, displacement, and exclusion that continue to multiply in the "world storm"—as Díaz's Oscar Wao would say, what could be "more sci-fi" than the lives of diasporic people, refugees from storms political, economic, and climatic?—are cognitively estranging referents *par excellence*.[67] And money itself, as a "near-intangible substance" used "to compel or convince others to do some of your work for you," continues to exercise its alienating, estranging, fantastic power.[68]

In 1993, to describe the breakdown of the Kuhnian model of scientific progress as a history of "normal science" punctuated by "revolutions," philosophers Silvio O. Funtowicz and Jerome R. Ravetz coined the term "post-normal science."[69] Some futurists have seized on this to name our present moment "post-normal times."[70] A post-normal fantastic, a cloud of evaporated genre unmoored from its customary epistemological and social relations to the normal, will need to be understood differently. In this, a darkening of not the darkest of timelines, as the familiar vanishes and the strange engulfs what we formerly knew as the world, we must rethink the relationship between the minor and the universal dimensions of the fantastic. The place of the fantastic as an imaginary structuring the public sphere requires, as Farah Mendlesohn and Brian Stableford have notably suggested, that we attend closely to its *rhetoric*, the doing of things with words and images, and particularly in Kenneth Burke's sense, to the way it serves to "frame" reality in terms of acceptance and/or rejection.[71] We need to investigate the ways in which the fantastic is mobilized to demand the impossible and set bounds to the possible, to repeat the unrepeatable and rehearse for the unimaginable, to confront the monstrous and represent the unrepresented real.

12

Acting Supernaturally (with Notes on the Monster of Anarchy)

> Those who automatically deny magic's reality position themselves as superior in understanding to indigenous peoples' ways of knowing the world. That's an untenable position to take when you're writing from the viewpoint of said indigenes.
> —Nisi Shawl, "Hope and Vengeance in Post-Apocalyptic Sudan"

The Return(s) of Magic

Recall that it was Alexei, the railway porter from the country, who reached for Abba Gordin's modern fairytale with the quickest interest, asking: "Where is the 'Kingdom of Anarchism'?" The same sort of ostensibly backward folks in Spain entered the ranks of the anarchist movements there, as Murray Bookchin recounts. Whereas "Spanish socialism built its agrarian program around the Marxist tenet that the peasantry and its social forms could have no lasting revolutionary value until they were 'proletarianized' and 'industrialized,'" he argues, "Spanish anarchism . . . sought out the precapitalist collectivist traditions of the village, [and] nourished what was living and vital in them."[1] By a historical conjuncture that is more than a coincidence, according to Thomas Disch, during the period when science fiction was emerging as a genre, "a large part of the American urban lower classes, from which the sf audience was drawn, were recent immigrants from what is commonly called the Old Country—that is to say, from the place where folk tales were still a living tradition." Hence, among the several persistent "embarrassments of science fiction" is "its close resemblance, often bordering on identity, with myth, legend, and fairy tales."[2] Without necessarily equating these two audiences for modern fairytales, we might note that both the greenhorns of New York or San Francisco and the urbanized peasants of Moscow or Barcelona were people looking to the future with the highest of expectations.

An ambiguity, if not an embarrassment, haunts both Marxism and anarchism. On the one hand, these two European radicalisms were formed in opposition to the power of the Church in a period when Christianity formed one of the ideological bulwarks of Capital and the State—so much so that for Proudhon, these three institutions were, if not identical, at least "analogous" to one another.[3] Materialist atheism became a central plank in the doctrines of both the Marxist and Bakuninist fractions of the First International, outlasting the organization itself. Yet both have also repeatedly been drawn to recognize their own place in a line of descent from premodern religious heresies, messianic and millenarian movements, and schools of mysticism. And although it was Bakunin who called for a revolt against "the domination of life by science," both anarchists and Marxists, at times, have felt their commitment to "science" as an obstacle and a burden, a brake on radical imagination or worse: for Adorno and Horkheimer, likewise, the scientist "behaves toward things as a dictator toward men. He knows them in so far as he can manipulate them."[4] Thus, Paul-François Tremlett speaks of a "romantic" as well as an "enlightenment" discourse about religion circulating within the anarchist tradition, and Paul Breines likewise speaks of an "antagonism between Romantic and Enlightenment-Utilitarian moments" in Marxism.[5]

In a strangely parallel manner, a longstanding argument about the S in SF is that it masks a lack, a zero: what passes for "science" is really magic in disguise, a wish-fulfillment fantasy with a veneer (thick or thin) of scientistic rhetoric (i.e., mimicry).[6] Much as Georg Lukács had declared that the novel was the epic of a world after the death of God, SF could be seen as a replacement for—or a perpetuation of—myth in a disenchanted world.[7] The possibility that $S = 0$ has been seen by some as an "embarrassment," a threat to the dignity of SF, its claim to Enlightened maturity: insufficiently rigorous thinking undermined SF's claim to produce genuinely "*cognitive* estrangement."[8] However, the residue of myth has also sometimes been read as the very mark of SF's dignity, a sign that it stands in a line of descent from the oldest human dreams and struggles. And another line of argument might ask whether it is magic or rhetoric that one seeks to banish from SF and whether any such purification is possible or even desirable.

Worries about "anticognitive" tendencies in science fiction, threats to its genre identity, have been perennial. They motivated John W. Campbell Jr.'s insistence on the tragic ending to "The Cold Equations," and a scant generation later, we can see them resurfacing in Lester Del Rey's complaint against Stanley Kubrick's *2001: A Space Odyssey*: "The real message, of course, is . . . intelligence is perhaps evil and certainly useless . . . Men can only be saved by some vague and unknown mystic experience by aliens."[9] Even the astute critic Peter Nicholls raises concerns about the decadence of SF, the "Monster of Anarchy" stalking the New Wave works of J. G. Ballard and Brian Aldiss:

> The Monster of Anarchy is characterised above all by his cool. He saunters laconically through landscapes pitted and scarred by the stigmata of self-destruction, resting occasionally in the Garden of Gethsemane, asking and expecting no advice or assistance from any quarter, human or godly. He fucks a lot, and does so most cheerfully when surrounded by the detritus of destruction, the crashed Car, the dully gleaming carbine slung over the shoulder, the empty syringe lying on the toilet floor. This Monster worries me, not least because it may be he is not so much a product of literature as of life itself.[10]

This specter of a destructively pessimistic irrationality appears in a strikingly similar manner in the late works of Murray Bookchin, who spent the last decade of his life in a campaign against what he saw as "a nihilistic reaction to the failures imputed to Enlightenment ideals of reason, science, and progress" that vitiated any "serious project for social change."[11] Among the signs of this reaction were the renewed circulation of "pagan, magical, and generically mystical notions" as well as "the myth of the primitive."[12]

The Manufacture of the Plausible

In "Why There Is (Almost) No Such Thing as Science Fiction," Brian Stableford offers a novel theory about the magical residue in SF's "scientific" cognition. He proposes that the "magic" in SF is a side effect of rhetoric (i.e., "the set of techniques by which plausibility is manufactured"), which manufactures "psychological plausibility" as against "rational plausibility":

> The real problem with science fiction does not arise from the imaginative seductiveness of its facilitating devices or any accident of history; it arises from the fundamental relationship between rational and psychological plausibility,

and the manner in which they have drawn further and further apart over the last two centuries, resulting in a situation in which they are implicitly—and perhaps irrevocably—antipathetic to one another.

The existence, since time immemorial, of an extremely rich and various store of fantastic stories not only demonstrates that entities and events impossible or highly unlikely in the world of experience can be forcefully endowed with literary plausibility, but also illustrates the remarkable fact that such narrative devices often seem far more psychologically plausible than everyday events: an awareness summarized in the dictum that "truth is stranger than fiction." Age-old as it is, though, this situation is not unchanging, and the simple fact is that what passes for truth nowadays is much stranger than it used to be, and is, in consequence, so much stranger than fiction that it beggars belief.[13]

Here opens up an entire domain of the "cognitively estranging objects" that Seo-Young Chu describes: not only the utterly counterintuitive realm of post-Einsteinian physics but the discoveries of neurobiology, which, so far from locating a single Cartesian seat of the soul, a unified stream of consciousness, or a compact topology of id, ego, and superego, have given us decentered, distributed, emergent, and extended models of consciousness which resist representation in models or metaphors.[14] Here, the psychological plausibility offered by Descartes' "ghost-in-the-machine" model of mind—"a sham," as Stableford remarks, but nonetheless "really ... how we appear to ourselves when we attempt to examine ourselves from within"—not only outweighs its rational implausibility, it lends its psychological plausibility to "numerous corollaries blessed with elementary psychological plausibility."[15] Such scenarios include most of the *novae* of the TV series *Sense8* (2015–2018)—metempsychosis, telepathy, psychic parasitism, etc.—all of them too easily squared with our sense of our selves as somehow transcendent, immaterial, independent of the flesh.[16] The rhetoric of scientificity (conveyed in *Sense8* by vague appeals to evolutionary theory and allusions to CIA programs) only serves to smooth over the feeble objections of reason.

The observation that, in SF, "science" exists only as a set of persuasive devices, as rhetoric manufacturing psychological plausibility, is given an additional dimension by certain studies of the history of rhetoric and the natural sciences. For William Covino and Ioan Couliano, rhetoric and science, respectively, are the historical products of magic. That is to say: "before about 1700," according to Covino, rhetoric (as the study of the available means of persuasion) and magic (as the study of the manipulation of symbols to achieve material effects) were "complementary and in some ways identical," and while

Couliano stops short of claiming that "the method of magic has something to do with the method of our natural sciences," he notes that such founding figures of modern astronomy and physics as Johannes Kepler and Isaac Newton were themselves investigators of "Pythagorean astral music" and the "occult sciences" of alchemy.[17] Both argue that the modern *epistème* emerged from a process of textual purification that effaced the magical roots of Western thought. Such histories rather strikingly bear out the Gordin Brothers' arguments that science entails a mythic element and that the application of scientific method to the study of social relations as a pseudo-object inevitably produces a social science that is also a "sociomagic."

The question is: what does this recognition of kinship between the rational and the magical entail for our understanding of the strangely twinned practices of radical social theory and speculative fiction—that is, for what we have been calling the radical imagination? Does this imply a terrifying descent, reminiscent of the climactic chapters of Umberto Eco's *Foucault's Pendulum*, into irrationality, what Carl Sagan called the "demon-haunted dark" of premodernity? Conversely, might it mean abandoning crafting fictions and actions that respond to the harsh reality of climate change, for instance, and retreating instead into a world of New Age daydreams? Does this all betoken a return to the Sixties of "the monster of Anarchy" and attempts to levitate the Pentagon?[18] What, in fact, *are* the implications of the radical imagination's reenchantment? The answers may hinge on how we understand the relationship between the magical and the political.

The Magic of Politics

"Magic is the continuation of politics by other means," writes the anthropologist Nils Bubandt—the continuation of politics outside of the public sphere, where whispers and murmurs circulate.[19] Yet according to Yvonne P. Chireau, the kinds of "Conjure practices" we saw opposed to Lovecraftian magic—manipulations of herbs, roots, and cemetery dust, bones and stones, hair and nails—constitute a "material rhetoric": rhetoric, the very hallmark of political life.[20] And the entanglement between politics and magic doesn't end there. "It is the peculiar feature of political life," writes David Graeber,

> that within it, behavior that could only otherwise be considered insane is perfectly effective. If you managed to convince everyone on earth that you can breathe under water, it won't make any difference: if you try it, you will still

drown. On the other hand, if you could convince everyone in the entire world that you were King of France, then you would actually be the King of France. (In fact, it would probably work just to convince a substantial portion of the French civil service and military.)

This is the essence of politics. Politics is that dimension of social life in which things really do become true if enough people believe them. The problem is that in order to play the game effectively, one can never acknowledge its essence....

In this sense politics is very similar to magic...[21]

In other words, politics is the domain *par excellence* of rhetoric and performative language, of the "order-word" and the "incorporeal transformation" it effects, magically turning an ordinary human body into the mystical body of the King of France (or, as in *Bush v. Gore*, President of the United States), or magically turning millions of people into stateless noncitizens, and of the vast networks of belief and action required for such transformations to take effect.[22] Indeed, Graeber argues, "magical action" can be considered "one of the purest forms of political action, because it can only have effects on others in so far as someone hears about it, or otherwise learns that it has happened." That is to say:

Unless you happen to believe that, say, secretly rubbing bits of wood on a man's picture really can cause him to fall madly in love with you, it is clear that such actions can only have effects in so far as word gets around that you have done so. In fact... magical action is the only kind that might be said to consist of a null set; it does nothing in the physical context of its enactment, but only in so far as it enters a broader, more political context of narration, discussion, and report.[23]

Here opens a controversy. Is Graeber then discounting the possibility that magical actions have effects *for the reasons their practitioners claim they do*? What is the ontological status of magic? Is there any way to suggest that magic requires belief without, by the same gesture, setting its reality to $M = 0$?

Marcio Goldman seems to argue that the answer is necessarily *no*. Goldman reads Graeber's essay on fetishism among the Merina of Madagascar as a doomed "effort to save the Marxist conception of fetishism"—i.e., an account of fetishism as the delusory attribution of human or even superhuman powers to mere things—and at the same time to "rescue the Africans," i.e., to grant a measure of respect to *their* account of what they are doing in creating the kinds of objects that Europeans called "fetishes."[24] To interpret fetish-making as an act of *social* creativity rather than anything truly magical—assuming that "Lunkanka cannot really tie anyone's intestines into knots; Ravololona cannot really prevent hail from falling on anyone's crops"—means writing off Merina understandings of magic as operating in and on the *natural* order (conceived as separate from the social order).[25]

"The implicit assumption," remarks Alf Hornborg, "is that if objects are perceived as subjects, then *who are we* to suggest that it is an illusion?"[26] In his forceful critiques of the "ontological turn" in anthropology, Hornborg argues that dropping the subject/object, nature/culture, and human/nonhuman distinctions—in short, Latour's Modern Constitution—means simply giving up the ability to criticize capitalism as a purveyor of fetishistic illusions.[27] Magic isn't real; fetishism, "the attribution of autonomous agency to inanimate or abiotic things," is a false consciousness of the world, no matter who engages in it.[28] At most, "non-living objects . . . can impact on their surroundings (that is, have consequences for them)," but this never amounts to any "agency" since these objects are devoid of purpose.[29]

Several considerations make this a persuasive position. It is difficult for me, as a person thoroughly acculturated to secular, modern assumptions, to really believe in the ontology of a person acculturated among the Merina; perhaps it is finally just not something I am willing to do, even if it were possible for me to re-acculturate, although I acknowledge that it may be perfectly in accord with their common sense and that mine is no better. With apologies to the Gordin Brothers, the explanatory and predictive power of the natural sciences is too great to ignore, even if they are perpetually in danger of being reified into a scientism, and they are made urgently necessary both by their entanglement in the global ecological crisis *and* by their importance for any plausible attempts to come to grips with it. Then there is the fact of my mutual entanglement with the Merina, however distant and indirect, via the circuits of a global economy that makes it more difficult than ever to erect boundaries between one ontology and another, for better or worse.[30] The anthropocentric materialist ontology of Hornborg, Malm, and Colombo, for whom "nature sends the rain" while "culture prints the dollars," is convincing, not only because it offers a readymade framework for ideology critique—just unmask the appearance of "naturalness" to show that X is a cultural construct!—but because it is, *for me*, so intuitive, so "natural," that I can easily forget that it, too, is a cultural construct, an imaginary that has existed on this planet for the blink of a historical eye.[31]

It seems to me that David Graeber, Arturo Escobar, and the pragmatists offer, if not a way out of the dilemma, a kind of compromise position, a way to hold to a certain concept of "a common world," the Zapatistas' "world within which many worlds might fit," which is precisely what Escobar calls "the pluriverse."³² The pluriverse, that is, the Earth itself ("we all live within the Earth as pluriverse"), is not a shared *object* but "a shared *project* based on a multiplicity of worlds."³³ To share a project, we need not to dominate but to cooperate and translate back and forth. If, while I don't want to accept a Merina ontology, I don't want to impose a Western ontology on the Merina either, then it seems to me that the best I might be able to do is to offer a *pluriversal* description of Merina magic, from the transitional standpoint of *umet'* or *mētis* (pragmatist doing-as-knowing), as symbolic action taking place through human agencies: in other words, *as sociomagic*. To speak of their practices *metafetishistically* would be to translate them into terms that make sense to me, leaving open the possibility of translating my own practices into terms that would make sense to them. This would mean giving a pluriversal, sociomagical account of my own society.

A positive Sociomagic

Philippe Pignarre and Isabelle Stengers attempt to do just this in *Capitalist Sorcery: Breaking the Spell*. To return to the two examples of sociotechnical disaster that we began with, both in various ways the work of "minions . . . believing they are doing the 'right thing,'" robotically following procedures and disavowing their own agency: what Hornborg might describe as human subjects pretending, in effect, to be mere passive objects, Stengers and Pignarre describe instead as a kind of *ensorcellment*, the object world created by human action returning to exercise a baleful influence over human subjects in an "operation of capture."³⁴ Inspired by the role taken by activists such as Starhawk in the Seattle protests of 1999, they propose "to try to take seriously those Americans who have dubbed themselves 'neopagan witches' and have cultivated the heritage of techniques of nonviolent action in a mode that combines 'resisting capture' and 'learning to give thanks.'"³⁵ *Giving thanks* means the creation of new networks of interdependence and mutual aid; *resisting capture* means what the Gordin Brothers called a "positive biotechnics," i.e., the ecological arts of living off the industrial capitalist grid, as well as an "inventive sociotechnics," the creation and practice of what I have elsewhere called a resistance culture.³⁶

What is the difference between pagans "call[ing] on the elements of nature to sustain us" at the World Trade Organization protest in 1999 and Harry Potter fans yelling "Expelliarmus!" at police during Occupy in 2011?³⁷ An outside observer might assume that in the second case, no one shouting a made-up magic word expects any physical result from it; rather, it is uttered in a spirit of play, perhaps as an attempt to disrupt the deadly seriousness of a confrontation with armed and armored bullies, whereas in the first case, the words must be uttered or chanted in a spirit of absolute seriousness. This does not necessarily accord with the pagan practitioners' self-understanding, however, as Starhawk underscores:

> A pagan ritual incorporates touch, sensuality, and humor. Anything we truly revere is also something that we can ridicule respectfully. The elements of laughter and play keep us from getting stuck on one level of power or developing an inflated sense of self-importance. Humor keeps kicking us onward, to go deeper.
>
> We break eggs over the initiate's body as she lies in the ritual bath. The eggs are symbols of her rebirth. She is a woman who loves calm, peace, order.
>
> "This must be eggs-scruciating!" says a priestess.
>
> "Oh shut up. This is a serious business."
>
> "Eggs-zactly!"
>
> "Would you stop it?"
>
> "Don't egg her on."
>
> Her coven becomes embroiled in a staged fight. They are pushing her to her limits—and past them. Soon everyone is laughing, close to tears, at silly puns. But under the laughter is the power—shining, throbbing. When we fall silent, as she is dried, and laid on towels, and stroked with feathers, and told a sacred story, we feel the power rise, strong because the laughter has pushed down our barriers.[38]

In this way, the sociomagical process can also be metafetishistic: ironically self-aware in a manner quite distinct from the cynical disavowal of commodity fetishism, capable of taking a critical distance from its own symbolic practices while also tapping into the "power"—intersubjective *social* power, on my pragmatist reading of it—that these practices afford. Stephen Duncombe notes that this metafetishistic stance is also present in many forms of fandom:

> My first immersion in SF came through the original Star Trek series, which aired while I was in college. Every weeknight, at 11 PM, I would gather with my friends, many of whom were also my activist comrades, around the TV set to lose ourselves in the adventures of the USS Enterprise. To be sure, we were captivated by the foreign places and strange societies that Kirk and his crew visited, and longed for a similar world where money had no meaning and racism and sexism had been (sort of) left behind, but what held our interest and kept

us coming back night after night was something else entirely. We reveled in the melodramatic dialogue, the blocky staging, and William Shatner's overacting. We looked for the seams in the sets and snickered at the low-rent costumes of the aliens. And we all knew that once a crew member donned a red shirt for a landing party their days were numbered. We loved Star Trek, but we also loved laughing at it....

Our political imagination was inspired by the presentation of Roddenberry and company's SF scenarios, and Star Trek took us on a voyage lightyears from the unsatisfactory present we knew too well, yet the sheer campiness of the series kept us from accepting the future it presented as a real possibility, or, rather, a valid fantasy. Refused satisfaction in the alternative futures provided by Star Trek, we were forced to imagine and act upon our own ideas and ideals of an alternative.[39]

This "satirical" element within utopian imaginaries, Duncombe argues, is what prevents them from closing in on themselves and becoming the kind of rigid schemas that Russell Jacoby calls "blueprint utopias."[40] In this way, SF and fantasy fan practices present a model for the kind of activism that Duncombe has been calling for since his *Dream: Reimagining Progressive Politics in an Age of Fantasy* (2007; recently updated as *Dream or Nightmare: Reimagining Politics in an Age of Fantasy*, 2019). Fandoms can be excellent schools for the production of Duncombe's "ethical spectacles," which he defines as

- *participatory*: "involv[ing] all participants in the creation of the spectacle"
- *open*: "responsive and adaptive to shifting contexts and the ideas of participants"
- *transparent*: "a spectacle that presents itself as a spectacle, without seeking to trick or deceive"
- *realistic*: "us[ing] fantasy to illuminate and dramatize real-world power dynamics"
- *utopian*: "a dream that we know is just a dream, yet motivates us nonetheless."[41]

As such, ethical spectacles enact a kind of counterspell against the illusions of the status quo, "mobilizing counterfantasies that might offer a space for their audience to imagine for themselves."[42] Without erasing the infinity of differences between all of these different forms of sociomagical action, I want to suggest that all of them are capable of mobilizing political imagination and social power against the phantasmagoria of late capitalism as well as the fantasies of pollution and purification summoned by its far-right guardians.

I have argued that any sufficiently advanced sociotechnics is indistinguishable from sociomagic, that every effective mode of social relationship entails a kind of fetishistic practice. As many have argued, it is from those cultures that the Great Sorting Machine has falsely relegated to "the past" that we may have the most to learn about the arts of collective creation, both biotechnical and sociotechnical. And this learning process, as Brandon Benallie of the K'é Infoshop reminds us, has been underway for centuries: colonial accounts of Native American societies and their practices of mutual aid and democratic confederation, such as the *Jesuit Relations* (1632–1673), Cadwallader Colden's *History of the Five Nations* (1747), and Lewis Henry Morgan's *The League of the Ho-dé-no-sau-nee or Iroquois* (1851), read by settlers and Europeans, nourished the development not only of liberal democracy as practiced in the United States, but also of Marxist and anarchist theory, which today are challenged to remember and recognize their forgotten affinities with Indigenous peoples, worlds, and struggles.[43] The K'é project belongs not to any "dustbin of [settler] history" but to what Theresa Warburton, following Eve Tuck and K. Wayne Yang, calls "indigenous futurity"—a future, perhaps, within which many futures might fit.[44]

It feels risky to speak of *any* futurity right now, but it seems to me that this just makes it all the more urgently necessary. The science-fictionalization of the public imaginary, the "capitalist realist" perception of the present *as* the (bad old) future, perversely threatens to foreclose the imagining of other futures, other possibilities; we are met at every turn with rhetorics of impossibility, the sociomagic of despair. White supremacy feeds on this despair; the cackling fascist goons whom the President has just asked to "stand by" (Sept. 29, 2020) are delighted by the atmosphere of apocalyptic drama, eager for the "boogaloo" (the fantasy of a "generalized disaster," as Susan Sontag put it, "that . . . releases one from normal obligations," a magical circle within which the normal rules are suspended).[45] Against this background, it is striking to see how often the New Black Fantastic mobilizes the rhetoric of magic. In an incisive observation about the discourse around Nnedi Okorafor's "africanfuturist" novel, *Who Fears Death*, Nisi Shawl writes:

> Once again we confront the question of genre purity . . . Many works cited in my history of black science fiction defy classification as SF or fantasy. *Who Fears Death* often gets described as "science fantasy," a hybrid term, because though indisputably taking place in a technological future, it features shapeshifters and diviners among its water distilling machines and pocket video cams. Such boundary enforcement annoys me . . .[46]

Magic, in this context, is a force that obeys rules and patterns of its own (recall Jemisin's corollary to Clarke's Third Law: "any sufficiently systematized magic is indistinguishable from science"), but which nonetheless permits an expression of desire that jams the "boundary enforcement" of the Great Sorting Machine and overflows the categories of the Modern Constitution. Such powers of the imagination pose a challenge both to white supremacy and to capitalist and statist realism, the ideological faces of the forces currently throttling the planet. They suggest that the end of *one* world may yet be the beginning of others.

APPENDIX

A Provisional Manifesto for Nonmodern Anarchisms

1. The two major institutions of global modernity—capitalism and the State—are making the world increasingly unlivable; to say "modernity" is to say genocide and ecocide.

2. Modernity raised hopes in some quarters that all the traditional forms of oppression associated with patriarchal "state-civilization" (Abdullah Öcalan) could be erased by capitalist progress and its secular institutions; it largely defaulted on such promises, bequeathing instead "division—of people, spheres of life, and knowledge" (Nazan Üstündağ). Perhaps most fundamentally, it divided humanity from the nonhuman, society from nature, and life from the nonliving world, calling this division "Enlightenment." To be nonmodern is not necessarily to invert or undo all that modernity did and was; it is to fundamentally question and challenge the modern world order.

3. Anarchism, as developed by European thinkers and movements, was a child of Enlightenment modernity, albeit fundamentally incompatible with the modern world system. It drew from modernity its finest tools for thought and struggle against the statist, capitalist, colonial world order that constituted modernity but also inherited some of modernity's eugenic logic. "In our own time, these legacies are intertwined like strands and subvert the clear patterns that existed in the past until the language of freedom becomes interchangeable with that of domination" (Murray Bookchin).

4. Nonmodern anarchisms refuse the "normative developmentalism" of modernity, in both its capitalist and anti-capitalist guises (Glen Coulthard), instead seeking pathways toward "a world within which many worlds fit" (EZLN).

5. Capitalist modernity has no use for traditions that obstruct it, much as it is avid to commodify some traditions (and to invent new ones). Far-right calls to "reject modernity, embrace tradition" cannot be taken at face value: the "traditions" they refer to are either stereotyped tableaus of Eisenhower-era conformity or nationalist wish-fulfillment dreams, and the "modernity" they reject consists of rights with ancient roots: women's rights, the right to dress and have sex as one likes, and so on. Nonmodern anarchisms do not regard traditions as inherently oppressive or emancipatory per se, nor does it assume that traditional societies are incapable of an immanent critique. "I am not one to believe in traditions simply because they are traditions. I believe in the legacies that multiply human freedom, and not in those that cage it" (Eduardo Galeano).

6. Nonmodern anarchisms are not a license for white anarchists to appropriate Indigenous traditions. They do not express a wish to become Indigenous: "This is not possible . . . The issue is not for an individual to become something they are not, but rather for a centering of other ways of being in the world that may make our final years on this planet—however many or few there may be—ones worth living" (Bettina Escauriza). They may provide meeting grounds for sincere seekers of an exit from the increasingly ruinous modern world system and its oppressive institutions.

7. Nonmodern anarchisms are relations woven to support the thought and struggles of modernity's excluded, notably Indigenous, Black, queer, disabled, trans, nonbinary, Two Spirit, and migratory or diasporic peoples, among the many others who failed to match the normative models that favored whiteness, cismasculinity, able-bodiedness, heterosexuality, and citizenship. They are formed in explicit opposition to the pernicious system of violent classification and stratification that is capitalism. As such, they can have nothing to do with the fascist aspirations of so-called "national anarchism."

8. Modernity, in its liberal guise, encourages its excluded to pursue inclusion within the modern world system, to seek recognition within one or another of its institutions and regimes of rights while still perpetuating its profitable hierarchies. The Western anarchist tradition rejected this conception of progress as a cruel illusion. Nonmodern anarchisms express the will of the excluded to go beyond the politics of recognition and its institutions.

9. From the Western anarchist tradition, nonmodern anarchisms draw a certain codification of practices already found in nonmodern cultural

repertoires, practices which anarchism imperfectly but usefully translated into the conceptual idiom of modernity: mutual aid, communism, affinity groups, direct democracy, consensus, direct action.

10. Nonmodern anarchisms may take on as many forms and as many idioms as needed. They will be reinvented in this process of translation.

11. The atheist commitments of the Western anarchist tradition, while never completely uncontested, were in relation to particular religious institutions, of which the Catholic Church of the nineteenth and early twentieth centuries was the paradigm. Nonmodern anarchisms are no longer committed to atheism but are free to pursue an immanent critique of the legacies of domination within the many traditions from which they draw their strength.

12. Nonmodern anarchisms are no longer committed to scientism but know many sciences, old and new; they are no longer committed to modern technics as a whole but practice many convivial technics, old and new.

13. Nonmodern anarchisms refuse to separate humanity and nonhumanity, society and nature, life and nonlife.

14. One thing that Indigenous knowledge can impart to nonmodern anarchisms is the lived experience of democratic confederation—how to establish effective unity between disparate social worlds without the imposition of uniformity, after the manner of the Three Fires confederation linking the three Anishinaabe peoples, the Haudenosaunee *Kaienerekowa* or "Great Peace," and many others. In place of colonialism's practices of universality, nonmodern anarchisms seek and practice pluriversality, a pragmatist Great Peace that allows both for coordination of action and pluriversal truths.

Notes

Introduction

1. Chantelle Gray, *Anarchism After Deleuze and Guattari: Fabulating Futures* (London: Zero Books, 2022), 14.
2. Kathryn Hume, *Fantasy and Mimesis: Responses to Reality in Western Literature* (New York: Routledge, 2014), 92.

1. Predictive Analytics for Modern Times

1. Deborah Lupton, *Digital Sociology* (Abingdon, Oxon: Routledge, 2014), 140.
2. Melissa Ames, *Small Screen, Big Feels: Television and Cultural Anxiety in the Twenty-First Century* (University Press of Kentucky, 2020), 7; Slavoj Žižek, "A Permanent Economic Emergency," *New Left Review*, no. 64 (August 2010), https://newleftreview.org/issues/II64/articles/slavoj-zizek-a-permanent-economic-emergency.
3. Brian McHale, *Postmodernist Fiction* (London: Routledge, 2003), 253. I would argue that McHale has it backward: the overwhelming obsession of the theory and literature called "postmodern" was with epistemology, and particularly with a relativist epistemology tied to the way that texts are open to indefinitely many interpretations. It is in this sense that, by shifting to "properly ontological" grounds, Deleuze and Guattari place their poststructuralist philosophy outside of postmodernism and the question of relativism (Michael Hardt, "Anti-Oedipus (1)," *Reading Notes on Deleuze and Guattari*, Capitalism & Schizophrenia, Duke University, December 18, 2012, https://people.duke.edu/~hardt/ao1.htm).
4. McHale, *Postmodernist Fiction*, 253.
5. Cory Doctorow, "Cold Equations and Moral Hazard," *Locus Online* (blog), March 2, 2014, https://locusmag.com/2014/03/cory-doctorow-cold-equations-and-moral-hazard/. The image of the "toolkit," however, might lead us into a falsely instrumentalist picture of things, reinstating the active subject/passive object dichotomy, unless we think of our tools as actants acting on *us*. See Ch. 3, "Methodological Metafetishism."
6. Seo-Young Chu, *Do Metaphors Dream of Literal Sleep?: A Science-Fictional Theory of Representation* (Cambridge, Mass: Harvard University Press, 2010), 68.
7. Chu, Do Metaphors Dream of Literal Sleep, 9. Last but not least on her list of such cognitively estranging referents: the possibility of "robot rights" or the really-emerging-but-difficult-to-imagine ethical relationship between human beings and our increasingly lifelike machines.
8. Bruno Latour, *We Have Never Been Modern*, translated by Catherine Porter (Cambridge, Mass.: Harvard University Press, 1993), 13.

9. As Arturo Escobar writes, catastrophic climate change marks not only a crisis of capitalism, the expansion of the market hitting a biospheric limit, but a much deeper "crisis of a particular civilizational model," "a crisis within modernity . . . [as] a particular ontology or mode of being in the world" (Arturo, *Pluriversal Politics: The Real and the Possible*, translated by David Frye [Durham, NC: Duke University Press, 2020], xii).

10. Matt Bernico, "Anthropodicy: An Anarchism of Things," *Anarchist Developments in Cultural Studies*, nos. 1 & 2 (November 24, 2015), https://journals.uvic.ca/index.php/adcs/article/view/17181, 74.

11. Zakiyyah Iman Jackson, *Becoming Human: Matter and Meaning in an Antiblack World* (New York: NYU Press, 2020), 1; Aimé Césaire, *Discourse on Colonialism*, translated by Joan Pinkham (New York: Monthly Review Press, 2000), 42.

12. Travis Linnemann, *The Horror of Police* [Minneapolis: University of Minnesota Press, 2022], 64.

13. Greg Tate quoted in Joy Sanchez-Taylor, *Diverse Futures: Science Fiction and Authors of Color* (Columbus, Oh.: Ohio State University Press, 2021), 7; Sanchez-Taylor, *Diverse Futures*, 7.

14. For Canavan, "despite her protests, most of Chu's excellent work really amounts to Suvinian/Jamesonian allegory sailing under the flag of metaphor" (Gerry Canavan, "Do Metaphors Dream of Literal Sleep? A Science-Fictional Theory of Representation (review)," *Comparative Literature Studies* 49, no. 4 [2012]: 618). Wishing, perhaps, to avoid damning by faint praise, he suggests a different way in which Chu fulfills her promise to "turn Suvin inside out": "Where Suvinian criticism has tended to focus on science fiction's dimension of estrangement, taking the cognition as read, Chu puts estrangement to one side and focuses instead on the principle of cognition." Nonetheless, he insists, it's all the same allegorical reading strategy in the end (Canavan, "Do Metaphors Dream of Literal Sleep," 619). On the contrary: Chu *begins* with estrangement, with the strangeness of the world, and that difference from Suvin makes all the difference.

15. Mark Bould, "The Ships Landed Long Ago: Afrofuturism and Black SF," *Science Fiction Studies* 34, no. 2 (2007): 177–86; Grace L. Dillon, "Imagining Indigenous Futurisms," in Grace L. Dillon, ed. *Walking the Clouds: An Anthology of Indigenous Science Fiction* (Tucson: The University of Arizona Press, 2012), 8.

16. N. K. Jemisin, *The Stone Sky* (New York: Orbit, 2017), 300.

17. N. K. Jemisin and Elizabeth Flock, "How the Protests in Ferguson Helped Inspire This Fantasy Novel," PBS NewsHour, June 12, 2019, https://www.pbs.org/newshour/arts/how-the-protests-in-ferguson-helped-inspire-this-fantasy-novel.

18. McKenzie Wark, *Molecular Red: Theory for the Anthropocene* (New York: Verso Books, 2015).

19. N. K. Jemisin, *The Fifth Season* (New York: Orbit, 2015), 460.

20. Even if the most ancient of the personal names we learn—the forty-thousand-year-old names of the Stone Eaters, when they were still human—are vaguely reminiscent of names from the Korean Peninsula, the shape of which, writ large, the Stillness loosely resembles, Jemisin has disavowed any such intention (Jemisin, "Once Again"), and there really is no compelling reason to superimpose our maps on hers.

21. *The Fifth Season*, 459, 462–63.

22. *The Fifth Season*, 459.

23. Richard J. F. Day and Adam Lewis, "Radical Subjectivity and the *N*-Row Wampum: A General Model for Autonomous Relations Against and Beyond the Dominant Global Order?" in *Subjectivity in the Twenty-First Century: Psychological, Sociological, and Political Perspectives*, edited by Romin W. Tafarodi (Cambridge University Press, 2013, pp. 169–89), 169–70, italics mine.

24. Amal de Chickera, Laura van Waas, and Zahra Albarazi, "The World's Stateless: a new report on why size does and doesn't matter," European Network on Statelessness (December 15,

2014), https://www.statelessness.eu/blog/world%E2%80%99s-stateless-new-report-why-size-does-and-doesn%E2%80%99t-matter; OECD, *States of Fragility* 2018 (OECD, 2018), https://doi.org/10.1787/9789264302075-en.

25. Hannah Arendt quoted in Daniel Colson, *Trois Essais de Philosophie Anarchiste: Islam, Histoire, Monadologie* (Paris: Léo Scheer, 2004), 174.

26. Žižek, "A Permanent Economic Emergency," 95.

27. Jemisin, *The Stone Sky*, 309.

28. Jessica Hurley and N. K. Jemisin, "An Apocalypse Is a Relative Thing: An Interview with N. K. Jemisin," *ASAP/Journal* 3, no. 3 (December 21, 2018): 469.

Jemisin, *The Stone Sky*, 321.

29. Jemisin, *The Fifth Season*, 124.

30. N. K. Jemisin, *The Obelisk Gate* (New York: Orbit, 2016), 180.

31. Jemisin, *The Stone Sky*, 6.

32. Jemisin, *The Obelisk Gate*, 198.

33. Jemisin, *The Stone Sky*, 240.

34. Alan Moore and Kevin O'Neill, *The League of Extraordinary Gentlemen. Volume 1* (United States: DC Comics, 2017), 6.

35. Bruce Sterling, "My Rihla," *Science Fiction EYE* 2, no. 2 (August 1990): 12–16, 12.

36. Noel Carroll, *The Philosophy of Horror: Or, Paradoxes of the Heart* (New York: Routledge, 2003), 176, 23.

37. Edward Ingebretsen, *At Stake: Monsters and the Rhetoric of Fear in Public Culture* (Chicago: University of Chicago Press, 2001), 4.

38. Jemisin, *The Stone Sky*, 313.

39. Jemisin, *The Fifth Season*, 98; *The Stone Sky*, 216.

40. Jemisin, *The Obelisk Gate* 230; *The Stone Sky* 245; Slavoj Žižek, *Looking Awry: An Introduction to Jacques Lacan through Popular Culture* (Cambridge, Mass.: MIT Press, 1992), 35.

41. Jacques Lacan, *The Seminar of Jacques Lacan, Book III: The Psychoses, 1955-1956*, edited by Jacques-Alain Miller, translated by Russell Grigg (New York: Routledge, 1993), 9.

42. Žižek, *Looking Awry*, 35.

43. Toward the end of his life, Lacan turned to other figures of thought to reimagine his schema of three "orders" or "registers" of existence in terms of entanglement, visualizing their interrelationship in terms of the Borromean Knot. The three rings of this "knot" are, however, self-contained structures; their knotting presents us only with a relationship of interdependence between orders, a mutual implication, while maintaining their "profoundly heterogeneous" identities (Dylan Evans, *An Introductory Dictionary of Lacanian Psychoanalysis* [London: Routledge, 1996], 135). Indeed, Lacan was scandalized by Chomsky's suggestion that human language use had a biological basis, that the symbolic order might arise from material processes within the real itself (Lacan, *The Seminar of Jacques Lacan, Book XXIII: The Sinthome*, edited by Jacques-Alain Miller, translated by A. R. Price [Polity Press, 2016], 28; Turkle 244–45).

44. Danette Gerald and Kati Haycock, *Engines of Inequality: Diminishing Equity in the Nation's Premier Public Universities* (Washington D.C.: Education Trust, 2006), http://www.edtrust.org/dc/publication/engines-of-inequality-diminishing-equity-in-the-nation's-premier-public-universities, 3.

45. Diane Reay, "Universities and the Reproduction of Inequalities," in *A Manifesto for the Public University*, edited by John Holmwood, 112–26 (London, New York: Bloomsbury, 2011), 114; Gerald and Haycock, 8; Russell W. Rumberger, "Education and the Reproduction of Economic Inequality in the United States: An Empirical Investigation," *Economics of Education Review*, Special Issue in Honor of Henry M. Levin, 29, no. 2 (April 1, 2010): 246–54, https://doi.org/10.1016/j.econedurev.2009.07.006, 253.

46. Clark Kerr, *The Great Transformation in Higher Education, 1960-1980* (Albany, New York: SUNY Press, 1991), 58.

47. Abigail Welles, "Recovery," in *Economic Inequality, Neoliberalism, and the American Community College*, edited by Patrick Sullivan, 101–10 (Cham: Palgrave Macmillan, 2017), 102–3.

48. Leanne Weber, *Policing Non-Citizens* (London: Routledge, 2013), 182, 101.

49. Evelyn Alsultany, *Arabs and Muslims in the Media: Race and Representation After 9/11* (New York: NYU Press, 2012), 123.

50. Lennard J. Davis, *Enforcing Normalcy: Disability, Deafness, and the Body* (New York: Verso, 1995), 24.

51. Theresia Degener, "Disabled Persons and Human Rights: The Legal Framework," in *Human Rights and Disabled Persons: Essays and Relevant Human Rights Instruments*, edited by Yolan Koster-Dreese and Theresia Degener, 9–39 (Dordrecht: Martinus Nijhoff Publishers, 1995), 9.

52. Judith Butler, *Frames of War: When Is Life Grievable?* (London; New York: Verso, 2016), 34–35.

53. Elizabeth A. Povinelli, *Geontologies: A Requiem to Late Liberalism* (Durham: Duke University Press, 2016).

54. Calvin L. Warren, *Ontological Terror: Blackness, Nihilism, and Emancipation* (Durham: Duke University Press, 2018), 12–13.

55. Donna Haraway, *Simians, Cyborgs, and Women: The Reinvention of Nature* (New York: Routledge, 1991), 204.

56. Daniel Colson, *A Little Philosophical Lexicon of Anarchism from Proudhon to Deleuze*, translated by Jesse Cohn (Colchester; New York; Port Watson: Minor Compositions, 2019), 60.

57. Latour, *We Have Never Been Modern*, 11. How to oppose the Great Sorting Machine of modernity if it is, in effect, both itself (purification) and its opposite (hybridization)? I suggest an analogy with Cornelius Castoriadis' analysis of capitalism: the "profound tendency" of capital is to turn human beings into obedient robots ("reification"), but it cannot fully realize this goal because it also relies fundamentally on the spontaneous, intelligent, creative activity of those same human beings; at the same time, it cannot fully allow this spontaneous creativity of workers to be fully realized, because that would render the owners and managers redundant (Cornelius Castoriadis, *The Imaginary Institution of Society*, translated by Kathleen Blamey [Cambridge, Mass.: MIT Press, 1997], 16, 80). Similarly, it seems to me that the Great Sorting Machine is locked into its genocidal and ecocidal *program*, as it were, of purification, although this strategy is suicidal as well, even if it also relies on tactics of hybridization which, if allowed to flourish, would spell *its* doom but *our* survival.

58. Hurley and Jemisin, "An Apocalypse Is a Relative Thing," 469.

59. Alfred Bester, "5,271,009," *Magazine of Fantasy & Science Fiction*, March 1954, 30.

60. Hurley and Jemisin, "An Apocalypse Is a Relative Thing," 469–70.

61. Marion Long, "Paradise Tossed: Personal Utopias of Elie Wiesel, Tammy Faye Bakker, Max Headroom, Hans Kung, Coretta Scott King, Stephen Jay Gould, Jesse Jackson, Grace Slick, David Rockefeller, Kurt Vonnegut, and Others," *Omni*, April 1988, 108.

62. Ryan McGee, "How to Approach Peak TV in a Post-Trump Era," *Pajiba*, January 3, 2017, https://www.pajiba.com/think_pieces/how-to-approach-peak-tv-in-a-posttrump-era.php.

63. Bernice M. Murphy, *Key Concepts in Contemporary Popular Fiction* (Edinburgh: Edinburgh University Press, 2017), 49.

64. Geoff Klock, *How to Read Superhero Comics and Why* (London: A&C Black, 2002), 27, 91.

65. Deborah Mellamphy, "Dead Eye: The Spectacle of Torture Porn in *Dead Rising*," in *Game On, Hollywood!: Essays on the Intersection of Video Games and Cinema*, edited by Gretchen Papazian and Joseph Michael Sommers (Jefferson, NC: McFarland, 2013), 35–36.

66. Campbell qtd. in Eric Solstein and Gregory Moosnick, "John W. Campbell's Golden Age of Science Fiction: Text Supplement to the DVD," Digital Media Zone, 2002, http://www.sfcenter.ku.edu/JWC_Study_Supplement.pdf, 72.

67. Justine Larbalestier, *The Battle of the Sexes in Science Fiction* (Middletown, Conn.: Wesleyan University Press, 2002), 117; John G. Russell, "Darkies Never Dream: Race, Racism, and the Black Imagination in Science Fiction," *CR: The New Centennial Review* 18, no. 3 (Winter 2018): 261–62; Solstein and Moosnick, "John W. Campbell's Golden Age of Science Fiction," 28–31.

68. "She should've been a hero,/But she was just a girl,/And sex trumps logic everywhere—/At least in this cold world . . . //Men may claim to pull their weight—/'Authority is hell'—/But women always pay the freight,/And take the blame as well" (Jack O'Brien, "Classics of Science Fiction: 'The Cold Equations,'" *Asimov's Science Fiction* 32, no. 3 (Mar. 2008): 107, 5–8, 21–24).

69. Campbell quoted in John Bangsund, ed., *John W. Campbell: An Australian Tribute* (Canberra: Ronald E. Graham and John Bangsund, 1972), 86.

70. Godwin, "The Cold Equations," 81–82.

71. Godwin, "The Cold Equations," 69.

72. Godwin, "The Cold Equations," 84.

73. Godwin, "The Cold Equations," 64, 76, 72.

74. Godwin, "The Cold Equations," 68, 76.

75. Godwin, "The Cold Equations," 66; John Huntington, "Hard-Core Science Fiction and the Illusion of Science," in *Hard Science Fiction*, edited by George E. Slusser and Eric S. Rabkin, 45–57 (Carbondale, Ill.: Southern Illinois University Press, 1986), 52; Mark Bould and Sherryl Vint, "There Is No Such Thing as Science Fiction," in *Reading Science Fiction*, edited by James Gunn, Marleen Barr, and Matthew Candelaria, 43–51 (New York: Palgrave Macmillan, 2009), 47; Doctorow, "Cold Equations and Moral Hazard."

76. Bruno Latour, *Science in Action: How to Follow Scientists and Engineers Through Society* (Cambridge, Mass.: Harvard University Press, 1987), 258.

77. Bruno Latour, *Reassembling the Social: An Introduction to Actor-Network-Theory* (Oxford: OUP Oxford, 2005), 63.

78. Roland Barthes, *Mythologies*, translated by Annette Lavers (New York: Farrar, Strauss, and Giroux, 1991), 107, 140–41.

79. Godwin, "The Cold Equations" 76.

80. Žižek, *Looking Awry*, 34.

81. Huntington, "Hard-Core Science Fiction," 54–55.

82. Steven Shaviro, *The Universe of Things: On Speculative Realism*, Posthumanities (Minneapolis: University of Minnesota Press, 2014), 48.

83. Godwin, "The Cold Equations" 76.

84. Hilgartner seems to use the term "sociotechnical" to emphasize the way in which social and technological systems are intertwined. As we shall see, however, his statement also reads as true in the special sense that Abba and Wolf Gordin gave to the word *sociotechnics* a century ago, referring less to the actual and more to a potential or emergent field of practice dedicated to "the endless and inexhaustible variety of forms of social construction" ("Manifest Pananarkhistov" 219). The *absence* of such a sociotechnics is at the root of many "natural" disasters.

85. John A. Powell, Hasan Kwame Jeffries, Daniel W. Newhart, and Eric Stiens, "Toward a Transformative View of Race: The Crisis and Opportunity of Katrina," in *There Is No Such Thing as a Natural Disaster: Race, Class, and Hurricane Katrina*, edited by Gregory D. Squires and Chester W. Hartmann, 59–84 (New York: Routledge, 2006), 63.

86. Nancy Tuana, "Viscous Porosity: Witnessing Katrina," in *Material Feminisms*, edited by Stacy Alaimo and Susan Hekman, 188–213 (Bloomington, Ind.: Indiana University Press, 2008), 193.

87. Noel King, "I Agree With DHS Overcrowding Report 100%, Border Agents' President Says," *Morning Edition, National Public Radio*, 3 July 2019, www.npr.org, https://www.npr.org/2019/07/03/738401304/i-agree-with-dhs-overcrowding-report-100-border-agents-president-says.

88. "I Agree With DHS Overcrowding Report 100%, Border Agents' President Says," Interview, *Morning Edition*. National Public Radio, July 3, 2019, https://www.npr.org/2019/07/03/738401304/i-agree-with-dhs-overcrowding-report-100-border-agents-president-says.

89. Godwin, "The Cold Equations," 64.

90. Huntington, "Hard-Core Science Fiction," 52–53.

91. Chu, *Do Metaphors Dream of Literal Sleep*, 68.

92. Brian Attebery, *Decoding Gender in Science Fiction* (London: Routledge, 2002), 180.

93. Godwin, "The Cold Equations," 65.

94. Attebery, *Decoding Gender*, 182–83.

95. Leslie A. Fiedler, "Introduction," in *In Dreams Awake : A Historical-Critical Anthology of Science Fiction*, 11–24 (New York: Dell, 1975), 19; Thomas M. Disch, "The Embarrassments of Science Fiction," in *On SF*, 3–15 (Ann Arbor: University of Michigan Press, 2005), 143; James E. Gunn, "A Touch of Stone," in *The Road to Science Fiction: From Heinlein to Here*, 3:213–15 (Lanham, MD: Scarecrow Press, 2002), 244.

96. Godwin, "The Cold Equations," 74, 73.

97. Hugo Gernsback, "A New Sort of Magazine," *Amazing Stories*, April 1926, 3.

98. Roger Luckhurst, "Bruno Latour's Scientifiction: Networks, Assemblages and Tangled Objects," *Science Fiction Studies* 33, no. 1 (March 2006): 4; Bould and Vint, "There Is No Such Thing as Science Fiction," 50. Yet SF hybridity is still partly visible in "The Cold Equations," which, to produce its backdrop of interstellar colonization, cheats the lightspeed limit with a "hyperspace drive"—this, on the same page that announces that basic Newtonian mechanics is "the law, and there could be no appeal" (Godwin, "The Cold Equations," 63).

99. Bould and Vint, "There is No Such Thing as Science Fiction," 49.

100. Gunn, "A Touch of Stone," 244; Attebery, *Decoding Gender*, 48.

101. Larbalestier, *The Battle of the Sexes in Science Fiction*, 104.

102. Rebekah Sheldon, "Reading for Transgression: Queering Genres," in *After Queer Studies*, edited by Tyler Bradway and E. L. McCallum, 171–87 (Cambridge: Cambridge University Press, 2019), 179.

103. John Clute, "Fantastika: Or, the Sacred Grove," *Fantastika Journal* 1, no. 1 (April 2017): 16.

104. John Rieder, *Colonialism and the Emergence of Science Fiction* (Middletown, CT: Wesleyan University Press, 2007), 2–3.

105. Sylvia Wynter, "A Utopia from the Semi-Periphery: Spain, Modernization, and the Enlightenment," edited by Stelio Cro, *Science Fiction Studies* 6, no. 1 (1979): 106.

106. Robert M. Philmus, "Science Fiction and Alienation," *Science Fiction Studies* 9, no. 1 (1982): 91; Attebery, *Decoding Gender*, 180.

107. A nod with some reservations. For one thing, Ellis was documenting some cultural phenomena that were very much of a particular moment, and I don't want to imply that *this* moment is in any way the same. Nonetheless, if history doesn't repeat, sometimes it rhymes: when Ellis (1989) quotes Anthony Davis' declaration that "I had to fight to overcome in the institution and transform it; make a white institution serve a black person" (Trey Ellis, *Platitudes* [Ann Arbor, MI: UPNE, 2003], 197), this seems to me to speak pretty directly to the kind of situation that Nalo Hopkinson describes in "A Reluctant Ambassador from the Planet of Midnight" (2010), documenting the pain of working within a community of writers and readers that is alternately inclusive and exclusive, nourishing and alienating, in unequal measure (Nalo Hopkinson, "A Reluctant Ambassador from the Planet of Midnight," *Journal of the Fantastic in the Arts* 21, no. 3 [2010]: 348).

108. If the audience for the direct-to-video film treatment of Hopkinson's *Brown Girl In the Ring* (2017) was limited, Marvel's *Black Panther* (2018), whatever its failings, brought an Afrofuturist aesthetic to a truly massive audience for the first time, producing a spinoff comic series, *Wakanda Forever* (2018), written by Nnedi Okorafor, whose *Who Fears Death* and *Binti* are under development for TV series (release dates unknown). DC Comics' *Green Lantern* series has more recently spawned *Far Sector* (2019–2020), an explicitly science-fictional series written by N. K. Jemisin. Barry Jenkins' TV adaptation of Colson Whitehead's *The Underground Railroad* won critical acclaim (2021), and Paramount beat out Lucasfilm for the rights to Tomi Adeyemi's *Children of Blood and Bone*. The close of 2022 will see an adaptation of Butler's *Kindred*.

109. A. O. Scott quoted in Marleen S. Barr, *Afro-Future Females: Black Writers Chart Science Fiction's Newest New-Wave Trajectory* (Columbus: Ohio State University Press, 2008), ix.

2. Sublime Machines: Valves of the Heart

1. Isaac Asimov, "The Caves of Steel (1)," *Galaxy Science Fiction* 7, no. 1 (Oct. 1953): 18.
2. Asimov, "The Caves of Steel (2)," *Galaxy Science Fiction* 7, no. 2 (Nov. 1953): 109.
3. Asimov, "The Caves of Steel (2)," 109–10.
4. A number of critics have traced the blackout motif not only to individual *noir* authors' struggles with alcoholism and post-traumatic stress disorder (e.g., Rita Elizabeth Rippetoe, *Booze and the Private Eye: Alcohol in the Hard-Boiled Novel* [Jefferson, NC: McFarland, 2004], 64, 66) but to the more widespread generational trauma of returning veterans, producing "a narrative structure . . . in which the protagonist has to account for a missing period of his life, when he was outside the world in which the film is set, and in which things happened to him which set him at a distance from that world and its inhabitants" (Richard Maltby, "Film Noir: The Politics of the Maladjusted Text," *Journal of American Studies* 18, no. 1 [Apr. 1984]: 63). If Lije Baley represents an identifiable reality, it is perhaps not a generic truth of the human confronting its robotic Other so much as—once or twice refracted through fictions—the dislocated men of 1941–1945 and 1950–1953. Perhaps this places him among other Korean-War-era protagonists of Asimov's fiction (James Gunn, *Isaac Asimov: The Foundations of Science Fiction* [Lanham, Md.: Scarecrow Press, 1996], 81) and science fiction film (Steffen Hantke, *Monsters in the Machine: Science Fiction Film and the Militarization of America after World War II* [Jackson: Univ. Press of Mississippi, 2016], 96–97).
5. Asimov, "The Caves of Steel (1)," 5–6.
6. C. Namwali Serpell, "A Heap of Cliché," in *Critique and Postcritique*, edited by Elizabeth S. Anker and Rita Felski (Durham: Duke University Press, 2017), 161. Reading cliché as mechanical doesn't preclude the recognition that it operates "like a magic formula": "Magical behaviour—if taken idealtypically—is programmed behaviour, not least because it is believed that magical power, *mana*, resides in repetitiveness" (Anton C. Zijderveld, *On Clichés: The Supersedure of Meaning by Function in Modernity* [London; Boston: Routledge and Kegan Paul, 1979], 16, 22). Let it be noted that Zijderveld's progressivist distinction between magical incantation (mere automatism) and religious belief (conscious and conscientious) is subverted in advance by this connection between the magical and the mechanical, the "primitive" and the modern (22–23). Here once again we find a hinge between disenchanted modernity and the world of enchantment—the word "enchantment" itself stemming from another kind of mechanical repetition: chanting—that it had claimed to eliminate forever.
7. "Cliché, n. and Adj.," in *OED Online*, Oxford University Press, accessed 22 Oct. 2020. http://www.oed.com/view/Entry/34264.

8. Herbert L. Gold, "That's Life, Etc.," *Galaxy Science Fiction* 7, no. 2 (November 1953): 4.

9. Gold, "That's Life, Etc.," 4.

10. Kenneth Burke, *Attitudes Toward History*. 3rd ed. (Berkeley: University of California Press, 1984), 140-41n; Gold, "That's Life, Etc.," 4.

11. Burke, *Attitudes Toward History*, 331.

12. Philip K. Dick, "The Electric Ant," *The Magazine of Fantasy and Science Fiction* (October 1969): 101.

13. Dick, "The Electric Ant," 104.

14. Max Stirner. *The Ego and Its Own*, edited by David Leopold, translated by Steven T. Byington (Cambridge: Cambridge University Press, 2000), 43.

15. Stirner, *The Ego and Its Own*, 43; Jacques Lacan, *Ecrits: The First Complete Edition in English*, translated by Bruce Fink (New York: W. W. Norton & Company, 2006), 413-14.

16. Lacan, *Ecrits*, 414-15.

17. Gilles Deleuze and Félix Guattari, *A Thousand Plateaus: Capitalism and Schizophrenia*, translated by Brian Massumi (Minneapolis: University of Minnesota Press, 1987), 71.

18. William Wordsworth, *The Prelude, 1799, 1805, 1850: Authoritative Texts, Context and Reception, Recent Critical Essays*, edited by Jonathan Wordsworth et al. (New York: W. W. Norton, 1979), book 6, 524-28.

19. George Eliot, *Middlemarch: An Authoritative Text, Backgrounds, Criticism*, edited by Bert G. Hornback (New York: W. W. Norton & Company, 2000), 92.

20. Eliot, *Middlemarch*, 40, 12.

21. Eliot, *Middlemarch*, 105.

22. David Graeber, *The Utopia of Rules: On Technology, Stupidity, and the Secret Joys of Bureaucracy* (Brooklyn: Melville House, 2015), 79-80.

23. Julien Offray de la Mettrie. *Man a Machine; And, Man a Plant* (Indianapolis: Hackett Publishing, 1994), 76.

24. Gilles Deleuze and Félix Guattari, *L'Anti-Oedipe: Capitalisme et Schizophrénie* (Paris: Minuit, 1972), 7.

25. Gilles Deleuze, *Spinoza: Practical Philosophy*, translated by Robert Hurley (San Francisco: City Lights Books, 1988), 122; Baruch Spinoza, *Ethics: With the Treatise on the Emendation of the Intellect and Selected Letters*, edited by Seymour Feldman, translated by Samuel Shirley, Hackett Publishing, 1992), 157.

26. Alan M. Turing, "Computing Machinery and Intelligence," *Mind* 59, no. 236 (Oct. 1950): 444.

27. Turing, "Computing Machinery and Intelligence," 442.

28. Turing, "Computing Machinery and Intelligence," 433-34.

29. Judith Halberstam, "Automating Gender: Postmodern Feminism in the Age of the Intelligent Machine," *Feminist Studies*, no. 3 (1991): 443.

30. John R. Searle, "The Case for a Traditional Liberal Education," *The Journal of Blacks in Higher Education*, no. 13 (Autumn 1996): 93-94.

31. John R. Searle, "Minds, Brains, and Programs," *The Behavioral and Brain Sciences*, vol. 3 (1980): 420, 422.

32. Searle, "Minds, Brains, and Programs," 417-18.

33. Long Bui, "Asian Roboticism: Connecting Mechanized Labor to the Automation of Work," *Perspectives on Global Development and Technology* 19, no. 1-2 (March 2020): 111.

34. Mel Chen suggests that something similarly suspicious is at work in the work of Searle's precursor, John L. Austin, when Austin proposes, as a "non-serious" variant of the speech-act that marries two human beings, "a marriage with a monkey" (quoted in Mel Y. Chen, *Animacies: Biopolitics, Racial Mattering, and Queer Affect* [Durham: Duke University Press, 2012], 94): "I read the 'exemplary ridiculousness' of Austin's example as indicating a wider anxiety about the legitimacy of exchange between properly animated figures . . . [combining] intimations of

sexual oddity with racial nonwhiteness and figural blackness" (Chen, *Animacies*, 14). The "intuition pumps" of philosophy tap into deep wells of sociogenic aversion and revulsion.

35. Kathryn Hume, *Fantasy and Mimesis: Responses to Reality in Western Literature* (New York; Routledge, 2014), 83.

36. Martin Heidegger, *The Fundamental Concepts of Metaphysics: World, Finitude, Solitude*, translated by William McNeill and Nicholas Walker (Bloomington, In.: Indiana University Press, 1995), 177.

37. Iwona Janicka, "Who Can Speak? Rancière, Latour and the Question of Articulation," *Humanities* 9, no. 4 (December 2020): 7, emphasis mine.

38. Colson, *Lexicon*, 93.

39. Horst Ruthrof, *Language and Imaginability* (Cambridge: Cambridge Scholars Publishing, 2014), 21; Janicka, "Who Can Speak," 11.

40. Mikhail Aleksandrovich Bakunin, *Œuvres*, vol. 4 (Paris: P-V. Stock, 1895), 269.

41. Edwin Hutchins, *Cognition in the Wild* (Cambridge, Mass.: MIT Press, 1995), 362; N. Katherine Hayles, *How We Became Posthuman: Virtual Bodies in Cybernetics Literature and Informatics* (Chicago: University of Chicago Press, 1999), 289.

42. Karl Marx, in Karl Marx and Friedrich Engels, *Marx Engels Collected Works* vol. 29 (London: Lawrence & Wishart, 1975), 92; Pierre-Joseph Proudhon, *Oeuvres complètes* vol. 11, edited by Célestin Charles Alfred Bouglé and Henri Moysset (Paris: Slatkine, 1982), 261.

43. Searle, "Minds, Brains, and Programs," 419.

44. Slavoj Žižek, *Iraq: The Borrowed Kettle* (New York: Verso, 2005), 9.

45. Gregory Bateson, *Steps to an Ecology of Mind* (New York: Ballantine Books, 1972), 136.

46. Searle, "Minds, Brains, and Programs," 419–20.

47. David J. Chalmers, *The Conscious Mind: In Search of a Fundamental Theory* (New York: OUP USA, 1996), 251.

48. Anatoliy Dneprov, "Igra," *Znanie-Sila*, no. 5 (1961): 39–42; Center for Consciousness Studies, "A Russian Chinese Room Story Antedating Searle's 1980 Discussion," *Center for Consciousness Studies, Philosophy Department of Moscow State University* (blog), June 15, 2018. http://www.hardproblem.ru/en/posts/Events/a-russian-chinese-room-story-antedating-searle-s-1980-discussion/index.php.

49. Ned Block, "Troubles With Functionalism," *Minnesota Studies in the Philosophy of Science* 9 (1978): 279.

50. Chalmers, *The Conscious Mind*, 251.

51. Center for Consciousness Studies, "A Russian Chinese Room Story."

52. John R. Searle, "The Failures of Computationalism," *Think*, no. 2 (June 1993): 68.

53. Daniel C. Dennett, *Consciousness Explained* (Boston: Little, Brown and Co., 1991), 12, 17.

54. Willy Ley, "Homemade Pseudo-Cells," *Galaxy Science Fiction* (Dec. 1953): 52–53.

55. Alfonso L. Herrera, in Henderson James Cleaves et al., eds., *Herrera's 'Plasmogenia' and Other Collected Works: Early Writings on the Experimental Study of the Origin of Life*, translated by Henderson James Cleaves and Fabiola Barraclough (New York: Springer, 2014), 163.

56. Herrera, in Cleaves et al., eds., *Herrera's 'Plasmogenia,'* 79, 9.

57. Jorge Quintana-Navarrete, "Biopolítica y vida inorgánica: La plasmogenia de Alfonso Herrera," *Revista Hispánica Moderna* 72, no. 1, 2019): 81; Herrera qtd. in Quintana-Navarrete, "Biopolítica y vida inorgánica," 88.

58. Ley, "Homemade Pseudo-Cells," 52–53.

59. A chemical recreation still routinely marketed to preteen boys in similar publications a generation later, along with "Sea Monkeys"—those disappointingly inert packets of brine shrimp that altogether failed to demonstrate the intelligent behavior promised between the pages of *Superboy* and *The Amazing Spider-Man*.

60. Ley, "Homemade Pseudo-Cells," 54.

61. Ley, "Homemade Pseudo-Cells," 52; Miriam Allen deFord, "Frustrated Frankenstein: Alfonso Herrera and His Colpoids," *The Magazine of Fantasy and Science Fiction* (Aug. 1954): 55–56, 59; Ley, "Homemade Pseudo-Cells," 53; A. I. Oparin, *The Origin of Life on the Earth*, trans. Ann Synge (New York: Academic Press, 1957), 89.

62. Foucault, *Order of Things*, 19.

63. Sir James George Frazer, *The Golden Bough, Part 1: The Magic Art and the Evolution of Kings* (Macmillan and Co., Ltd., 1921), 51.

64. Mary Shelley, *Frankenstein: The 1818 Text Contexts Criticism*, Second edition, ed. J. Paul Hunter (New York: W. W. Norton, 2012), 22.

65. M. Shelley, *Frankenstein*, 91. It is interesting that even now, efforts to bridge the gap between social and natural histories, to study the "more-than-human sociality" of mushrooms or rivers, are "commonly dismissed as 'mere description'" (Anna Tsing, "More-Than-Human Sociality: A Call for Critical Description," in *Anthropology and Nature*, pp. 37–52 [New York: Routledge, 2013], 27). A scientific aversion to "description," to resemblances and analogies, continues to structure University Discourse.

66. Oparin, *The Origin of Life on the Earth*, 91,

67. Sidney W. Fox and Nicols Fox. *The Emergence of Life* (New York: Basic Books, 1988), 35–36.

68. Fox and Fox, *Emergence of Life*, 35.

69. Alfonso L. Herrera, "La aspiración gigantesca de la plasmogenia," *Nervio* 1, no. 3 (July 1931), 17; Herrera, in Henderson James Cleaves et al., eds., *Herrera's 'Plasmogenia' and Other Collected Works*, 184.

70. DeFord, "Frustrated Frankenstein," 58.

71. Alfonso L. Herrera, "Discurso en honor de Metchnikoff," *Boletín de la Dirección de Estudios Biológicos* 2, no. 67 (1917): 77. If the appeal to "ether" seems impossibly dated to us now—wasn't its existence disproved by the famous Michelson-Morley experiment in 1887?—Massimiliano Badino and Jaume Navarro remind us that theories of ether survived well into the 20th century (Massimiliano Badino and Jaume Navarro, "Introduction: Ether: The Multiple Lives of a Resilient Concept," in *Ether and Modernity: The Recalcitrance of an Epistemic Object in the Early Twentieth Century*, edited by Jaume Navarro [Oxford: Oxford University Press, 2018], 1–2). It has also enjoyed a strange revival as a philosophical concept, as Alexander Weheliye's appeal to the reality of black "flesh" as "the ether that holds together the world of Man while at the same time forming the condition of possibility for this world's demise" (Alexander G. Weheliye, *Habeas Viscus: Racializing Assemblages, Biopolitics, and Black Feminist Theories of the Human* [Durham: Duke University Press, 2014], 40)—in J. Kameron Carter and Sarah Jane Cervenak's formulation, a "black ether" (J. Kameron Carter and Sarah Jane Cervenak, "Black Ether," *CR: The New Centennial Review* 16, no. 2 [2016]: 203–24).

72. Alfonso L. Herrera, "Nueva Teoría Cosmogónica del Proto-Rayo: Una Tempestad en la Nada." *Estudios* 13, no. 141 (May 1935): 27.

73. Fox and Fox, *Emergence of Life*, 35.

74. Herrera, in *Herrera's 'Plasmogenia' and Other Collected Works*, 187.

75. quoted in Comisión Nacional para el Conocimiento y Uso de la Biodiversidad, "Alfonso Luis Herrera, 1868-1942," *Curiosos y Comprometidos*, https://www.biodiversidad.gob.mx/biodiversidad/curiosos/sXIX/AlfonsoLHerrera.php, accessed 9 Oct. 2020.

76. Ley, "Homemade Pseudo-Cells," 52.

77. DeFord, "Frustrated Frankenstein," 55.

78. DeFord, "Frustrated Frankenstein," 55–56, 58, italics mine.

79. DeFord, "Frustrated Frankenstein," 58. Certainly, nothing can be "more dubious" than discourse that reminds scientists that the sciences were once called *natural philosophy*—unless, of course, it is to remind scientists of their magical and alchemical ancestry.

80. Anthony Enns, "Pseudoscience," *The Oxford Handbook of Science Fiction*, edited by Rob Latham (Oxford: Oxford University Press, 2014), 498.

81. Fox and Fox, *Emergence of Life*, 27, 35–36.

82. Alfonso L. Herrera, "La sordera mística de la ciencia," *Estudios* 12, no. 130 (June 1934): 10.

83. Alfonso L. Herrera, "Mi Labor Revolucionaria en la Enseñanza," *Crisol* 73 (Jan. 1935): 58.

84. Luis J. Zalce y Rodríguez, *Apuntes para la historia de la masoneria en México (de mis lecturas y mis recuerdos)* (Distrito Federal, México: Talleres Tipográficos de la Penitenciaría del Distrito Federal, 1950), 466–67.

85. Alfonso L. Herrera, quoted in J. Estour, "Plasmogénie," *Encyclopédie anarchiste*, edited by Sébastien Faure, vol. 4 (La Librairie internationale, 1934), 2061; Moura quoted in Nabylla Fiori de Lima, *Maria Lacerda de Moura Na Revista Estudios (1930–1936): Anarquismo Individualista e Filosofia da Natureza* (Universidade Tecnológica Federal do Paraná, 2016), 143.

86. André Blavier, *Les fous littéraires* (Paris: Editions des Cendres, 2000), 286.

87. Israel Castellanos and Alfonso L. Herrera, *La plasmogenia* (Habana: Imprenta y papelería de Rambla, Bouza y ca., 1921), 123.

88. Konstantin Tsiolkovskii, quoted in Asif A. Siddiqi, "Imagining the Cosmos: Utopians, Mystics, and the Popular Culture of Spaceflight in Revolutionary Russia," *Osiris* 23 (2008): 268.

89. Alfonso L. Herrera, "La aspiración gigantesca de la plasmogenia," 19.

90. Quintana-Navarrete, "Biopolítica y vida inorgánica," 91.

91. Alfonso L. Herrera, "Origen de la sociedad humana: I," *Crisol: revista de critica* 4, no. 20 (1930): 146.

92. Alfonso L. Herrera, "Origen de la sociedad humana: I," 146; "Filosofía Comparada: El Animal y El Salvaje" *Memorias de La Sociedad Científica Antonio Alzate*, no. 9 (1896–1895): 91–94; *La vie sur les hauts plateaux: Influence de la pression barométrique sur la constitution et le développement des êtres organisés. Traitement climatérique de la tuberculose. I* (Escalante, 1899), 23–24, 139, 307; "Origen de la sociedad humana: VI," *Crisol* 6, no. 33 (1930): 228.

93. Nadia Ledesma Prietto, "Apuntes sobre la eugenesia y la libertad sexual en el discurso de dos médicos anarquistas. Argentina, 1930-1940," *Nomadías* no. 16 (2012): 75.

94. Un Medico Rural [Isaac Puente], "Ideales Redentores," *Estudios* 10, no. 101 (1932): 5–6.

95. Jorge Quintana-Navarrete, "Biopolítica y vida inorgánica: La plasmogenia de Alfonso Herrera," *Revista Hispánica Moderna* 72, no. 1 (2019): 94, 92.

96. M. Shelley, *Frankenstein*, 32.

97. Quintana-Navarrete, "Biopolítica y vida inorgánica," 91.

98. M. Shelley, *Frankenstein*, 36

99. Samuel Taylor Coleridge, "Frost at Midnight," in *The Broadview Anthology of Literature of the Revolutionary Period 1770–1832*, edited by D. L. Macdonald and Anne McWhir (Peterborough, Canada: Broadview Press, 2010), 17-20.

100. Coleridge, "Frost at Midnight," 31; Sigmund Freud, *The Standard Edition of the Complete Psychological Works of Sigmund Freud*, translated by James Strachey and Anna Freud, vol. 17 (London: Hogarth Press, 1971), 220, 245.

101. Massimo Canevacci, in Renata Lemos Morais and Massimo Canevacci, "Digital Ubiquity in the Anthropocene: The Non-Anthropocentric Anthropology of Massimo Canevacci," *Antrocom: Online Journal of Anthropology* 12, no. 1 (2016): 7.

102. Immanuel Kant, *Lectures on Ethics*, edited by Peter Heath and J. B. Schneewind, translated by Peter Heath (Cambridge: Cambridge University Press, 2001), 70, 147.

103. Jonathan Culler, *Theory of the Lyric* (Cambridge, Mass.: Harvard University Press, 2015), 16; Culler, "Apostrophe," in *The Pursuit of Signs: Semiotics, Literature, Deconstruction* (Ithaca: Cornell University Press, 2002), 137.

104. William Wordsworth, "The World Is Too Much With Us; Late and Soon," in *The Poems, Volume One* (London: Penguin, 1990), 9–14.

105. Theodor W. Adorno and Max Horkheimer, *Dialectic of Enlightenment*, translated by John Cumming (New York: Verso, 1997), 3; Immanuel Kant, *Groundwork of the Metaphysics of Morals*, translated by Mary Gregor (Cambridge: Cambridge University Press, 1998), 41–42.

106. Charlotte Perkins Stetson, "The Yellow Wall-Paper," *The New England Magazine* 5, no. 5 (January 1892): 648.

107. Colson, *Lexicon*, 186–87.

108. Stetson, "The Yellow Wall-Paper," 649–50.

109. Stetson, "The Yellow Wall-Paper," 651.

110. Stetson, "The Yellow Wall-Paper," 647.

3. Methodological Metafetishism

1. Bruno Latour, "Agency at the Time of the Anthropocene," *New Literary History* 45, no. 1 (April 2014): 7; Eva Marie Garroutte in Eva Marie Garroutte and Kathleen Delores Westcott, "The Story Is a Living Being: Companionship with Stories in Anishinaabe Studies" in *Centering Anishinaabeg Studies: Understanding the World through Stories*, edited by Jill Doerfler et al. (East Lansing, MI: MSU Press, 2013), 68.

2. Arjun Appadurai, "Introduction: Commodities and the Politics of Value" in *The Social Life of Things: Commodities in Cultural Perspective* (Cambridge: Cambridge University Press, 1986), 5.

3. Alejandro de Acosta, "Two Undecidable Questions for Thinking in Which Anything Goes," in *Contemporary Anarchist Studies: An Introductory Anthology of Anarchy in the Academy*, edited by Randall Amster et al. (New York: Routledge, 2009), 32.

4. Alejandro de Acosta, "Two Undecidable Questions for Thinking in Which Anything Goes," in *Contemporary Anarchist Studies: An Introductory Anthology of Anarchy in the Academy*, edited by Randall Amster et al. (New York: Routledge, 2009), 32.

5. Robinson, *Ministry for the Future*, 538.

6. Kim Stanley Robinson, *The Ministry for the Future* (New York: Orbit Books, 2021), 60.

7. Joshua Clover, *The Matrix* (London: BFI, 2004), 71.

8. Slavoj Žižek, "Cyberspace, or the Unbearable Closure of Being," in *Endless Night: Cinema and Psychoanalysis, Parallel Histories*, edited by Janet Bergstrom (University of California Press, 1999), 107.

9. Theodor W. Adorno, *Minima Moralia: Reflections from Damaged Life*, translated by E. F. N. Jephcott (New York: Verso, 1999), 197.

10. Bruno Latour and T. Hugh Crawford, "An Interview with Bruno Latour," *Configurations* 1, no. 2 (May 1993), 254–55, 259.

11. Arjun Jayadev, and Samuel Bowles, "Guard Labor," *Journal of Development Economics* 79, no. 2 (April 2006), 337.

12. Mike Michael, "The Power-Persuasion-Identity Nexus: Anarchism and Actor Networks," *Anarchist Studies*, 2, no. 1 (Spring 1994): 27.

13. Gustav Landauer quoted in Martin Buber, *Paths in Utopia* (Syracuse, NY: Syracuse University Press, 1996), 46.

14. Michael, "The Power-Persuasion-Identity Nexus," 30–32.

15. Bakunin, *God and the State* (New York: Dover, 1970) 32.

16. Michael, "The Power-Persuasion-Identity Nexus," 40. Or perhaps the question (and its answer) are merely paranoid, in Eve Kosofsky Sedgwick's sense—that is, premised on a "disavowal" of anything that might seem too naive, too utopian, too hopeful (Eve Kosofsky Sedgwick, "Paranoid Reading and Reparative Reading, or, You're So Paranoid, You Probably

Think This Essay Is About You," in *Touching Feeling: Affect, Performativity, Pedagogy* [Durham: Duke University Press, 2003], 144).

17. Michael, "The Power-Persuasion-Identity Nexus," 32.

18. How we respond to the experience of betrayal is crucial. I wonder if Latour's diffidence towards any kind of left-wing politics isn't the effect of any political centrism that would be inherent to his theory but a rejection of the sort of reflexive leftism that left French intellectuals allied with the Communist Party in such humiliating positions after the crushing of the Hungarian revolt of 1956 or the Prague Spring in 1968. In the 1993 interview, Latour remarks that after the fall of the Berlin Wall, "we cannot believe a word of socialism," but reflects that "I might be biased by our recent French political history" (Bruno Latour and T. Hugh Crawford, "An Interview with Bruno Latour," *Configurations* 1, no. 2 [1993]: 260, 254). Perhaps Latour himself is a paranoid reader?

19. Robert Damien, "Transport ferroviaire et ordre politique: Proudhon, une pensée philosophique des réseaux?" in *Penser les réseaux*, edited by Daniel Parrochia (Seyssel, France: Champ Vallon, 2001), 219–20.

20. Pierre-Joseph Proudhon quoted in Damien, "Transport ferroviaire et ordre politique," 226–27.

21. Damien, "Transport ferroviaire et ordre politique," 228.

22. Colin Ward, *Anarchy in Action* (London: George Allen & Unwin, 1973), 26; Todd May, *The Political Philosophy of Poststructuralist Anarchism* (University Park, Pa.: Pennsylvania State U.P., 1994), 52.

23. Jacques Derrida, *Writing and Difference*, translated by Alan Bass (Chicago: University of Chicago Press, 1978), 279; Proudhon quoted in Damien, "Transport ferroviaire et ordre politique," 227.

24. Eduardo Colombo, "À propos du *Petit lexique philosophique de l'anarchisme* de Daniel Colson," *Réfractions* no. 8 (Printemps-été 2002): 130.

25. Colombo, "À propos du *Petit lexique*," 130–31.

26. John D. Rockefeller quoted in Richard Hofstadter, *Social Darwinism in American Thought: 1860–1915* (Philadelphia: University of Pennsylvania Press, 1944), 31.

27. Colombo, "À propos du *Petit lexique*," 140.

28. Michel Foucault quoted in Colson, *Lexicon*, 32.

29. Daniel Colson, "Réponse de Daniel Colson à Eduardo Colombo," *Réfractions* no. 8 (Printemps-été 2002): 143–44.

30. Jerome Groopman, "Can Brain Science Help Us Break Bad Habits?" *The New Yorker*, Oct. 2019, https://www.newyorker.com/magazine/2019/10/28/can-brain-science-help-us-break-bad-habits.

31. Pierre-Joseph Proudhon, *System of Economical Contradictions: Or, The Philosophy of Misery*, translated by Benjamin R. Tucker (Boston: Benjamin R. Tucker, 1888), 198.

32. Karl Marx, *Grundrisse: Foundations of the Critique of Political Economy* (New York: Vintage Books, 1973), 702.

33. Karl Marx and Friedrich Engels, *Marx Engels Collected Works* vol. 35 (London: Lawrence & Wishart, 1975), 81.

34. Pierre-Joseph Proudhon, *Oeuvres complètes* vol. 10, edited by Célestin Charles Alfred Bouglé and Henri Moysset (Paris: Slatkine, 1982), 366.

35. China Miéville, "Marxism and Fantasy: Editorial Introduction," *Historical Materialism* 10, no. 4 (2002): 42, emphasis mine.

36. Karl Marx, in *Marx Engels Collected Works* vol. 35 (London: Lawrence & Wishart, 1975), 81–83.

37. David Graeber, "Fetishism as Social Creativity: Or, Fetishes Are Gods in the Process of Construction," *Anthropological Theory* 5, no. 4 (Dec. 2005), 411.

38. William Pietz, "The Problem of the Fetish, II: The Origin of the Fetish," *RES: Journal of Anthropology and Aesthetics* no. 13 (Mar. 1987): 30–31.

39. Andrew Feenberg, *Lukács, Marx, and the Sources of Critical Theory* (New York: Oxford University Press, 1986), 61.

40. Jackson, *Becoming Human*, 111, 113–14.

41. Jackson, *Becoming Human*, 34.

42. Kevin Young, *The Grey Album: On the Blackness of Blackness* (Minneapolis: Graywolf Press, 2012), 36–37, emphasis mine.

43. Graeber, "Fetishism as Social Creativity," 424.

44. David Graeber, "Radical Alterity Is Just Another Way of Saying 'Reality': A Reply to Eduardo Viveiros de Castro," *HAU: Journal of Ethnographic Theory* 5, no. 2 (Sept. 2015): 3–4.

45. Angela M. Haas, "Wampum as Hypertext: An American Indian Intellectual Tradition of Multimedia Theory and Practice," *Studies in American Indian Literatures* 19, no. 4 (2007): 78–80; Day and Lewis, "Radical Subjectivity and the *N*-Row Wampum," 172–73.

46. Ruth Kinna, Alex Prichard, and Thomas Swann, "Occupy and the Constitution of Anarchy," *Global Constitutionalism* 8, no. 2 (July 2019): 385.

47. Gerald R. Alfred and Taiaiake Alfred, *Peace, Power, Righteousness: An Indigenous Manifesto* (Oxford: Oxford University Press, 2009), 76; Penelope Myrtle Kelsey, *Reading the Wampum: Essays on Hodinöhsö:ni' Visual Code and Epistemological Recovery* (Syracuse, NY: Syracuse University Press, 2014), 2.

48. Alfred and Alfred, *Peace, Power, Righteousness*, 76; Day and Lewis, "Radical Subjectivity and the *N*-Row Wampum," 186.

49. Abbey Lolcano, "I Teach Social Movements at My Job . . ." *Facebook*, 2 June 2020, https://www.facebook.com/abbeyvolcano/posts/10221550293817910.

50. Karl Marx, in Karl Marx and Friedrich Engels, *Marx Engels Collected Works* vol. 35, 83.

51. Karl Marx, in Karl Marx and Friedrich Engels, *Marx Engels Collected Works* vol. 35, 163–64, emphasis mine.

52. China Miéville, *Perdido Street Station* (New York: Del Rey Books, 2000), 391.

53. Miéville, *Perdido Street Station*, 392.

54. Miéville, *Perdido Street Station*, 392, 396.

55. Hugh Dalziel Duncan in Kenneth Burke, *Permanence and Change: An Anatomy of Purpose* (Indianapolis: Bobbs-Merrill Educational Publishing, 1965), xxviii–xxix.

56. Colson, *Lexicon*, 163–64.

57. Philip Winn, editor, *Dictionary of Biological Psychology* (New York: Routledge, 2003), 1612.

58. Mark Bracher, *Lacan, Discourse, and Social Change: A Psychoanalytic Cultural Criticism* (Ithaca: Cornell University Press, 2018), 55–57.

59. Bracher, *Lacan, Discourse, and Social Change*, 58-59; Immanuel Kant, *Political Writings*, edited by Hans Reiss, translated by H. B. Nisbet (Cambridge: Cambridge University Press, 1991), 55, 59.

60. Latour, "On Interobjectivity." Mind, Culture, and Activity, vol. 3, no. 4, Routledge, Oct. 1996, pp. 228–45; Shaviro, *Universe*, 48.

61. Jay Driskell, "A Strong Statement, but It Doesn't Go Nearly Far Enough . . ." *Facebook*, July 15, 2020. https://www.facebook.com/jay.driskell/posts/pfbid024x3pNiXRtjXATP1SizXFx PVYEM3Aj2396nn3pCbjv7G5gWrq5dYLz6x7rYBTgLDQl.

62. John M. Barry, "The Pandemic Could Get Much, Much Worse. We Must Act Now," *The New York Times*, July 14, 2020, sec. Opinion, https://www.nytimes.com/2020/07/14/opinion/coronavirus-shutdown.html.

63. John M. Barry, "The Pandemic Could Get Much, Much Worse"

64. Kant, *Political Writings*, 55.

65. Michel Serres, *The Natural Contract* (Ann Arbor: University of Michigan Press, 1995).
66. Latour, "Agency at the Time of the Anthropocene," 18.
67. Latour, "Agency at the Time of the Anthropocene," 13.

4. Unbarring the Other

1. Rudy Rucker, "Alien Contact," in Stephen Leigh, *Alien Tongue* (New York: Bantam Spectra, 1991), 301–2.
2. Jeannette Ng, *Under the Pendulum Sun: A Novel of the Fae* (Nottingham, UK: Angry Robot 2017), 10.
3. René Descartes, *Discourse on Method and Meditations on First Philosophy*, translated by Donald A. Cress, Fourth ed. (Indianapolis: Hackett Publishing, 1999), 22.
4. Ng, *Under the Pendulum Sun*, 180–81.
5. Brianna DaSilva and Jeannette Ng, *Othering in Fantasy with Jeannette Ng*, 2019. soundcloud.com, https://soundcloud.com/femalesinfantasy/othering-in-fantasy-with-jeannette-ng.
6. William Dean, *The China Mission: Embracing a History of the Various Missions of All Denominations Among the Chinese* (New York: Sheldon, 1859).
7. Ng, *Under the Pendulum Sun*, 91.
8. DaSilva and Ng, *Othering in Fantasy with Jeannette Ng*.
9. Patrick Parrinder, "The Alien Encounter: Or, Ms Brown and Mrs Le Guin," *Science Fiction Studies* 6, no. 1 (1979), 52.
10. Stanisław Lem, *His Master's Voice* (San Diego: Harcourt Brace Jovanovich, 1983), 22.
11. Jackson, *Becoming Human*, 128–29.
12. Liu Cixin, *The Dark Forest* (New York: Tom Doherty Associates, 2015), 9–16.
13. Liu Cixin, *The Three-Body Problem* (New YorkNew York: Tom Doherty Associates, 2014), 329.
14. Liu, *Three-Body Problem*, 383.
15. Liu, *Dark Forest*, 484.
16. Ursula K. Le Guin, "'The Author of the Acacia Seeds' and Other Extracts from the Journal of the Association of Therolinguistics," in *Fellowship of the Stars: Nine Science Fiction Stories*, edited by Terry Carr (New York: Simon and Schuster, 1974), 170–71.
17. Le Guin, "The Author of the Acacia Seeds," 171.
18. Peter Davis, *Hearts and Minds*, Warner Bros., 1974.
19. Sylvia Wynter, "Unsettling the Coloniality of Being/Power/Truth/Freedom: Towards the Human, after Man, Its Overrepresentation—An Argument," *CR: The New Centennial Review* 3, no. 3 (2003): 260.
20. Ryner, *The Human Ant*, translated by Brian Stableford (Tarzana, CA: Black Coat Press, 2014), 25–26.
21. Ryner, *The Human Ant*, 42–44.
22. Thomas Nagel, "What Is It Like to Be a Bat?" *The Philosophical Review* 83, no. 4 (1974): 438.
23. Kai Heron, "[Review] Lacan and Deleuze: A Disjunctive Synthesis," *Psychoanalysis, Culture & Society* 24, no. 2 (June 2019), 230.
24. Ted Chiang, *Stories of Your Life and Others* (New York: Vintage Books, 2016), 140. The Sapir-Whorf Hypothesis has itself come under increasing criticism, as studies show that the underlying "hardware" of the human brain enables shared perceptions of color spectra, for instance, despite profound differences in the way those spectra are encoded by human languages. Nonetheless, the data seem to support a "weak" version of the hypothesis in which language *focuses* perception, attention, memory, etc., without completely determining them. Like many other obsolete bits of science, however, "strong" Whorfianism makes for excellent science fiction.

25. Le Guin, "The Author of the Acacia Seeds," 173–78.

26. Le Guin, "The Author of the Acacia Seeds," 178.

27. Chalmers, *The Conscious Mind*, 251; Sara Marielle Gaup Beaska, *Gulahallat Eatnamiin / WE SPEAK EARTH*, 2015, https://www.youtube.com/watch?v=VzBmDpDAskQ.

28. Jemisin, *Stone Sky*, 47, 44.

29. Clayton Koelb, "The Language of Presence in Varley's 'The Persistence of Vision,'" *Science Fiction Studies* 11, no. 2 (1984), 154; Jemisin, *Stone Sky*, 47–48.

30. Slavoj Žižek, *Absolute Recoil: Towards A New Foundation Of Dialectical Materialism* (New York: Verso, 2014), 12.

31. Žižek, *Absolute Recoil*, 12.

32. Bakunin, *God and the State*, 12.

33. Pierre-Joseph Proudhon, *Oeuvres*, vol. 8, part 3, 267; Svyatogor quoted in Ekaterina Lebadeva, "Svyatogor about Volcanism as a True Art," *Constructivists, Biocosmists, and the New Human: Technology and Philosophy in Petrograd and Leningrad* (28 May 2020) techpeterburg.wixsite.com, https://techpeterburg.wixsite.com/mysite/blog/categories/cosmism/, modifications mine; Brat'ia Gordiny [A. L. Gordin and V. L. Gordin], *Anarkhiia v Mechte: Strana Anarkhiia: Utopiia-poema* [*Anarchy in a Dream: The Land of Anarchy: Utopia-Poem*] (Moskva: Izd. Pervogo tsentral'nogo sotsiotekhnikuma, Vremennyi tekhnikum propagandy i agitatsii, 1919), 186; Adrián del Valle, "The World as a Plurality," *Inter-América* 8, no. 2 (Dec. 1924), pp. 99–113.

34. Tracy McNulty, "Speculative Fetishism," *Konturen* 8, no. 0 (October 9, 2015): 99; Colson, *Lexicon*, 133–34.

35. Colson, *Lexicon*, 133.

36. Colson, *Lexicon*, 165.

37. David Graeber, "Consumption," *Current Anthropology* 52, no. 4 (Aug. 2011), 494.

38. Eduardo Colombo, *L'Espace politique de l'anarchie: Esquisses pour une philosophie politique de l'anarchisme* (Lyon: Atelier de création libertaire, 2008), 145; Félix Martí Ibáñez, *Psicoanálisis de La Revolución Social Española* (Barcelona: Tierra y Libertad, 1937), 28, 36.

39. P. Djèli Clark, *Ring Shout, or, Hunting Ku Kluxes In the End Times* (New York: Tom Doherty Associates, 2020).

40. Madhu Dubey, "Becoming Animal in Black Women's Science Fiction," in *Afro-Future Females: Black Writers Chart Science Fiction's Newest New-Wave Trajectory*, edited by Marleen S. Barr (Ohio State University Press, 2008), 34.

41. (N. K. Jemisin in Ezra Klein and N. K. Jemisin, "Transcript: Ezra Klein Interviews N. K. Jemisin," *The New York Times*, October 18, 2022, sec. Podcasts, https://www.nytimes.com/2022/10/18/podcasts/ezra-klein-interviews-nk-jemisin.html.

42. Stephen Hawking quoted in Ki Mae Heussner, "Stephen Hawking: Alien Contact Could Be Risky," *ABC News*, accessed 26 July 2020, https://abcnews.go.com/Technology/Space/stephen-hawking-alien-contact-risky/story?id=10478157.

43. Christopher L. Robinson, "Teratonymy: The Weird and Monstrous Names of H. P. Lovecraft," *Names* 58, no. 3 (2010): 132; H. P. Lovecraft quoted in Robinson, "Teratonymy," 128.

44. Lovecraft quoted in Robinson, "Teratonymy," 128.

45. Graham Harman, *Weird Realism: Lovecraft and Philosophy* (Winchester, UK: Zero Books, 2012), 4.

46. Some of Lovecraft's progressive readers resist such attempts to dissolve the negativity of his vision. Benjamin Noys, for instance, dismisses Grant Morrison's "Lovecraft In Heaven" (in which it is suggested that perhaps chaos and bodily intermingling are not bad but good things) as a well-intentioned bowdlerization; "The body that transforms itself is the body that conforms to the chaos of market capitalism, which demands a malleable body open to new desires and new patterns of consumption" (15-- must be feared and defended against: is this not precisely what Eve Kosofsky Sedgwick called the "paranoid" style of theorizing?

47. Leonid Andrenko, "Les Habitants des Astres," *L'En dehors* 16, no. 284–85 (juillet-août 1935): 17–18.

48. Leonid Andrenko, "La Diversidad del Principio Vital en el Universo," *Estudios*, vol. 7, no. 65 (Enero 1929): 56-57.

49. Andrenko, "La Diversidad del Principio Vital en el Universo," 57-58.

50. Leonid Andrenko, "Les habitants des astres (fin)," *L'En dehors* 16, no. 286 (Sept. 1935): 34.

51. Difficult as it is to believe, Andrenko seems never to have been persecuted for this flirtation with a forbidden ideology; rather, in February of 1931, he was arrested by the OGPU, charged with "participating in a counterrevolutionary group," the Russian Society of Enthusiasts for the Study of the World (Russkoe Obshchestvo Liubitelei Mirovedeniia, or ROLM) and sentenced to 3 years of internal exile (Natsional'na Akademiya 25n1; Siddiqi, *Red Rockets' Glare* 31). After spending most of the year in Irkutsk, Siberia, he was released on the request of his mentor, Nikolai Morozov, but apparently was still subjected to "humiliating persecution from fellow astronomers for his own 'scientific views' and 'original inventions,' to the point of attempts to completely ban his activities as pseudoscientific" (Natsional'na Akademiya 25n1). Following a mysterious period during World War II—as one historian remarks, "the data on this period are contradictory" (Shkuratov 222n373), but the Andrenkos may have been captured by the Nazis after the fall of Odessa in 1941, passing then through work assignments at observatories in Cracow and Sigmaringen (Walter 272; Natsional'na Akademiya 25–26n1)—the Andrenkos emigrated to the United States as Displaced Persons in 1951.

52. Walt Whitman, "Song of Myself," *Leaves of Grass: Authoritative Texts, Prefaces, Whitman on His Art, Criticism*, edited by Sculley Bradley and Harold William Blodgett (New York: Norton, 1973), 1325, 568.

53. Daniel Colson, *Trois essais*, 42–43.

54. Peter Kropotkin, "Anarchist Morality (Continued)," *Mother Earth* 11, no. 10 (Dec. 1916): 716; see also Peter Kropotkin, *Mutual Aid: A Factor of Evolution* (New York: McClure, Philips & Co., 1902).

55. Murray Bookchin, *The Ecology of Freedom: The Emergence and Dissolution of Hierarchy* (Montréal: Black Rose Books, 1991), 232.

5. Notes toward a Rhetoric of the Fantastic

1. James O. Bailey, *Pilgrims Through Space and Time: Trends and Patterns in Scientific and Utopian Fiction* (Westport, Conn.: Greenwood Press, 1977), 203.

2. Ray Lynn Anderson, "Persuasive Functions of Science Fiction: A Study in the Rhetoric of Science," Dissertation, University of Minnesota (1968), 304.

3. Aristotle, *Aristotle: The Art of Rhetoric* (Cambridge, Mass.: Harvard University Press, 1967), 168; John Poulakos, "Toward a Sophistic Definition of Rhetoric," Philosophy & Rhetoric 16, no. 1 1983): 43. Christine Brooke-Rose's *Rhetoric of the Unreal* and Farah Mendlesohn's *Rhetorics of Fantasy* serve as an inspiration here, although neither really attempts the kind of catalog of persuasive means (however incomplete) that I'm trying to construct.

4. Darko Suvin, *Metamorphoses of Science Fiction: On the Poetics and History of a Literary Genre* (New Haven: Yale University Press, 1979), 4, 64.

5. Arrianna Planey, "Thoughts on Margaret Atwood's 'A Handmaid's Tale': Part 2," *Arrianna Planey's Blog* (blog), December 30, 2013, https://arriannaplaney.wordpress.com/2013/12/30/thoughts-on-margaret-atwoods-a-handmaids-tale-part-2/; Suvin, *Metamorphoses*, 71.

6. Freud, *Standard Edition* vol. 17, 245.

7. Aaron Worth, "Introduction," in Arthur Machen, *The Great God Pan and Other Horror Stories* (Oxford: Oxford University Press, 2018), xiv.

8. Jemisin, *The Fifth Season*, 7.

9. Walidah Imarisha, "Walidah Imarisha on Visionary Sci Fi," presented at the Are the Gods Afraid of Black Sexuality? Conference, Columbia University, March 3, 2015, https://www.youtube.com/watch?v=iYnR9_2UgLE.

10. Frank L. Cioffi, *Formula Fiction?: An Anatomy of American Science Fiction, 1930-1940* (Westport, Conn.: Greenwood Press, 1982), 128; Farah Mendlesohn, *Rhetorics of Fantasy* (Middletown, Conn.: Wesleyan University Press, 2008), xiv.

11. Christine Brooke-Rose, *A Rhetoric of the Unreal: Studies in Narrative and Structure, Especially of the Fantastic* (Cambridge: Cambridge University Press, 1988), 234.

12. Barthes, *Mythologies*, 10.

13. Damien Broderick, *Reading by Starlight: Postmodern Science Fiction* (New York: Routledge, 2005), 85. Another example: the title sequence for the HBO series *True Blood* (2008–2014), set in contemporary rural Louisiana, the premise of which is that vampires are real and have "come out of the coffin," revealing themselves to humanity and demanding rights as one more minority among others. This sequence splices together bits of documentary or pseudo-documentary footage signifying "the South"—ministers preaching, racist police and citizens beating Civil Rights demonstrators in the 1960s, grainy footage of shotgun shacks and lewd barroom dances—with just one major counterfactual inserted: a shot of a church marquee bearing the words GOD HATES FANGS. The factual and the counterfactual blend smoothly into one montage, placing the outlandish scenario in continuity with what viewers believe they know about "the South," its cultural and historical contradictions.

14. Darcie Little Badger, "Nkásht Íí," *Strange Horizons* (December 15, 2014), http://strangehorizons.com/fiction/nksht/.

15. Mircea Eliade, *The Sacred and the Profane: The Nature of Religion* (New York: Houghton Mifflin Harcourt, 1959), 9.

16. Linnemann, *The Horror of Police*, 16.

17. Comey quoted in Linneman, *The Horror of Police*, 166.

18. Susan Stryker, "Performing Transgender Rage: My Words to Victor Frankenstein Above the Village of Chamounix," *GLQ* 1, no. 3 (1994): 238; Mary Daly, *Gyn/Ecology: The Metaethics of Radical Feminism* (Boston: Beacon Press, 1978), 71.

19. Carroll, *Philosophy of Horror*, 100–101.

20. Octavio Paz, *Sunstone = Piedra de sol*, translated by Eliot Weinberger (New York: New Directions Pub. Corp., 1991), 37; Stanisław Lem, *Microworlds*, edited by Franz Rottensteiner (New York: Harcourt Brace Jovanovich, 1986), 34.

21. Caesar Grant, "All God's Chillen Had Wings," in *The Book of Negro Folklore*, edited by Langston Hughes and Arna Wendell Bontemps (New York: Dodd, Mead, 1958), 62–63.

22. Grant, "All God's Chillen," 64.

23. Nnedi Okorafor, *The Book of Phoenix* (New York: DAW Books, 2017), 63–64.

24. Chu, *Do Metaphors Dream of Literal Sleep*, 68.

25. Chu, *Do Metaphors Dream of Literal Sleep*, 3.

26. Bailey, *Pilgrims Through Space and Time*, 203.

27. H. G. Wells, *The Complete Science Fiction Treasury of H. G. Wells* (New York: Crown Publishers, 1978), iii.

28. Ben Wetherbee thinks of this in terms of a *topos* or "common place" of dystopian film rhetoric, which he names "the *technological swamp*, or the expansive bricolage of high-tech detail, excess, and waste that implies decay and neglect" in contrast to an opposing *topos*, "the *clinic*, or the overlit surveillance site that implies oppressive, panoptic oversight" (Ben Wetherbee, "Dystopoi of Memory and Invention: The Rhetorical 'Places' of Postmodern Dystopian Film," *Journal of Multimodal Rhetorics* 2, no. 2 [Fall 2018]: 117). While this schema of "*dystopoi*"

or "*bad places*" (Wetherbee, "Dystopoi," 124) brilliantly explicates a particular cluster of genre films, I want to place this analysis of the way the known and new function within the fantastic on a broader basis.

29. Brooke-Rose, *A Rhetoric of the Unreal*, 234.

30. Margaret Scammell, *Consumer Democracy: The Marketing of Politics* (Cambridge: Cambridge University Press, 2014), 158.

31. Barthes, *Mythologies*, 41.

32. Fisher, *Capitalist Realism*, 2.

33. John W. Campbell, Jr., "The Science of Science-Fiction Writing," in *Of Worlds Beyond: The Science of Science-Fiction Writing*, edited by Lloyd Arthur Eshbach (Reading, Penn.: Fantasy Press, 1947), 86.

34. Isaac Asimov, "Introduction," in *More Soviet Science Fiction* (New York: Collier Books, 1962), 8, 10; Wells, *Complete Science Fiction Treasury*, iv.

35. James O'Toole, "Santorum: No Apology," *Post-Gazette*. April 23, 2003, https://old.post-gazette.com/nation/20030423santorum0423p1.asp.

36. Joanna Russ, *To Write Like a Woman: Essays in Feminism and Science Fiction* (Bloomington: Indiana University Press, 1995), 18.

37. Rieder, *Colonialism and the Emergence of Science Fiction*, 69

38. Anderson, "Persuasive Functions of Science Fiction," 279.

39. Jean-Paul Sartre quoted in Peter Brooks, *Reading for the Plot: Design and Intention in Narrative* (Cambridge, Mass.: Harvard University Press, 1992), 93.

40. Émile Pataud and Émile Pouget, *Syndicalism and the Co-Operative Commonwealth: How We Shall Bring about the Revolution*, translated by Charlotte Charles and Frédéric Charles (Oxford: The New International Publishing Company, 1913), xviii.

41. Puget Sound Anarchists, "A Letter of Solidarity From the Year 3017," November 30, 2017, https://pugetsoundanarchists.org/a-letter-of-solidarity-from-the-year-3017/.

42. Farah Mendlesohn, *Rhetorics of Fantasy* (Middletown, Conn.: Wesleyan University Press, 2008), 1.

43. Damion Kareem Scott, "Afrofuturism and Black Futurism: Some Ontological and Semantic Considerations," in *Critical Black Futures: Speculative Theories and Explorations*, edited by Philip Butler (Singapore: Springer, 2021), 159.

44. Arthur C. Clarke, *Profiles of the Future* (New York: Harper & Row, 1973), 39n1.

45. Anderson, "Persuasive Functions of Science Fiction," 302.

46. Joshua Clover, "Retcon: Value and Temporality in Poetics." *Representations* 126, no. 1 (2014): 9–30.

47. Brian Stableford, "Why There Is (Almost) No Such Thing as Science Fiction: Observations on Rhetoric and Plausibility in Science and Science Fiction," in *Narrative Strategies in Science Fiction and Other Essays on Imaginative Fiction* (Rockville, MD: Borgo Press/Wildside Press, 2010), 10–11; Chu, *Do Metaphors Dream of Literal Sleep*, 68.

48. W. R. Irwin, *The Game of the Impossible: A Rhetoric of Fantasy* (Urbana, Ill.: University of Illinois Press, 1976), 69; Franz Kafka, *The Metamorphosis, in the Penal Colony and Other Stories: The Great Short Works of Franz Kafka*, translated by Joachim Neugroschel (New York: Simon and Schuster, 2010), 117.

49. Jonathan Chait, "'We Do That All the Time, Get Over It.' Mulvaney Boasts About Ukraine Plot," *New York Magazine* (Oct. 2019), https://nymag.com/intelligencer/2019/10/mulvaney-ukraine-get-over-it.html.

50. Broderick, *Reading By Starlight*, xi.

51. Latour, *Science In Action*, 80.

52. Anderson, "Persuasive Functions of Science Fiction," 296–97.

53. David Roth, "The Man Who Was Upset: Making Sense of Donald Trump's Petulant Reign," *The New Republic* (June 2019), https://newrepublic.com/article/154100/making-sense-donald-trump-petulant-presidency; Vivien A. Schmidt, "Britain-out and Trump-in: A Discursive Institutionalist Analysis of the British Referendum on the EU and the US Presidential Election," *Review of International Political Economy* 24, no. 2, (Mar. 2017): 261; Anne Applebaum, "You're Not Supposed to Understand the Rumors About Biden," *The Atlantic* (Oct. 2020), https://www.theatlantic.com/ideas/archive/2020/10/smears-against-biden-dont-need-make-any-sense/616824/.

54. Kevin Young, *Bunk: The Rise of Hoaxes, Humbug, Plagiarists, Phonies, Post-Facts, and Fake News* (Minneapolis: Graywolf Press, 2018), 22–23.

55. Edward James, *Science Fiction in the Twentieth Century* (Oxford: Oxford University Press, 1994), 148.

56. Dick, "The Electric Ant," 101.

57. Stanisław Lem, *The Futurological Congress (from the Memoirs of Ijon Tichy)*, translated by Michael Kandel (San Diego: Harcourt Brace Jovanovich, 1985), 109.

58. Dick, "The Electric Ant," 100–101, 107.

59. Dick, "The Electric Ant," 108.

60. Marc Angenot, "The Absent Paradigm: An Introduction to the Semiotics of Science Fiction," *Science Fiction Studies* 6, no. 1 (1979): 15; Dick, "The Electric Ant," 101.

61. Frank Herbert, *Dune* (New York: Penguin, 2005), 8.

62. Godwin, "Cold Equations," 63.

6. Rhetorics of the Impossible

1. Umberto Eco, *The Role of the Reader: Explorations in the Semiotics of Texts* (Bloomington, Ind.: Indiana University Press, 1984), 110–11

2. Wolfgang Max Faust and R. Baird Shuman, "Comics and How to Read Them," *The Journal of Popular Culture* 5, no. 1 (1971), 198.

3. Gerard Hynes, "Locations and Borders," in *The Routledge Companion to Imaginary Worlds*, edited by Mark J. P. Wolf (New York: Routledge, 2017), 4; Andrew J. Friedenthal, *The World of DC Comics* (New York: Routledge, Taylor & Francis Group, 2019), n.p.

4. Brooks, *Reading for the Plot*, 103; David Graeber, *The Utopia of Rules: On Technology, Stupidity, and the Secret Joys of Bureaucracy* (Brooklyn: Melville House, 2015), 211.

5. Richard P. Feynman, *QED: The Strange Theory of Light and Matter* (Princeton, NJ: Princeton University Press, 2014), 8.

6. Samuel R. Delany, *The American Shore: Meditations on a Tale of Science Fiction by Thomas M. Disch—"Angouleme"* (Middletown, Conn.: Wesleyan University Press, 2014), xvi–xvii.

7. Samuel R. Delany, *The Jewel-Hinged Jaw: Notes on the Language of Science Fiction* (Middletown, Conn.: Wesleyan University Press, 2011), 11.

8. Dean Spade, *Normal Life: Administrative Violence, Critical Trans Politics, and the Limits of Law* (Durham: Duke University Press, 2015), 41.

9. Rosemary Jackson, *Fantasy: The Literature of Subversion* (New York: Routledge, 2013), 26.

10. Adorno and Horkheimer, *Dialectic of Enlightenment*, 258; John Barth, *Lost in the Funhouse: Fiction for Print, Tape, Live Voice* (New York: Anchor Press, 1988), 115.

11. Hume, *Fantasy and Mimesis*, 91.

12. Jackson, *Fantasy*, 78, italics mine.

13. Ursula K. Le Guin, *The Dispossessed* (New York: Harper, 2003), 28–29.

14. Franz Kafka, *The Complete Stories*, edited by Nahum N. Glatzer (New York: Schocken Books, 1995), 5. Just the opposite, as we shall see, is the shrinking space of the Gordin Brothers' Russian utopia *Anarkhiia v Mechte: Strana Anarkhiia* [*Anarchy In a Dream: The Land of Anarchy*, 1919]: as our five protagonists join one another on the road to freedom, "oddly enough, the path became even easier, the road underfoot seemed to curl up, shorten, and space began to lose its length!" (*Strana* 87). Paradoxically, the logic of contracting spaces, like the elegant pinch of Madeleine L'Engle's "tesseract" or the Alcubierre Drive, often seems to be associated with a Romantic logic of escape, whereas the expanding space of the "Five and a Half Minute Hallway" in *House of Leaves*, like the extra dimension opening behind Gilman's yellow wallpaper or the hidden spaces of the Court in Kafka's *Trial*, seems to embody a Gothic logic of confinement and trauma (Alan Gibbs, *Contemporary American Trauma Narratives* [Edinburgh: Edinburgh University Press, 2014], 106–7).

15. J. G. Ballard, "Build-Up," *New Worlds Science Fiction* 19, no. 55 (Jan. 1957), 70.

16. Ballard, "Build-Up," 64–65.

17. Ballard, "Build-Up," 70.

18. Josephine Saxton, "The Wall," *Science Fantasy* 24, no. 78 (Nov. 1965), 76–77.

19. Saxton, "The Wall," 72, 75.

20. Saxton, "The Wall," 78.

21. Freud, *Standard Edition* vol 17, 101; Joan Copjec, *Supposing the Subject* (New York: Verso, 1994), 21.

22. Ng, *Under the Pendulum Sun*, 114.

23. Ng, *Under the Pendulum Sun*, 118, italics mine.

24. W. H. Auden, *The Dyer's Hand and Other Essays* (New York: Random House, 1962), 222.

25. Slavoj Žižek in John Milbank, Slavoj Žižek, and Creston Davis, *Paul's New Moment: Continental Philosophy and the Future of Christian Theology* (Grand Rapids, Mich: Brazos), 180.

26. Melinda Hall, "Horrible Heroes: Liberating Alternative Visions of Disability in Horror," *Disability Studies Quarterly* 36, no. 1 (Mar. 2016): n.p.

27. Carroll, *The Philosophy of Horror*, 184.

28. Howard P. Lovecraft, *The Annotated H. P. Lovecraft*, edited by Sunand T. Joshi (New York: Dell, 1997), 323.

29. Lisa Yaszek, "The Domestic SF Parabola," in *Parabolas of Science Fiction*, edited by Brian Attebery and Veronica Hollinger (Middletown, Conn.: Wesleyan University Press, 2013), 106.

30. Darko Suvin, "Preface," *Other Worlds, Other Seas: Science-Fiction Stories from Socialist Countries* (New York: Random House, 1970), 11.

31. Herbert George Wells, *The Sleeper Awakes* (London: Collins, 1921), 5.

32. Lauren Berlant, *Cruel Optimism* (Durham: Duke University Press, 2011), 73.

33. Sigmund Freud, *The Interpretation of Dreams*, translated by James Strachey (New York: Basic Books, 2010), 159–60.

34. Isaac Butler, "V For Vile," *The Hooded Utilitarian* (17 Sept. 2012), https://www.hoodedutilitarian.com/2012/09/v-for-vile/.

35. Alan Moore and David Lloyd, *V For Vendetta* (New York: DC Comics, 1989), 172.

36. Robert Silverberg, "What We Learned from This Morning's Newspaper," in *Infinity Four*, edited by Robert Hoskins (New York: Lancer Books, 1972), 28, 30, 34–35.

37. Isaac Asimov, *Foundation; Foundation and Empire; Second Foundation* (London: Everyman's Library, 2010), 29.

38. Asimov, *Foundation; Foundation and Empire; Second Foundation*, 30.

39. Asimov, *Foundation; Foundation and Empire; Second Foundation*, 67–69.

40. Constance Penley, "Time Travel, Primal Scene, and the Critical Dystopia," *Camera Obscura: Feminism, Culture, and Media Studies* 5, no. 3 (1986): 72.

41. Colombo, *L'Espace Politique*, 72.

42. David T. Mitchell and Sharon L. Snyder, *Narrative Prosthesis: Disability and the Dependencies of Discourse* (Ann Arbor: University of Michigan Press, 2000), 47-48; Liat Ben-Moshe, "Infusing Disability in the Curriculum: The Case of Saramago's 'Blindness,'" *Disability Studies Quarterly* 26, no. 2 (Mar. 2006): n.p.

43. Ria Cheyne, "'She Was Born a Thing': Disability, the Cyborg and the Posthuman in Anne McCaffrey's *The Ship Who Sang*," *Journal of Modern Literature* 36, no. 3 (July 2013): 142.

44. Mitchell and Snyder, *Narrative Prosthesis*, 63.

45. Le Guin, "Introduction," *The Norton Book of Science Fiction: North American Science Fiction, 1960–1990*, edited by Ursula K. Le Guin and Brian Attebery (New York: W. W. Norton & Co., 1993), 30.

46. Moshe, "Infusing Disability."

47. Palmer Rampell, "The Science Fiction of Roe v. Wade," *ELH* 85, no. 1, (Mar. 2018): 228.

48. Koelb, "The Language of Presence," 154.

49. Colson, *Lexicon*, 27. Travis Linnemann makes a strikingly similar kind of claim in his theorization, following Eugene Thacker's *Philosophy of Horror*, of a "world-for-us," "the Cartesian world of order" surrounded by a "thin blue line" of police power, beyond which lies a "world-in-itself" that we can only imagine as "horizons of disorder—primitivism, anarchism," and finally a completely unthinkable "world-without-us" or "world-without-police" (Travis Linnemann, "Bad Cops and True Detectives: The Horror of Police and the Unthinkable World," *Theoretical Criminology* 23, no. 3 [2019]: 356–57). To my mind, these formulations owe a little too much to the Lacanian concept of the Real, risking *ontologizing* "the horror of police" (Linnemann, "Bad Cops and True Detectives," 357) treating it as a permanent fixture of the universe rather than a contingent historical institution. Nonetheless, Linneman offers a distinct and persuasive way to use horror fiction as a particularly sharp instrument for dealing with political reality in the era of Blue Lives Matter.

50. Berlant, *Cruel Optimism*, 2; Judith Butler, *The Psychic Life of Power: Theories in Subjection* (Stanford, CA: Stanford University Press, 1997), 57; Berlant, *Cruel Optimism*, 194.

51. P. A. Zoline, "The Heat-Death of the Universe," *New Worlds Speculative Fiction* (July 1967): 33.

52. Zoline, "The Heat-Death of the Universe," 34–35.

53. John Ronald Reuel Tolkien, *The Lord of the Rings* 50th anniversary ed. (London: Houghton Mifflin Company, 2004), 717–18.

54. Thomas Aquinas, *Summa Theologiae: Volume 8, Creation, Variety and Evil*, edited by Thomas Gilby (Cambridge: Cambridge University Press, 2006), 145, 147.

55. Valerie Rohy, "On Fairy Stories," *MFS: Modern Fiction Studies* 50, no. 4 (2004): 931.

56. Rohy, "On Fairy Stories," 933.

57. Bensusan, "Polemos" 67, italics mine.

58. Thomas Pynchon, *The Crying of Lot 49* (New York: Harper & Row, 1986), 21-22.

59. Clarke, *Profiles*, 39n1.

60. N. K. Jemisin, "Writing The Fantastic In 2017," *Science Friday, National Public Radio* (13 Oct. 2017), https://www.sciencefriday.com/segments/writing-the-fantastic-in-2017/.

61. Jemisin, *Fifth Season*, 462; *Obelisk Gate*, 102.

62. Claude Lévi-Strauss, "The Structural Study of Myth," *The Journal of American Folklore* 68, no. 270 (Dec. 1955), 434; Jemisin, *Fifth Season*, 167.

63. Jemisin in Jason Kehe, "WIRED Book Club: Fantasy Writer N. K. Jemisin on the Weird Dreams That Fuel Her Stories," *Wired* (June 2016), *www.wired.com*, https://www.wired.com/2016/06/wired-book-club-nk-jemisin/.

64. Jemisin, interview with DaSilva.

65. Frank Wilderson III, *Afropessimism* (New York: W. W. Norton & Co., 2020), 228.

66. Calvin L. Warren, *Ontological Terror: Blackness, Nihilism, and Emancipation* (Durham: Duke University Press, 2018), 37; Warren, *Onticide: Afropessimism, Queer Theory, & Ethics*, 2017, https://illwilleditions.noblogs.org/files/2015/09/Warren-Onticide-Afropessimism-Queer-Theory-and-Ethics-READ.pdf, 6

67. Ebony Elizabeth Thomas, *The Dark Fantastic: Race and the Imagination from Harry Potter to the Hunger Games* (New York: New York University Press, 2020), 151.

68. Warren, *Ontological Terror*, 5.

69. Colson, *Trois Essais*, 21; Walter Benjamin, *Selected Writings* (Cambridge, Mass.: Harvard University Press, 2004), 697; James Martel, *Textual Conspiracies: Walter Benjamin, Idolatry, and Political Theory* (Ann Arbor: University of Michigan Press, 2011), 92.

70. Chanda Prescod-Weinstein, "Making Black Women Scientists under White Empiricism: The Racialization of Epistemology in Physics," *Signs: Journal of Women in Culture and Society* 45, no. 2 (Jan. 2020): 421–47.

71. Colson Whitehead, *The Intuitionist* (New York: Anchor Books, 1999), 239.

72. Warren, *Ontological Terror*, 145, 19–20; Martel, *Textual Conspiracies*, 92

73. Suvin, *Metamorphoses*, 78.

74. Rieder, *Colonialism*, 15.

75. Wark, *Molecular Red*; Warren, *Onticide*, 7.

76. N. K. Jemisin, *The City We Became* (New York: Orbit, 2020), 28.

77. N. K. Jemisin, *The Inheritance Trilogy*, 4; *Fifth Season*, 42.

78. Judith Halberstam in Fred Moten and Stefano Harney, *The Undercommons: Fugitive Planning & Black Study* (Wivenhoe: Minor Compositions, 2013), 5.

79. Jemisin, *Stone Sky*, 387.

80. Ytasha Womack, *Afrofuturism: The World of Black Sci-Fi and Fantasy Culture* (Chicago: Chicago Review Press, 2013), 24. Perhaps it is even—to entertain the Demon of Analogy a moment longer—an exercise in Gerald Vizenor's "survivance."

81. Jemisin, *The Stone Sky*, 283.

82. Jemisin, *The Stone Sky*, 299.

83. Wilderson, *Afropessimism* 158; Warren, *Ontological Terror*, 152.

84. Warren, *Ontological Terror*, 152.

85. Wilderson, *Afropessimism*, 313; "Anarkata: A Statement" (October 19, 2019) https://drive.google.com/file/d/1XU9P2PwO2wECh vsRdvu51hQBQnIiTbzG/view, 20.

86. Wilderson, *Afropessimism*, 41–42.

87. Jemisin, *The Stone Sky*, 313.

88. Colson, *Lexicon*, 76–77.

89. Jemisin, *Obelisk Gate*, 267-69.

90. Jemisin, *Obelisk Gate*, 390-91.

91. Colson, *Lexicon*, 77.

92. Jemisin, *Fifth Season*, 76.

93. Jemisin, *Fifth Season*, 296.

94. Jemisin, *Fifth Season*, 358–59.

95. Jemisin, *Fifth Season*, 372.

96. Delany, *Jewel-Hinged Jaw*, 150–52.

97. Lilian S. Robison, *Wonder Women: Feminisms and Superheroes* (New York: Routledge, 2004), 118.

98. Jemisin, *Fifth Season*, 371.

99. Jemisin, *Obelisk Gate*, 319–20.

100. Jemisin, *Stone Sky*, 283.

101. Jemisin, *Obelisk Gate*, 156–57.
102. Jemisin, *Obelisk Gate*, 180.
103. Jemisin, *Stone Sky*, 92.
104. Jemisin, *The Stone Sky*, 387.
105. Juan Duchesne Winter, "Literary Communism: A Manifesto of the Rearguard," *Journal of Latin American Cultural Studies: Travesia* 19, no. 3 (2010): 229.
106. Sharae Deckard, "Trains, Stone, and Energetics: African Resource Culture and the Neoliberal World-Ecology," in *World Literature, Neoliberalism, and the Culture of Discontent*, edited by Sharae Deckard and Stephen Shapiro, (Cham: Palgrave Macmillan, 2019), 258–59.
107. Colson, *Lexicon*, 25.
108. Nisi Shawl, *Everfair* (New York: Tor Books, 2016), 171.
109. Jasmine A. Moore, *Sankofa: Framing Afrofuturistic Dialectical Utopias in N. K. Jemisin's "The Fifth Season," Nisi Shawl's "Everfair," and Nnedi Okorafor's "Binti."* The University of Alabama in Huntsville, 2018), 43–44.
110. Nisi Shawl, in Kelly Lynn Thomas and Nisi Shawl, "Spitting in the Face of Empire: *The Millions* Interviews Nisi Shawl," *The Millions*, 2 Oct. 2017, https://themillions.com/2017/10/spitting-in-the-face-of-empire-the-millions-interviews-nisi-shawl.html.
111. Nisi Shawl in G. G. Silverman and Nisi Shawl, "Women in Speculative Fiction: Interview with Nisi Shawl, Author of EVERFAIR," G. G. Silverman (1 Feb. 2017), http://www.ggsilverman.com/women-in-speculative-fiction-interview-with-nisi-shawl-author-of-everfair/.
112. See Javier Padilla, "Science Fiction as Theory Fiction," *Modernism/Modernity Print Plus* (May 16, 2022), https://modernismmodernity.org/forums/posts/padilla-science-fiction-theory-fiction.
113. Terry Eagleton, *Criticism and Ideology: A Study in Marxist Literary Theory* (London, New York: Verso, 2006), 43.
114. Barbara Christian, "The Race for Theory," *Feminist Studies* 14, no. 1 (1988): 69-71; Madelyn Jablon, *Black Metafiction: Self-Consciousness in African American Literature* (Iowa City: University of Iowa Press, 1997), 1–2.

7. Rhetorical Sovereignty

1. Grace L. Dillon, "Imagining Indigenous Futurisms," *Walking the Clouds: An Anthology of Indigenous Science Fiction*, edited by Grace L Dillon (Tucson: University of Arizona Press, 2012), 2.
2. Theresa Warburton, *Other Worlds Here: Honoring Native Women's Writing in Contemporary Anarchist Movements* (Evanston, Ill.: Northwestern University Press, 2021), 34.
3. Suvin, *Metamorphoses*, 7–8.
4. Pyle, Kai Minosh. "How to Survive the Apocalypse for Native Girls." *Love After the End: An Anthology of Two-Spirit and Indigiqueer Speculative Fiction*, edited by Joshua Whitehead (Vancouver: Arsenal Pulp Press, 2020), 85.
5. Hilan Bensusan and Tomás Ribeiro Cardoso, "Por Uma Metafísica de Tramas: O Mundo Sem Arché," *Kriterion* 53, no. 125 (June 2012): 287.
6. David Graeber, *Fragments of an Anarchist Anthropology* (Chicago: Prickly Paradigm Press, 2004), 51–53.
7. Scott Richard Lyons, *X-Marks: Native Signatures of Assent* (U of Minnesota Press, 2010), 452.
8. Lyons, *X-Marks*, 449.
9. Adam Garnet Jones, "History of the New World," *Love After the End: An Anthology of Two-Spirit and Indigiqueer Speculative Fiction*, edited by Joshua Whitehead (Vancouver: Arsenal Pulp Press, 2020), 41.

10. Blaire Topash-Caldwell, "Sovereign Futures in Neshnabé Speculative Fiction," *Borderlands Journal* 19, no. 2 (Mar. 2021), 46.

11. Suvin, *Metamorphoses*, 8–9.

12. Farah Mendlesohn, *Rhetorics of Fantasy* (Middletown, Conn.: Wesleyan University Press, 2008), 99.

13. Kristina Baudemann, "'I HAVE SEEN THE FUTURE AND I WON'T GO': The Comic Vision of Craig Strete's Science Fiction Stories," *Studies in American Indian Literatures* 29, no. 4 (2017): 84.

14. Drew Hayden Taylor, *Take Us to Your Chief: And Other Stories: Classic Science-Fiction with a Contemporary First Nations Outlook* (Madeira Park, BC: D & M Publishers, 2016), 140.

15. Taylor, *Take Us to Your Chief*, 142–43.

16. Taylor, *Take Us to Your Chief*, 144–46.

17. David Brin, "The Dogma of Otherness," in David Brin, *Otherness* (New York: Bantam Books, 1994), 91.

18. Daniel Heath Justice, "Indigenous Wonderworks and the Settler-Colonial Imaginary," *Apex Magazine* (10 Aug. 2017), https://apex-magazine.com/indigenous-wonderworks-and-the-settler-colonial-imaginary/; Aileen Moreton-Robinson, *The White Possessive: Property, Power, and Indigenous Sovereignty* (University of Minnesota Press, 2015), xiii.

19. Taylor, *Take Us to Your Chief*, 139.

20. Taylor, *Take Us to Your Chief*, 144.

21. Taylor, *Take Us to Your Chief*, 139.

22. Gerald Robert Vizenor, Fugitive Poses: Native American Indian Scenes of Absence and Presence (Lincoln, Neb.: U of Nebraska Press, 2000), 15.

23. Craig Strete, "A Horse of a Different Technicolor," *Galaxy* 36, no. 1 (Jan. 1975): 77, 82.

24. Strete, "Horse," 77.

25. Vizenor, *Fugitive Poses*, 15; Baudemann, "I HAVE SEEN THE FUTURE AND I WON'T GO," 94.

26. Strete, "Horse," 81, 79, 78.

27. Strete, "Horse," 80.

28. Jean Baudrillard, *Simulacra and Simulation*, translated by Sheila Faria Glaser (Ann Arbor: University of Michigan Press, 1994), 1.

29. Bailey, *Pilgrims Through Space and Time*, 203.

30. Baudemann, "I HAVE SEEN THE FUTURE AND I WON'T GO," 77.

31. Philip Joseph Deloria, *Playing Indian* (New Haven: Yale University Press, 1998), 184.

32. See Acee Agoyo, "'The Elizabeth Warren of the Sci-Fi Set': Author Faces Criticism for Repeated Use of Tribal Traditions," *Indianz.com*, 24 June 2020, https://www.indianz.com/News/2020/06/24/the-elizabeth-warren-of-the-scifi-set-au.asp, and Elsa Ruth Klingensmith-Parnell, "Cultural Appropriation or Much-Needed Representation? On Rebecca Roanhorse's *Trail of Lightning*," *SFRA Review*, no. 330, Fall 2019, pp. 90–98.

33. Rebecca Roanhorse, "Welcome to Your Authentic Indian Experience," *Apex Magazine* (8 Aug. 2017), https://www.apex-magazine.com/welcome-to-your-authentic-indian-experience/.

34. Miriam C. Brown Spiers, "Reimagining Resistance: Achieving Sovereignty in Indigenous Science Fiction," *Transmotion* 2, nos. 1 & 2, (Nov. 2016): 53.

35. Dillon quoted in Chelsea May Vowel, *Where No Michif Has Gone Before: The Form and Function of Métis Futurisms* (Diss., University of Alberta, 2020), 6.

36. Richard Van Camp, "Aliens," in *Love Beyond Body, Space, and Time*, edited by Hope Nicholson (Winnipeg: Bedside Press, 2016), 20.

37. Van Camp, "Aliens," 21.

38. Justice, "Indigenous Wonderworks."

39. Irwin, *The Game of the Impossible*, 69.

40. Van Camp, "Aliens," 22–23.
41. Van Camp, "Aliens," 25.
42. Van Camp, "Aliens," 26.
43. Van Camp, "Aliens," 27–28.
44. Van Camp, "Aliens," 28, 30.
45. Little Badger, "Nkásht Íí."
46. Little Badger, "Nkásht Íí."
47. Pyle, "How to Survive the Apocalypse," 82.
48. Fisher, *Capitalist Realism*, 2.
49. Pyle, "How to Survive the Apocalypse," 89–90.
50. Pyle, "How to Survive the Apocalypse," 94.

8. The Rhetoric of Fictive Science, or the Academy of Outrageous Books

1. David Hume, *An Enquiry Concerning Human Understanding: With Hume's Abstract of A Treatise of Human Nature and A Letter from a Gentleman to His Friend in Edinburgh* (Indianapolis: Hackett Publishing, 1993), 114.

2. Gilles Deleuze, *Spinoza: Practical Philosophy*, translated by Robert Hurley (San Francisco: City Lights Books, 1988), 17–18.

3. Baruch Spinoza, *Ethics: With the Treatise on the Emendation of the Intellect and Selected Letters*, edited by Seymour Feldman, translated by Samuel Shirley (Indianapolis: Hackett Publishing, 1992) 105, italics mine.

4. Murray Bookchin, *Re-enchanting Humanity: Defense of the Human Spirit against Antihumanism, Misanthropy, Mysticism and Primitivism* (London: Cassell, 1995), 216.

5. Stephen Jay Gould, *Wonderful Life: The Burgess Shale and the Nature of History* (New York: W. W. Norton & Company, 1990), 274.

6. Andrew Ross, *Science Wars* (Durham: Duke University Press, 1996), 14; Paul R. Gross and Norman Levitt, *Higher Superstition: The Academic Left and Its Quarrels with Science* (Baltimore: JHU Press, 1997).

7. William James, *Pragmatism, a New Name for Some Old Ways of Thinking: Popular Lectures on Philosophy* (Longmans, Green, and Company, 1907), 54. Erica Lagalisse warns us that "the disciplinary 'ontological turn'" means that, while "the anthropologist (finally) grants the 'reality' of plants that think, clouds that have agendas, and spiritual animal protectors," this is "only by inventing multiple realities in the process: ontology (reality) becomes plural such that the white man can still enjoy his office without having to worry about the weather" (Erica Lagalisse, "Occult Features of Anarchism," *Essays in Anarchism and Religion*, edited by Alexandre Christoyannopoulos and Matthew S. Adams, vol. 2 [Stockholm: Stockholm University Press, 2018], 314n81). In other words, "When entire cosmologies are reified as 'proper' only to specific pre-ordained identities, we are effectively saying they are false to the extent that they do not apply across the cosmos whatsoever" (Lagalisse, "Occult Features," 314). I would rather say with Wolf Gordin that *all* cosmologies are false to some extent at which we can only guess and that the bridging of discrepancies between perspectives does not have to aim at the resolution of all such world-pictures into one, as in the grandiose dream of an E. O. Wilson; even pragmatism is not to be thought as an end but as a means to such bridgings, a transitional device. As Kafka knew too well from his upbringing in the cramped architecture of a Prague flat—his room was literally the passageway between other rooms—it isn't really possible to dwell in a corridor (Frederick Robert Karl, *Franz Kafka, Representative Man* [New York: Fromm International Publishing Corporation, 1993], 205).

8. Charles Richard Johnson, *Being & Race: Black Writing Since 1970* (Bloomington, IN: Indiana University Press, 1988), 32; @phillosoraptor, "Why Is Lord of the Flies Happening in an American City?" Twitter (31 July 2020), *twitter.com*, https://twitter.com/phillosoraptor/status/1289179367411597313.

9. Chiara Bottici, *A Philosophy of Political Myth* (Cambridge: Cambridge University Press, 2010), 7; Kenneth Burke, *The Philosophy of Literary Form: Studies in Symbolic Action* (Baton Rouge: Louisiana State University Press, 1967), 4.

10. A. M. Gittlitz, "Let Them Drink Blood," *The New Inquiry* (blog), December 27, 2016, https://thenewinquiry.com/let-them-drink-blood/.

11. Alan D. Sokal and Jean Bricmont, *Fashionable Nonsense: Postmodern Intellectuals' Abuse of Science* (New York: Picador, 1999), 25–26.

12. Donna J. Haraway, *Staying with the Trouble: Making Kin in the Chthulucene* (Durham: Duke University Press, 2016), 10.

13. Michel Foucault, *The Order of Things: An Archaeology of the Human Sciences* (London: Routledge, 2002), 62.

14. Karl Marx, in Karl Marx and Friedrich Engels, *Marx Engels Collected Works*, vol. 36 (Lawrence & Wishart, 1975), 31.

15. Karl Marx, in Karl Marx and Friedrich Engels, *Marx Engels Collected Works*, vol. 35 (Lawrence & Wishart, 1975), 221–22.

16. Marx, in Marx and Engels, *Marx Engels Collected Works*, vol. 35, 222.

17. James Arnt Aune, *Rhetoric and Marxism* (s.l.: Taylor & Francis Group, 2020).

18. Foucault, *Order of Things*, 79–80.

19. Marx, in Marx and Engels, *Marx Engels Collected Works*, vol. 35, 77, 750. Note that for David Graeber, this gives *Capital* a quasi-fictive status, as it presents not the full truth of industrial capitalism but a "utopian vision" of it, a *reductio ad absurdum* in which all of the ideal constructs of the capitalists themselves are supposed to be true but "the whole system will eventually destroy itself" (Graeber, *Debt*, 354).

20. Evgeniy Kuchinov, "Anarchy as Access to Space: Language, Imagination, Technics," Science Fiction and Communism Conference, Sofia, 25–27 May, accessed 26 July 2020, https://www.academia.edu/41807554/Anarchy_as_Access_to_Space_Language_Imagination_Technics_eng_, 2.

21. Fernando Tárrida del Mármol, *Problemas Trascendentales. Estudios de Sociología y Ciencia Moderna* (Paris: Sociedad de Ediciones Literarias y Artísticas, 1908), 47, 154.

22. Tárrida del Mármol, *Problemas Trascendentales*, 222.

23. Tárrida del Mármol, *Problemas Trascendentales*, 111.

24. Tárrida del Mármol, *Problemas Trascendentales*, 113–15. A few years after Tárrida's death, the anarchist who writes under the name "Ret Marut" (later to adopt the pen name "B. Traven," under which he will become an internationally famous but hermetic author) will denounce this abuse of *mathesis* ("there is no mathematical proof that leads to the truth") by promulgating his own mock-equation, "the Marcurve," for which he makes an increasingly grandiose, Bouvard-and-Pecuchet series of claims: "The Marcurve is the same curve in which the sun runs, in which the earth runs, in which the (more than 240 seen so far) planets of our planetary system run, in which all the celestial bodies run, in which every movement that I make, that an animal makes, runs"; "By means of the Marcurve, I can communicate with all people who live on any planet in the universe"; "no matter what the tempo of the movement of an atom, the atom always moves in the Marcurve"; "He who recognizes the Marcurve no longer sees all things of the world with the imperfect eye of man. He sees the things and events in the world as they really are" (Ret Marut, "Die Zerstörung unseres Welt-Systems durch die Markurve," *Der Ziegelbrenner* 4, no. 20–22 [Jan. 1920] 2, 28, 29, 41, 43). Perhaps this amounts a kind of Stirnerian exorcism of the "spook" of mathematics itself.

25. Tárrida del Mármol, *Problemas Trascendentales*, 112.

26. Tárrida del Mármol, *Problemas Trascendentales*, 109–10.

27. Benedict Richard O'Gorman Anderson, *Under Three Flags: Anarchism and the Anti-Colonial Imagination* (New York: Verso, 2005), 169, 171–72

28. Beobi, *Grammatika*, 23.

29. There are bibliographic records of at least three, although these all seem to be dictionaries of the first (1920) version of AO, rather than the mature version described by the Grammatika (1924): *AO-russkiy slovar' chelovechestvo-izobretatelya* (*AO-Russian Dictionary of Humanity-Inventor*), *Slovar' yazyka AO: AO-russkiy* (*Dictionary of the AO language: AO-Russian*), and *AO-russkiy grammaticheskiy slovar'* (*AO-Russian Grammatical Dictionary*). None of these appear in library holdings or antiquarian sales catalogs.

30. Beobi, *Grammatika*, 5.

31. Beobi, *Grammatika*, 1–2.

32. Christian Lundberg, *Lacan in Public: Psychoanalysis and the Science of Rhetoric* (Tuscaloosa, Ala.: University of Alabama Press, 2012), 15.

33. Lacan, *Ecrits*, 694.

34. Bruce Fink in Lacan, *Ecrits*, 132.

35. William Gibson and Bruce Sterling, *The Difference Engine* (Random House Publishing Group, 2011), 428–29; Greg Bear, *Queen of Angels* (Warner Books, 1991), 278, 358.

36. Lacan, *Ecrits*, 693.

37. Bruce Fink, *Lacan to the Letter: Reading Ecrits Closely* (U of Minnesota Press, 2004), 133.

38. Lacan, *Ecrits*, 696.

39. Fink, *Lacan to the Letter*, 134.

40. Ian Parker, *Psychology After Lacan: Connecting the Clinic and Research* (New York: Routledge, 2014), 28.

41. Haraway, *Staying With the Trouble*, 10.

42. Stephen Duncombe, "Does It Work?: The Æffect of Activist Art," *Social Research: An International Quarterly* 83, no. 1 (Spring 2016): 128.

43. Duncombe nonetheless insists that he *is* "advocating for ... a rigorous methodology that will allow activist artists, and their critics, to judge the æfficacy of their practice. In a word, I am making an argument for metrics" (130). Without wanting to say that the affective efficacy of art should never be assessed, nor even that quantitative measures are always irrelevant to art, I am skeptical of the practicality of such efforts, which strike me as likely to attribute "success" (S) falsely to artistic interventions in which unanticipatable external factors are really driving, for instance, how many people show up to a meeting—or, just as likely, to falsely lay blame for their "failure."

44. Haraway, *Staying With the Trouble*, 169n2.

45. Haraway, *Staying With the Trouble*, 14.

46. The alpha and omega of her equation were originally intended to mean *bios* and *zoe*, the qualified and unqualified kinds of life in Giorgio Agamben's schema of "biopolitics" (Donna J. Haraway, *When Species Meet* [U of Minnesota Press, 2013], 324n3).

47. Liza Featherstone, "Radical Academics for the Status Quo," *Jacobin* (Dec. 2019), https://jacobinmag.com/2019/12/radical-academics-judith-butler-kamala-harris-donation.

48. Raymond Tallis, *Not Saussure: A Critique of Post-Saussurean Literary Theory* (Basingstoke, UK: Palgrave, 1995), 147.

49. David Graeber, *Debt: The First 5,000 Years* (Brooklyn: Melville House, 2014), 354.

50. Horst Ruthrof, *Semantics and the Body: Meaning from Frege to the Postmodern* (Toronto: University of Toronto Press, 1997), 46–47; Robert E. Scholes, *Textual Power: Literary Theory and the Teaching of English* (New Haven: Yale University Press, 1985), 92.

51. Slavoj Žižek, *The Sublime Object of Ideology* (New York: Verso, 1989), 12.

52. Burke, *Philosophy of Literary Form*, 4. Indeed, Žižek acknowledges this necessary degree of fetishism at some points: e.g., in *Tarrying with the Negative*, when he speaks of "the delusion that we can dispense with this fetishistic split [between reality and fictions]" (Slavoj Žižek, *Tarrying with the Negative: Kant, Hegel, and the Critique of Ideology* [Durham: Duke University Press, 1993], 88). However, he does not draw the conclusion that his own antifetishism, i.e., his entire critique of ideology, is delusory or impracticable.

53. Beobi, *Gnoseologiya: Vvedeniye vo Vseizobretatel'stvo* (Moskva: Izdaniye Vseizobretal'ni, 1921), 20.

54. Bookchin, *Ecology of Freedom*, 34, 229, 25, 164.

9. How the Gordin Brothers Escaped Western Gravity

1. Bruno Latour, *War of the Worlds: What about Peace?* (Chicago: Prickly Paradigm Press, 2002), 7.

2. Latour, *War of the Worlds*, 8, 11–12.

3. Khaldoun Samman, *Clash of Modernities: The Making and Unmaking of the New Jew, Turk, and Arab and the Islamist Challenge* (New York: Routledge, 2015), 43.

4. Donna J. Haraway, "Teddy Bear Patriarchy: Taxidermy in the Garden of Eden, New York City, 1908–1936," *Social Text* no. 11 (1984), 55. Scott Richard Lyons argues that some of the Indigenous people who entered the reservations were not merely beaten into submission but also "making a choice to modernize and nationalize" (Lyons, *X-Marks*, 127), in which case we might also speak of Indigenous time machines. Survivance takes many paths, to be sure, but it is unclear to me whether such bargains weren't just as Faustian as those struck by Samman's modernizing Turks and Jews.

5. William M. Denevan, "The Pristine Myth: The Landscape of the Americas in 1492," *Annals of the Association of American Geographers* 82, no. 3 (1992): 369.

6. Patricia Leighten, "The White Peril and *L'Art Nègre*: Picasso, Primitivism, and Anticolonialism," *The Art Bulletin* 72, no. 4 (1990): 611.

7. David Pan, "Kafka as a Populist: Re-Reading 'In the Penal Colony,'" *Telos* no. 101 (Sept. 1994): 8; Peter Kropotkin, "The Morality of Nature," *The Nineteenth Century*, vol. 57 (Mar. 1905): 425. Perhaps the Great Sorting Machine is also Aimé Césaire's colonial "forgetting machine" (Césaire, *Discourse*, 52).

8. Colson, *Trois Essais*, 31.

9. Brat'ya Gordinii, *Anarkhiia v Mechte*, 183, 185.

10. Brat'ya Gordinii, *Anarkhiia v Mechte*, 186–88.

11. Brat'ya Gordinii, *Anarkhiia v Mechte*, 185–86.

12. Chu, *Do Metaphors Dream of Literal Sleep*, 2, 14.

13. Chu, *Do Metaphors Dream of Literal Sleep*, 14.

14. Pablo Neruda, "Oda a La Piedra/Ode to the Stone," in Pablo Neruda, *Fifty Odes*, translated by George Schade (Austin, Tex.: Host Publications, Inc., 1996), 169.

15. Chu, *Do Metaphors Dream of Literal Sleep*, 10–11; Jackson, *Fantasy*, 24; Brat'ya Gordinii, *Anarkhiia v Mechte*, 185.

16. Hallowell quoted in Graham Harvey, *Animism: Respecting the Living World* (New York: Columbia University Press, 2006), 33; Arthur Caswell Parker, *Seneca Myths and Folk Tales* (Buffalo, NY: Buffalo Historical Society, 1923), 97–100.

17. Suvin, *Metamorphoses*, 8; Stanislaw Lem, "On the Structural Analysis of Science Fiction" in Stanislaw Lem, *Microworlds*, edited by Franz Rottensteiner, translated by Franz Rottensteiner et al. (San Diego: Harcourt Brace Jovanovich, 1986), 33.

18. Kuchinov, "Anarchy as Access to Space."

19. The phenomenon of a stone singing, for the Gordin Brothers, is not the reflection of an intelligible "law" but a power manifested by the thing in action. Their ontology is, in this respect, similar to that of the Haudenosaunee/Wendat concept of immanent powers or *orén:na* (orenda)—encoding, according to the anthropologist J. N. B. Hewitt, both "the idea that things both are and are known by what they do" and, importantly, "song or voice or, more generally, the way things express themselves" (Scott L. Pratt, *Native Pragmatism: Rethinking the Roots of American Philosophy* [Bloomington, In.: Indiana University Press, 2002], 192–93; "Karén:na," in *Kanienkeha: An Open Source Endangered Language Initiative (Mohawk Dictionary)*, 17 Sept. 2015, https://kanienkeha.net/body-parts/karenna/).

20. Abba Gordin and Wolf Gordin, "Manifest Pananarkhistov," in Abba Gordin and Wolf Gordin, *Strana Anarkhiya (Utopii)*, edited by Evgeniy Kuchinov (Moscow: Common place, 2019), 218.

21. Gordin and Gordin, "Manifest Pananarkhistov," 217. This is incorrect, as the Inuit creation story centers on Raven. The Ainu people, who had a branch in Kamchatka, did believe that they descended from the polar bear, and were one of a number of peoples living within the Russian Empire with ritual practices involving bears, as recorded in the Sakhalin Island memoir of Anton Chekhov (Anton Chekhov, *Sakhalin Island* [Surrey, UK: Alma Books, 2018], 205), one of Wolf Gordin's favorite writers (Abba Gordin, *Draysik yor in Lite un Poyln: Oytobiografye* [Buenos Aires: Bukhgemaynshaft bay der yidisher ratsiyonalistisher gezelshaft, 1958], 361). It is possible that the Gordin Brothers are passing along some misremembered bit of ethnography.

22. Gordin and Gordin, "Manifest Pananarkhistov," 218.

23. Eugene Kuchinov, "Russkiy Mars [Russian Mars]." *KROT* no. 4 (2018), https://krot.me/articles/mars-is-ours.

24. Latour, *We Have Never Been Modern*, 139, 10.

25. Brat'ya Gordinii, *Anarkhiia v Mechte*, 95.

26. Colson, *Lexicon*, 93.

27. Harvey, *Animism*, 38.

28. Murray Bookchin, *Toward an Ecological Society* (Montréal: Black Rose Books, 1980), 277–78.

29. Bookchin, *Ecology of Freedom*, 43.

30. Bookchin, *Ecology of Freedom*, 233–34.

31. Gordin and Gordin, "Manifest Pananarkhistov," 279.

32. Adorno and Horkheimer, *Dialectic of Enlightenment*, 3.

33. Gordin and Gordin, *Anarkhiia v Mechte*, 105.

34. Gordin and Gordin, "Manifest Pananarkhistov," 284, 218–19, 233, 247, 279, 299.

35. Gordin and Gordin, "Manifest Pananarkhistov," 217–18.

36. Bookchin, *Ecology of Freedom*, 64, 98, 28.

37. Gordin and Gordin, "Manifest Pananarkhistov," 217, italics mine.

38. Gordin and Gordin, *Anarkhiia v Mechte*, 112, 107.

39. James C. Scott, *Seeing Like a State: How Certain Schemes to Improve the Human Condition Have Failed* (New Haven: Yale University Press, 1998), 6.

40. Gordin and Gordin, "Manifest Pananarkhistov," 284.

41. Bookchin, *Re-enchanting Humanity*, 127. Of course, in his "Theses on Feuerbach," Marx writes that "All mysteries which mislead theory into mysticism find their rational solution in human practice and in the comprehension of this practice" (Marx, "Theses on Feuerbach," in Karl Marx and Friedrich Engels, *Marx and Engels Collected Works*, vol. 5 [London: Lawrence & Wishart, 1979], 8). A radicalization of this eighth thesis would have to dispense with the automatic dismissal of "mysteries" and "mysticism" and forego the gesture that reinstates theory ("the *comprehension* of this practice") as the guardian of practice.

42. Lawrence William Gross, *Anishinaabe Ways of Knowing and Being* (London: Ashgate, 2014), 132–33.

43. Harvey, *Animism*, 19–20; Jerry Lee Rosiek, Jimmy Snyder, and Scott L. Pratt, "The New Materialisms and Indigenous Theories of Non-Human Agency: Making the Case for Respectful Anti-Colonial Engagement," *Qualitative Inquiry* 26, no. 3–4 (Mar. 2020): 337–38.

44. Quoted in Harvey, *Animism*, 33.

45. Gordin and Gordin, *Anarkhiia v Mechte*, 149, 187.

46. Gordin and Gordin, *Anarkhiia v Mechte*, 183.

47. Gordin and Gordin, *Anarkhiia v Mechte*, 121.

48. Gordin and Gordin, *Anarkhiia v Mechte*, 122.

49. Gordin and Gordin, *Anarkhiia v Mechte*, 145.

50. Gordin and Gordin, *Anarkhiia v Mechte*, 167, emphasis mine.

51. Vizenor, *Fugitive Poses*, 15.

52. Max Weber, *The Vocation Lectures*, edited by David S. Owen and Tracy B. Strong, translated by Rodney Livingstone (Indianapolis: Hackett Publishing, 2004), 13.

53. Gordin and Gordin, *Anarkhiia v Mechte*, 144, 163–64.

54. Gordin and Gordin, *Anarkhiia v Mechte*, 87.

55. Gordin and Gordin, *Anarkhiia v Mechte*, 95.

56. Philip K. Dick, *A Scanner Darkly* (New York: Vintage Books, 1991) 22–23; Gordin and Gordin, *Anarkhiia v Mechte*, 85.

57. A. Farsayt, "The Worker's Movement," *Smorgonie, District Vilna; Memorial Book and Testimony (Smarhon, Belarus)*, edited by Marc D. Hodies, translated by Janie Respitz, https://www.jewishgen.org/yizkor/smorgon/smo237.html, 237; Moshe Tzinovitz, "Rabbis of the Community of Smorgon," *Smorgonie, District Vilna; Memorial Book and Testimony (Smarhon, Belarus)*, edited by Marc D. Hodies, translated by Jerrold Landau, https://www.jewishgen.org/yizkor/smorgon/smo078.html, 96; Gordin and Gordin, *Anarkhiia v Mechte*, 167, 153. I can't resist a remark here about how delightfully this cuts against Darko Suvin's historical-materialist scientism, conservative as it sometimes is in spite of his attachment to the notion of the "novum" or "new thing" that marks science fiction's otherness. Perhaps *The Land of Anarchy* shows us the need for Aaron Worth's concept of the *antiquum*, the "old thing" that returns in the unfamiliar guise of the new. What if this, too, turned out to describe Native American writers' uses of the fantastic?

58. Leonid Heller, "Voyage au pays de l'anarchie: Un itinéraire: l'utopie," *Cahiers Du Monde Russe* 37, no. 3 (1996), 270, emphasis mine.

59. Colson, *Trois Essais*, 42–43.

60. Vizenor, *Fugitive Poses*, 97.

61. Jonathan Boyarin, "Europe's Indian, America's Jew: Modiano and Vizenor," *boundary 2* 19, no. 3 (1992): 197–98.

62. Vizenor, *Fugitive Poses*, 67–68, 61; Lea Speyer, "Israel Advocates at Columbia University to Hold 'Indigenous People Unite' Event to 'Reclaim Narrative' About Jewish Right to Native Land," Algemeiner.com, accessed 23 May 2020, https://www.algemeiner.com/2016/12/04/columbia-university-students-reclaiming-israels-narrative-event-aimed-countering-false-palestinian-claims-jewish-rights-land/.

63. Colson, *Lexicon*, 21, emphasis mine.

64. Vizenor, *Fugitive Poses*, 47.

65. Evgeniy Kuchinov, "Ot pananarkhizma k AOizmu i AIIZu: ocherk istorii i mifologii odnogo inoplanetnogo plemeni" *Etnograficheskoye obozreniye* no. 6, Dec. 2019, 34.

66. Abba Gordin and Wolf Gordin, *Sotsiomagiya i Sotsiotekhnika, ili Obshcheznakharstvo i Obshchestroitel'stvo* [*Sociomagic and Sociotechnics, or Generalized Quackery Versus Global Construction*] (Moskva: Pervyy tsentral'nyy sotsiotekhnikum, 1918), 31.

67. Gordin and Gordin, *Anarkhiia v Mechte*, 18.
68. Gordin and Gordin, *Anarkhiia v Mechte*, 23.
69. Gordin and Gordin, "Manifest Pananarkhistov," 242–43.
70. Gordin and Gordin, *Sefer el ha-golah*, quoted in Marc D. Hodies, ed., *Smorgonie, District Vilna; Memorial Book and Testimony*, translated by Jerrold Landau and Sara Mages (JewishGen, 2019), https://www.jewishgen.org/yizkor/smorgon/Smorgon.html, 212.
71. Deleuze and Guattari, A Thousand Plateaus, 158; Gordin and Gordin, *Sefer el ha-golah*, quoted in Hodies, ed., *Smorgonie, District Vilna*, 212.
72. Michael G. Smith, *Rockets and Revolution: A Cultural History of Early Spaceflight* (Lincoln: University of Nebraska Press, 2014), 131.
73. Sergei N. Kuznecov, "Linguistica Cosmica: Rozhdeniye 'Kosmicheskoy Paradigmy,'" *Sovremennaya Nauka*, no. 2 (2014): 225; Beobi, *Grammatika*, 16.
74. Beobi, *Grammatika*, 17, 5.
75. Friedrich Nietzsche, "Twilight of the Idols," *The Anti-Christ and Other Writings*, edited by Aaron Ridley, translated by Judith Norman (Cambridge: Cambridge University Press, 2005), 170; Beobi, *Grammatika*, 7
76. Smith, *Rockets and Revolution*, 131.
77. Jorge Luis Borges, *Other Inquisitions, 1937–1952*, translated by Ruth L. C. Simms (New York: Washington Square Press, 1966), 103.
78. Foucault, *The Order of Things*, xvi.
79. Foucault, *The Order of Things*, 64.
80. Galileo Galilei, *Discoveries and Opinions of Galileo*, translated by Stillman Drake (New York: Doubleday & Co., 1957), 237–38.
81. Galileo, *Discoveries and Opinions*, 276-77; Galileo quoted in Bottici, *Philosophy of Political Myth*, 66–67.
82. Latour, *We Have Never Been Modern*, 46.
83. Foucault, *Order of Things*, 422.
84. Foucault, *Order of Things*, 47. As Leonid Heller notes, the Gordin Brothers' "admiration for the word" reflected "the Russian corporeal 'philosophy of the name,' which considers the word as a force acting in the world, but for the Gordins, the word is even more objectified in the conscious mixture of words and things" (Leonid Heller, "Brat'ya Gordiny, Anarkhizm i Russkii Avangard," *The Russian Word in the Land of Israel, the Jewish Word in Russia*, edited by V. Khazan and W. Moskovich [Jerusalem: Hebrew University, Center for Slavic Languages and Literatures, 2006], 141–42).
85. Michael Hardt and Antonio Negri, *Empire* (Cambridge, Mass.: Harvard University Press, 2001), 91.
86. Christopher Alexander, Sara Ishikawa, Murray Silverstein, Max Jacobson, Ingrid F. King, and Shlomo Angel, *A Pattern Language: Towns, Buildings, Construction* (New York: Oxford University Press, 1977), 876–81.
87. Alexander, Ishikawa, Silverstein, Jacobson, King, and Angel, *A Pattern Language*, 72–74.
88. Jacques Rancière, *The Politics of Aesthetics*, Edited and translated by Gabriel Rockhill (London: A&C Black, 2013), 7–8.
89. Alexander, Ishikawa, Silverstein, Jacobson, King, and Angel, *A Pattern Language*, x.
90. Gordin and Gordin, *Anarkhiia v Mechte*, 27.
91. Gordin and Gordin, *Anarkhiia v Mechte*, 35.
92. Gordin and Gordin, *Anarkhiia v Mechte*, 64–65.
93. Gordin and Gordin, *Anarkhiia v Mechte*, 117.
94. Gordin and Gordin, *Anarkhiia v Mechte*, 118.
95. Gordin and Gordin, *Sotsiomagiya i Sotsiotekhnika*, 97, 102.

96. Burke, *Philosophy of Literary Form*, 4.
97. Gordin and Gordin, *Anarkhiia v Mechte*, 98–101.
98. Gordin and Gordin, *Anarkhiia v Mechte*, 101.
99. Graeber, *Fragments*, 25–26.
100. Ng, *Under the Pendulum Sun*, 212, 42, 198-200, 183, 318.
101. Ng, *Under the Pendulum Sun*, 128–29.
102. Gordin and Gordin, *Anarkhiia v Mechte*, 106.
103. Gordin and Gordin, *Anarkhiia v Mechte*, 130.
104. Gordin and Gordin, *Anarkhiia v Mechte*, 135.
105. Gordin and Gordin, *Anarkhiia v Mechte*, 152.
106. Deleuze and Guattari, *Anti-Oedipus* 8; Deleuze, *Spinoza*, 122.
107. Gordin and Gordin, "Manifest Pananarkhistov," 269; Marshall McLuhan, *The Gutenberg Galaxy: The Making of Typographic Man* (Toronto: University of Toronto Press, 1962), 4.
108. Technically—pun unavoidable—they only visit two of the departments, the Visual and the Auditory, before the narrative somewhat abruptly concludes. McLuhan might have seen this truncation not as resulting from the novel's serial character (it debuted in the pages of their newspaper, *Anarkhiia*) but from the relative centrality of eye and ear in human cultures, from the way in which orality and literacy reorganize the rest of the sensorium. Or perhaps beyond the last chapter of *Anarkhiia v Mechte* stands the unnameable otherwise to which John Varley's Kellerites go in "The Persistence of Vision."
109. Gordin and Gordin, *Anarkhiia v Mechte*, 166.
110. Gordin and Gordin, *Anarkhiia v Mechte*, 166.
111. Gordin and Gordin, *Anarkhiia v Mechte*, 167–68.
112. Gordin and Gordin, *Anarkhiia v Mechte*, 173, 177.
113. Evgeniy Kuchinov, "Fictio Audaciae: Po Tu Storonu Samosokhraneniya Prosveshcheniya," *Logos* 29, no. 4 (2019): 122–23.
114. Kuchinov, "Fictio Audaciae" 123; Meillassoux, *Science Fiction and Extro-Science Fiction*, 40.
115. Gordin and Gordin, *Anarkhiia v Mechte*, 176.
116. Gordin and Gordin, *Anarkhiia v Mechte*, 180–81.
117. Gordin and Gordin, *Anarkhiia v Mechte*, 114.
118. Is this manifestation of same-sex desire a reflection of the authors' subjectivities? At its extremes, the "intellectual-emotional intimacy" between the brothers verged on something incestuous: in his 1958 Yiddish memoir, Abba reflects that he and Wolf had enjoyed "a kind of platonic homosexuality," a "cohesive[ness]" that could manifest itself as "longing" (Abba Gordin, *Draysik Yor in Lite un Poyln: Oytabyagrafye* [Buenos Aires: Bukhgemaynshaft bay der Yidisher Ratsyonalistisher gezelshaft, 1958] 458).
119. Gordin and Gordin, *Anarkhiia v Mechte*, 199–201.
120. Jules Verne quoted in Andrew Milner, *Locating Science Fiction* (Oxford University Press, 2012), 32.
121. Wells, *Complete Science Fiction Treasury*, iii–iv.
122. Wells, *Complete Science Fiction Treasury*, iv.
123. Gordin and Gordin, *Manifest Pananarkhistov*, 218.
124. Gordin and Gordin, *Anarkhiia v Mechte*, 136.
125. Gordin and Gordin, *Anarkhiia v Mechte*, 138–39.
126. Slavoj Žižek, "The Matrix, Or, The Two Sides of Perversion," in William Irwin, ed., *The Matrix and Philosophy: Welcome to the Desert of the Real* (Chicago: Open Court Publishing, 2002), 265.
127. Žižek, "The Matrix," 264–65. Interestingly, this is not far from the project of the Gordin Brothers and the anarcho-biocosmists.

128. I confess that I grow tired of defending Lacan and his *École Freudienne* to undergraduates who wonder (justifiably) why Freud hasn't been canceled for his egregious abandonment of the so-called "seduction theory." It is difficult to suspend my disbelief when I really actively disbelieve in the fundamentals of Lacanianism, from the Death Drive to "lack" (see Andrew Robinson's "The Political Theory of Constitutive Lack: A Critique" and, with Simon Tormey, "A Ticklish Subject? Žižek and the Future of Left Radicalism").

129. Shmuel Spector and Geoffrey Wigoder, *The Encyclopedia of Jewish Life Before and During the Holocaust* (Jerusalem: Yad Vashem, 2001), 3.1204–5.

130. Joseph Nedava, "Abba Gordin: A Portrait of a Jewish Anarchist," *Soviet Jewish Affairs* 4, no. 2 (1974): 74; A. N. Garjavin and A. V. Pashkin, "Problemy Vospitaniya, Obrazovaniya i Prosveshcheniya v Trudakh A. L. i V. L. Gordinykh [Problems of Upbringing, Education and Enlightenment in the Works of A. L. and V. L. Gordin]," *Prosveshcheniye Na Rusi, v Rossii: Istoricheskiy Opyt: Materialy Devyatnadtsatoy Vserossiyskoy Zaochnyy Nauchnoy Konferentsii [Enlightenment in Russia, in Russia: Historical Experience: Materials of the Nineteenth All-Russian Absentee Scientific Conference]*, edited by Sergey Nikolaevich Poltorak ("Nestor," 2000), 101; Levin in Hodies, *Smorgonie*, 221; Abba Gordin and Wolf Gordin, *Podrazhatel'no-Ponimatel'nyy Metod' dlya Obucheniya Gramote* (Vilna: Novaya Pedagogika, 1909), 7.

131. Leo Tolstoy, "Yasnaya Polyana School," *The Novels and Other Works of Lyof N. Tolstoi: The Long Exile and Other Stories*, translated by Nathan H. Dole (New York: C. Scribner's Sons, 1911), 267.

132. Gordin and Gordin, *Podrazhatel'no-ponimatel'nyi metod*, 9–10.

133. Tolstoy, "Yasnaya Polyana School," 263–68.

134. Paulo Freire, *Pedagogy of the Oppressed*, translated by Myra Bergman Ramos (New York: Continuum, 1970), 70.

135. A. L. Gordin and V. L. Gordin, "Deklaratsiya Pervyy Tsentral'nyy Sotsiotekhnikum [Declaration of the First Central Sociotechnicum]," in *Anarkhisty. Dokumenty i Materialy. 1883–1935 Gg.*, vol. 2 (Moscow: ROSSPEN, 1999), 204.

136. Br. Gordin [Abba Gordin], *Pochemu? ili Kak muzhik popal v stranu 'Anarkhiia'* ["Why": Or, How a Peasant Arrived in the Land of Anarchy] (Moscow: Mosk. federatsiya anarkhist. grupp, 1918), 34–35.

137. Br. Gordin [Abba Gordin], *Pochemu?*, 49–50.

138. Br. Gordin, *Pochemu?*, 50.

139. Heller, "Voyage" 258. Heller compares the Gordin brothers' Pochemu to such well-known protagonists as Voltaire's Candide and that stock figure of Russian folktales, Ivan the Fool ("Brat'ya Gordiny" 134). The premise of visiting a counterfactual world where more enlightened norms prevail, serving as the basis for a similar anarchist propaganda-novel for children by Jean Grave (*Les Aventures de Nono*, 1901, translated into Spanish the following year), is, in fact, a classic technique of utopian writing. More broadly, the process of calling norms into question through a dialogue in which one party operates from the position of ignorance has a lineage dating back at least to Socrates.

140. Max Stirner, *The Ego and Its Own*, edited by David Leopold, translated by Steven T. Byington (Cambridge: Cambridge University Press, 2000), 13.

141. Stirner, *The Ego and Its Own*, 14–15.

142. Kuchinov, "Russkiy Mars."

143. Gordin and Gordin, *Sderot ha-Iledim*, quoted in Hodies, ed., *Smorgonie*, 220.

144. Deleuze and Guattari, *A Thousand Plateaus*, 76.

145. Paul Goodman, *Speaking and Language: Defence of Poetry* (New York: Random House, 1971), 21; David Graeber, *Debt: The First 5,000 Years* (Brooklyn: Melville House, 2014), 122–24.

146. John L. Austin, *How to Do Things With Words* (Oxford: Oxford University Press, 1962), 76–77.

147. Kenneth Burke, *Language as Symbolic Action: Essays on Life, Literature, and Method* (Berkeley: University of California Press, 1966), 10.

148. Goodman, *Speaking and Language*, 21.

149. Goodman and Goodman, *Anarkhiia v Mechte*, 120.

150. Gordin and Gordin, in Hodies, ed., *Smorgonie*, 210.

151. Michel Foucault, *Discipline and Punish: The Birth of the Prison*, translated by Alan Sheridan (New York: Vintage Books, 1977), 25; Karin A. Martin, "Becoming a Gendered Body: Practices of Preschools," *American Sociological Review* 63, no. 4 (1998): 502, italics mine.

152. Peter J. Schakel, *The Way into Narnia: A Reader's Guide* (Grand Rapids, Mich.: William B. Eerdmans Pub. Company, 2005) 142-43; C. S. Lewis, *The Silver Chair: Book 4 in the Chronicles of Narnia* (New York: Collier Books, 1976), 1.

153. Hanoch Levin in Hodies, ed., *Smorgonie*, 223.

154. Abba and Wolf Gordin in Hodies, ed., *Smorgonie*, 212.

155. Levin in Hodies, ed., *Smorgonie*, 224.

156. Austin, *How to Do Things With Words*, 4, 7.

157. Abba and Wolf Gordin in Hodies, ed., *Smorgonie*, 215.

158. Abba and Wolf Gordin in Hodies, ed., *Smorgonie*, 217–18, modifications mine.

159. Mikhail Ignatievich Popov, qtd. in Anton Ivanovich Pervushin, *Kosmonavty Stalina. Mezhplanetnyy Proryv Sovetskoy Imperii* (Moskva: Eksmo, 2005), 357.

160. Lev Efimovich Kolodnyi, *Moskva v Ulitsakh i Litsakh: Zamoskvoreche: Avtorskii Putevoditel'* (Moscow: Golos-press, 2002), 290.

161. Quoted in Pervushin, *Kosmonavty Stalina*, 357.

162. Vladimir Zotov, "Na Poroge Zvezdnogo Veka," *Ural'skiy Sledopyt*, no. 11 (1977): 53.

163. Sergei N. Kuznetsov, "Linguistica Cosmica: Rozhdeniye 'Kosmicheskoy Paradigmy,'" *Sovremennaya Nauka* no. 2 (2014): 47; translation after Fiks.

164. Amalia Viktorovna Arolovich, *Anarkhizm-universalizm v kontekste russkoy "kosmicheskoy paradigmy" nachala XX veka* [Anarchism-Universalism In the Context of the Russian "Cosmic Paradigm" of the Early Twentieth Century], Diss., Moscow State University, 2005), 147.

165. Kuznecov, "Linguistica Cosmica" 47, 59.

166. Kuznecov, "Linguistica Cosmica," 52.

167. Kolodnyi, *Moskva v Ulitsakh i Litsakh*, 290; N. A. Rynin, *Interplanetary Flight and Communication Vol. II, No.4: Rockets* (Washington, D.C.: National Aeronautics and Space Administration and the National Science Foundation, 1971), 202.

168. Asif A. Siddiqi, *The Red Rockets' Glare: Spaceflight and the Soviet Imagination, 1857-1957* (New York: Cambridge University Press, 2013), 92n54; Frank Winter, *Prelude to the Space Age: The Rocket Societies, 1924-1940* (Washington, D.C.: Smithsonian Institution Press, 1983), 29; Jean-Louis Carra, *Essai Sur La Nautique Aérienne, Contenant l'art de Diriger Les Ballons Aérostatiques à Volonté, & d'accélérer Leur Course Dans Les Plaines de l'air, Avec Le Précis de Deux Expériences Particulières de Météorologie à Faire, Lu à l'Académie Royale Des Sciences de Paris, Le 14 Janvier 1784* (1784), 3.

169. China Miéville, *October: The Story of the Russian Revolution* (New York: Verso, 2018), 315.

170. Irina Sirotkina and Marina Kokorina, "The Dialectics of Labour in a Psychiatric Ward: Work Therapy in the Kaschenko Hospital," in *Psychiatry in Communist Europe*, edited by Sarah Marks and Matt Savelli (Basingstoke, UK: Palgrave Macmillan, 2015), 35; Marx, in *Marx and Engels Collected Works vol. 3*,182.

171. Deleuze and Guattari, *Anti-Oedipus*, 10; Gilles Deleuze and Félix Guattari, *Kafka: Toward a Minor Literature*, translated by Dana Polan (Minneapolis: University of Minnesota Press, 1986), 60.

172. Wolf Gordin, "K Osvobozhdeniyu!" in V. V. Krivenkiy, ed. *Anarkhisty: dokumenty i materialy, 1883–1935 gg. v. 2: 1917–1935 gg.* (Moscow: ROSSPĖN, 1999) 61.
173. Wolf Gordin et al, "Ni Za Kogo!" in Krivenkiy, ed., *Anarkhisty*, 62.
174. Gordin and Gordin, "Sotsial'naya tekhnika," in *Strana Anarkhiya*, 26.
175. Gordin and Gordin, "Deklaratsiya," 199.
176. Vserossiyskaya sektsiya anarkhistov-universalistov [Abba Gordin et al], "Osnovnyye polozheniya anarkhistov-universalistov," in Krivenkiy, ed., *Anarkhisty*, 284.
177. Vserossiyskaya sektsiya anarkhistov-universalistov [Abba Gordin et al], "Osnovnyye polozheniya anarkhistov-universalistov," in Krivenkiy, ed., *Anarkhisty*, 284.
178. Abba Gordin et al, "Deklaratsiya moskovskoy organizatsii anarkho-universalistov (K vos'momu s"yezdu Sovetov)" in Krivenkiy, ed, *Anarkhisty*, 298.
179. Abba Gordin, Zikhroynes vol. 2, p. 366.
180. la paperson [K. Wayne Yang], *A Third University Is Possible* (Minneapolis: U of Minnesota Press, 2017), n.p.
181. Gordin and Gordin, "Deklaratsiya," 201.
182. Gordin and Gordin, "Deklaratsiya," 204.
183. Wolf Gordin quoted in Amalia Viktorovna Arolovich, *Anarkhizm-universalizm v kontekste russkoy "kosmicheskoy paradigmy" nachala XX veka* [*Anarchism-Universalism In the Context of the Russian "Cosmic Paradigm" of the Early Twentieth Century*] (Diss., Moscow State University, 2005), 146.
184. Quoted in Arolovich, *Anarkhizm-universalizm*, 148–49.
185. Victor Serge, "New Tendencies in Russian Anarchism," in *Anarchists Never Surrender: Essays, Polemics, and Correspondence on Anarchism, 1908–1938*, edited and translated by Mitchell Abidor (Oakland: PM Press, 2015), 186.
186. Sirotkina and Kokorina, "Dialectics of Labour," 36.
187. Allan Antliff, "Anarchy, Power, and Poststructuralism," *SubStance* 36, no. 2 (Aug. 2007): 62.

10. Unicorn Rhetoric

1. I am temporarily adopting something like the convention popularized by George Lakoff for the study of conceptual metaphors. For Lakoff, small capitals are used to identify a recurring figure of thought (e.g., the notion that TIME IS MONEY); I am using it in the same spirit, only without much of an attempt at linguistic rigor.
2. Graham Harman, *Bruno Latour: Reassembling the Political* (London: Pluto Press, 2014), viii–ix.
3. Harman, *Bruno Latour*, 105; Andreas Malm, *The Progress of This Storm: Nature and Society in a Warming World* (New York: Verso Books, 2018), n.p.
4. Harman, *Bruno Latour*, viii.
5. A Bro Tryna Catch His Unicorn, "Unicorn," *Urban Dictionary*, 12 Feb. 2012, https://www.urbandictionary.com/define.php?term=Unicorn.
6. Kathleen Smith, *The Fangirl Life: A Guide to Feeling All the Feels and Learning How to Deal* (New York: TarcherPerigee, 2016), 4–5.
7. "ecstasy, n.," OED Online, September 2022, Oxford University Press, https://www-oed-com.pnw.idm.oclc.org/view/Entry/59423?rskey=fTcHda&result=1&isAdvanced=false (accessed 3 Dec. 2022).
8. Smith, *Fangirl Life*, 4.
9. ThinkGeek quoted in David Pescovitz, "Unicorn Chaser Beverage from ThinkGeek!" *Boing Boing*, 1 Apr. 2009, https://boingboing.net/2009/04/01/unicorn-chaser-bever.html.

10. Charlie Wells, "Unicorns, Unicorns Everywhere," *The Wall Street Journal* (11 Mar. 2015): D1.

11. Arianna Davis, "What's Really Behind Unicorn Fever," *Refinery29*, https://www.refinery29.com/en-us/2017/05/152423/unicorn-trend-explanation-history.

12. "Save Me from This Epidemic of Unicorns," *Daily Telegraph (London)* (Jan. 2019): 21.

13. Elizabeth Balboa, "The Fabled Starbucks Unicorn Frappuccino: 'It's Like Watered-Down Cotton Candy,'" Benzinga, Apr. 2017, https://www.benzinga.com/general/restaurants/17/04/9319107/the-fabled-starbucks-unicorn-frappuccino-its-like-watered-down-cot.; Maura Judkis, "Starbucks' Unicorn Frappuccino Tastes like Sour Birthday Cake and Shame," *Washington Post* (19 Apr. 2017), https://www.washingtonpost.com/news/food/wp/2017/04/19/starbucks-unicorn-frappuccino-tastes-like-sour-birthday-cake-and-shame/?noredirect=on.

14. Paul Fussell, *Class: A Guide through the American Status System* (New York: Summit Books, 1983), 123–24; Florian Cramer, "Rosa Einhörner [Pink Unicorns]," *pleintekst.nl*, http://cramer.pleintekst.nl/essays/rosa_einhoerner/.

15. Wells, "Unicorns, Unicorns Everywhere," D1.

16. Tim Grierson, "'Unicorn Store': Toronto Review," *Screen Daily*, 13 Sept. 2017, https://www.screendaily.com/reviews/unicorn-store-toronto-review/5122310.article.

17. Camille Vigreux, quoted in Sophie Barel, "La symbolique de la licorne et sa place dans la sémiologie du web?" *Medium*, 29 Feb. 2016, https://medium.com/@SophieB./la-symbolique-de-la-licorne-et-sa-place-dans-la-s%C3%A9miologie-du-web-6ec33ef2745f., translation mine.

18. Cramer, "Rosa Einhörner."

19. Sokolowski qtd. in Mark Peters, "The Linguistic Sex Appeal of the Unicorn," *The Boston Globe*, 21 Feb. 2016, https://www.bostonglobe.com/ideas/2016/02/21/see-any-unicorns-lately/oCEhFoIgeaHw71hjbKSWVO/story.html.

20. Cramer, "Rosa Einhörner."

21. Davis, "What's Really Behind Unicorn Fever."

22. Cramer, "Rosa Einhörner"; Landyn Pan and Anna Moore, *The Gender Unicorn*, TSER: Trans Student Educational Resources, 2014, http://www.transstudent.org/gender.

23. Frank Lowe, "Editor's Note," *Raised by Unicorns: Stories from People with LGBTQ+ Parents*, edited by Frank Lowe (Jersey City, NJ: Red Wheel/Weiser, 2018), 3–4.

24. Tiffany Haddish, *The Last Black Unicorn* (New York: Simon and Schuster, 2019), 124.

25. Cramer, "Rosa Einhörner."

26. Devin Thorpe, "Successful African-American Silicon Valley Entrepreneur Feels 'Like A Black Unicorn,'" *Forbes*, Jan. 2018, https://www.forbes.com/sites/devinthorpe/2018/01/15/successful-african-american-silicon-valley-entrepreneur-feels-like-a-black-unicorn/#58220a6e29ab.

27. Team Unicorn, "About," *Team Unicorn FTW*, 17 July 2011, https://web.archive.org/web/20110717163826/http://teamunicornftw.com/about/.

28. Katie L. Kamelamela, "Tweet: 'For the Record . . .'" *Twitter*, 14 July 2019, https://twitter.com/kteabam/status/1150561729421516800?s=20.

29. Joreth, "Unicorn Hunter," *Urban Dictionary*, 8 Apr. 2010, https://www.urbandictionary.com/define.php?term=Unicorn%20Hunter.

30. Richard Dawkins, *The God Delusion* (Boston: Houghton Mifflin, 2006), 52-53; Cramer, "Rosa Einhörner."

31. Ian Chant, "North Korea Finds Ancient Unicorn Lair, Because Sure, Why Not," *The Mary Sue*, 30 Nov. 2012, https://www.themarysue.com/north-korea-unicorn/.

32. Ian Chant, "Bad News, Everyone—North Korean Unicorn Is Actually Just A Stupid Kirin," *The Mary Sue*, 7 Dec. 2012, https://www.themarysue.com/korean-unicorn-actually-a-kirin/.

33. Farhad Manjoo, "Fitting Label, Yet Unicorns Are No Myth," *The New York Times* (6 July 2015), B1, B4.

34. Roger Parloff, "This CEO's out for Blood." *Fortune* 169, no. 9 (June 2014): 66, Dan Primack, "Theranos Controversy Has Little to Do with 'Unicorns,'" *Fortune.com*, 22 Oct. 2015; Andrew Hill, "Fresh Blood: Why Everyone Fell for Theranos," *Financial Times*, 19 Mar. 2018, https://www.ft.com/content/a45ffdf0-2850-11e8-b27e-cc62a39d57a0.; "Herd of Unicorns," *Servant Financial News*, 18 Aug. 2016, https://www.servantfinancialnews.com/2016/08/herd-of-unicorns.html.

35. Rebecca Collard, "Why Syrian 'Safe Zones' Could Be Dangerous for Civilians and U.S. Policy," *Time*, July 2015, https://time.com/3979906/syria-safe-zones/; Ted Cruz in "Transcript: Republican Presidential Debate," *The New York Times*, 15 Dec. 2015, https://www.nytimes.com/2015/12/16/us/politics/transcript-main-republican-presidential-debate.html; "Peters: Obama Is 'Chasing a Unicorn' in Iraq and Syria," *Fox News Insider*, 2 Mar. 2015, https://insider.foxnews.com/2015/03/02/peters-sounds-after-us-backed-syrian-group-disbands; Jeffrey Goldberg in "Crisis in Syria," italics mine

36. Alan Rappeport, "On Budget, No Clarity, and So Far No 'Magic Unicorn,'" *New York Times*, 25 May 2017, p. A15.

37. Amy Deen Westbrook and David A. Westbrook, "Unicorns, Guardians, and the Concentration of the U. S. Equity Markets," *Nebraska Law Review* 96, no. 3 (2018): 697, 704–12.

38. Roger Cohen, "Robert Mueller in the Age of the Unicorn," *The New York Times*, 19 Apr. 2019, https://www.nytimes.com/2019/04/19/opinion/mueller-trump-russia-report.html.

39. John Nolte, "Benjamin Kunkel: The Lena Dunham of Literature," *Breitbart*, 12 Mar. 2014, https://www.breitbart.com/entertainment/2014/03/12/benjamin-kunkel-the-lena-dunham-of-literature/.

40. Cheryl K. Chumley, "AOC's Unicorn Flight of 'Green New Deal' Fancy," *The Washington Times*, 9 Feb. 2019, https://www.washingtontimes.com/news/2019/feb/9/alexandria-ocasio-cortezs-unicorn-flight-green-new/; Ben Shapiro, "AOC's Green New Deal Proposal Is One Of The Stupidest Documents Ever Written," *The Daily Wire*, 7 Feb. 2019, /news/43194/aocs-green-new-deal-proposal-one-stupidest-ben-shapiro; Katherine Timpf quoted in Katherine Hignett, "AOC Is in 'Bananaland': Fox News Contributor Slams Green New Deal in Unicorns-and-Cannibalism Tirade," *Newsweek*, 3 Mar. 2019, https://www.newsweek.com/alexandria-ocasio-cortez-cannibalism-green-new-deal-fox-news-katherine-timpf-1349999.

41. Matt McGrath, "Caution Urged over 'Carbon Unicorns,'" *BBC News*, 5 Oct. 2018. www.bbc.com, https://www.bbc.com/news/science-environment-45742191; Holmes à Court in Sabra Lane, "Carbon Capture a 'Unicorn' for Power Generation," interview with Simon Holmes à Court, *AM*, Australian Broadcasting Corporation, https://www.abc.net.au/am/content/2016/s4677884.htm.

42. Abbey Volcano, "Solidarity Unicornism and the Future of the Left: WSA's May Day Poster Explained," *Ideas and Action*, 25 Apr. 2012, http://ideasandaction.info/2012/04/solidarity-unicornism-and-the-future-of-the-left-wsas-may-day-poster-explained/.

43. Harold Beaver, "Homosexual Signs (In Memory of Roland Barthes)," *Critical Inquiry* 8, no. 1 (Oct. 1981): 106.

44. Baynard Woods, "How Unicorn Riot Covers the Alt-Right Without Giving Them a Platform," *Columbia Journalism Review*, 1 Nov. 2017, https://www.cjr.org/united_states_project/charlottesville-alt-right-unicorn-riot.php.

45. Angela Nagle, *Kill All Normies: Online Culture Wars From 4Chan and Tumblr to Trump and the Alt-Right* (Winchester, UK: Zero Books, 2017), 16–17.

46. Vivian Kane, "We Are in Love with the New She-Ra Design (And the Awful Men Hating on It Is Just a Bonus)," *The Mary Sue*, 16 July 2018, https://www.themarysue.com/new-she-ra-trolls-gtfo/.

47. Johan Huizinga, *Homo Ludens: A Study of the Play-Element in Culture* (London: Routledge & Kegan Paul, 1980), 10, 12, italics mine.

48. John Suler, "The Online Disinhibition Effect," *Cyberpsychology & Behavior: The Impact of the Internet, Multimedia and Virtual Reality on Behavior and Society* 7, no. 3 (July 2004): 321.

49. Suler, "The Online Disinhibition Effect," 321.

50. Brandon Morse, "My Little Pony to Children: Marxism Is Not Magic," *The Federalist*, 8 Apr. 2015, https://thefederalist.com/2015/04/08/my-little-pony-to-children-marxism-is-not-magic/.

51. Kathleen Richter, "My Little Homophobic, Racist, Smart-Shaming Pony," *Ms. Magazine*, 9 Dec. 2010, https://msmagazine.com/2010/12/09/my-little-homophobic-racist-smarts-shaming-pony/.

52. Peter Sloterdijk, *The Critique of Cynical Reason*, translated by Michael Eldred (Minneapolis: University of Minnesota Press, 1987), 320, 242.

11. The Fantastic from Counterpublic to Public Imaginary

1. Soumya Karlamangla, "I've Seen 'Contagion' Four Times. No, the Coronavirus Outbreak Isn't the Same," *Los Angeles Times*, 11 Mar. 2020, www.latimes.com, https://www.latimes.com/california/story/2020-03-11/coronavirus-contagion-outbreak-movie-comparison; Times Editorial Board, "Editorial: Panic Is Making Coronavirus Worse for Everyone," *Los Angeles Times*, 11 Mar. 2020, https://www.latimes.com/opinion/story/2020-03-11/panic-covid-19-worse-for-everyone; Katrin Park, "COVID-19: Life in Seoul Amid the Epidemic," *The Diplomat*, 11 Mar. 2020, thediplomat.com, https://thediplomat.com/2020/03/covid-19-life-in-seoul-amid-the-epidemic/.

2. Jason Kottke, "Decrease in Economic Activity Due to COVID-19 Reduced Air Pollution and Saved Lives," Kottke.Org: Home of Fine Hypertext Products, 11 Mar. 2020. kottke.org, https://kottke.org/20/03/covid-19-reduced-air-pollution.

3. Derek Beres, "Will the Future Be 'Mad Max' or 'Star Trek'? Coronavirus Offers Clues," *Big Think*, 11 Mar. 2020, bigthink.com, https://bigthink.com/politics-current-affairs/jordan-hall-coronavirus; Andrea Remke, "To panic or not to panic: A single mom's coronavirus worries," March 11, 2020, *Today Parenting Team*, https://community.today.com/parentingteam/post/to-panic-or-not-to-panic-single-moms-coronavirus-worries.

4. Istvan Csicsery-Ronay, *The Seven Beauties of Science Fiction* (Middletown, CT: Wesleyan University Press, 2012), 2.

5. @twakeonline, "Sarah Palin Singing Baby Got Back as the World Burns. This Is the Darkest Timeline," *Twitter*, 11 Mar. 2020, https://twitter.com/twakeonline/status/1237936502664916994. Lest we conclude too quickly that the density of fantastic scenarios and imagery swirling around that particular day in 2020 was solely due to the estranging influence of the pandemic, a momentary conjuncture, we might reflect, nearly three years on, on Kanye West's increasingly bizarre allegations about Elon Musk and Barack Obama being "clones" ("they probably made 10 to 30 Elons," he mused [*Newshub*, "Kanye West Claims Elon Musk Is a Clone in Latest Bizarre Rant," sec. *Entertainment*, accessed 7 Dec. 2022, https://www.newshub.co.nz/home/entertainment/2022/12/kanye-west-claims-elon-musk-is-a-clone-in-latest-bizarre-rant.html]) or the final weeks of Herschel Walker's 2022 campaign for a Senate seat in Georgia, a goal he came within less than a hundred thousand votes of obtaining, despite the fact that he had fantasized publicly about the preferability of being a werewolf to being a vampire (vampires, you see, lack "faith" [Graig Graziosi, "Herschel Walker Moans about Obama Mocking His

Vampire vs Werewolf Debate," *The Independent*, 5 Dec. 2022, https://www.independent.co.uk/news/world/americas/us-politics/herschel-walker-obama-werewolf-debate-b2239441.html]).

6. Gary K. Wolfe, *Evaporating Genres: Essays on Fantastic Literature* (Middletown, CT: Wesleyan University Press, 2011), viii; Broderick, *Reading By Starlight*, xi; Csicsery-Ronay, *The Seven Beauties of Science Fiction*, 2.

7. Wolfe, *Evaporating Genres*, 57; Bould and Vint, "There Is No Such Thing as Science Fiction," 50.

8. Wolfe, *Evaporating Genres*, 17; Chu, *Do Metaphors Dream of Literal Sleep*, 7. Let this expression, "the fantastic," serve here not to indicate any particular theorization of genre (certainly not Tzvetan Todorov's classic but prohibitively narrow category) but to shorthand a.) a massively hybrid transmedia "supergenre" encompassing science fiction, fantasy, and horror, the notoriously genre-straddling phenomena of the superhero narrative, RPGs of the tabletop, electronic, and live-action varieties, and more—in short, more or less the entity gestured at but left pointedly undefined by the International Association for the Fantastic in the Arts and its eponymous journal, along with b.) its accompanying "megatext" or "encyclopedia of tropes and enabling devices" (Broderick) and c.) the affective-cognitive aura surrounding this supergenre, an experiential quality that Chu, Ndalianis, Vint, and Csicsery-Ronay refer to as "the science-fictional" or "science-fictionality," but which could perhaps more accurately be called "fantasticality."

9. Chu, *Do Metaphors Dream of Literal Sleep*, 6-7; Brooks Landon, "Synthespians, Virtual Humans, and Hypermedia: Emerging Contours of Post-SF Film," in *Edging into the Future: Science Fiction and Contemporary Cultural Transformation*, edited by Veronica Hollinger and Joan Gordon (Philadelphia: University of Pennsylvania Press, 2002), 59.

10. John Clute, *Pardon This Intrusion* (London: Gateway).

11. Sherryl Vint, *Science Fiction: A Guide for the Perplexed* (London: A&C Black, 2014), 163.

12. Adam Vary, "The Year of the Geek," *Entertainment Weekly*, 21 Dec. 2007, https://ew.com/article/2007/12/21/year-geek/; "San Diego Comic-Con—Frequently Asked Questions," *San Diego Comic-Con Unofficial Blog*, 16 Dec. 2019, https://sdccblog.com/san-diego-comic-con-frequently-asked-questions/.

13. René Schallegger, "The Nightmares of Politicians: On the Rise of Fantasy Literature from Subcultural to Mass-Cultural Phenomenon," in *Collision of Realities: Establishing Research on the Fantastic in Europe*, edited by Lars Schmeink and Astrid Böger (Berlin: De Gruyter, 2012), 30; Adam Roberts, *The History of Science Fiction* (Basingstoke: Palgrave Macmillan, 2006), 499.

14. Michael Warner, *Publics and Counterpublics* (New York: Zone Books, 2005), 29.

15. Elena Machado Sáez, "Dictating Desire, Dictating Diaspora: Junot Díaz's 'The Brief Wondrous Life of Oscar Wao' as Foundational Romance," *Contemporary Literature* 52, no. 3, 2011): 522–23.

16. Bruce Sterling, "Preface," in William Gibson, *Burning Chrome* (New York: Ace Books, 1987), xi.

17. Samuel R. Delany, *Silent Interviews: On Language, Race, Sex, Science Fiction, and Some Comics—A Collection of Written Interviews* (Hanover, NH: Wesleyan University Press, 2011), 71.

18. Leslie A. Fiedler, "Introduction," in *In Dreams Awake: A Historical-Critical Anthology of Science Fiction* (New York: Dell, 1975), 18.

19. Sarah Brouillette, "Corporate Publishing and Canonization: *Neuromancer* and Science-Fiction Publishing in the 1970s and Early 1980s," *Book History* 5 (2002): 193.

20. Catherine Belsey, *Critical Practice* (London: Routledge, 2002), 62, 47–48.

21. Tom Frank, "Alternative to What?" *The Baffler*, no. 5 (1993): 5–14.

22. Wolfe, *Evaporating Genres*, xi, 51.

23. Landon, *Science Fiction*, 15; Wolfe, *Evaporating Genres*, 2.

24. Author Earnings, *2018 SFWA Nebula Conference Presentation*, 24 May 2018, https://web.archive.org/web/20180613083320/http://authorearnings.com/sfwa2018/.

25. Roberts, *History of Science Fiction*, 479-80.

26. Sam Otten, "DC v. Marvel Comic Sales Before Rebirth," *Comic and Screen*, 23 Jan. 2016, http://comicandscreen.blogspot.com/2016/01/dc-v-marvel-comic-sales-before-rebirth.html.

27. Christine Folch, "Why the West Loves Sci-Fi and Fantasy: A Cultural Explanation," *The Atlantic*, 13 June 2013, https://www.theatlantic.com/entertainment/archive/2013/06/why-the-west-loves-sci-fi-and-fantasy-a-cultural-explanation/276816/.

28. Peter Cullen Bryan, "Geeking Out and Hulking Out: Toward an Understanding of Marvel Fan Communities," in *Age of the Geek: Depictions of Nerds and Geeks in Popular Media*, edited by Kathryn E. Lane (Cham, Switzerland: Springer International Publishing, 2018), 153; Landon, *Science Fiction*, xv.

29. Kristina Busse, "Geek Hierarchies, Boundary Policing, and the Gendering of the Good Fan." Participations 10, no. 1 (2013): 77; Henry Jenkins, Sam Ford, and Joshua Green, *Spreadable Media: Creating Value and Meaning in a Networked Culture* (New York: NYU Press, 2018), 142.

30. Patton Oswalt, "Wake Up, Geek Culture. Time to Die," *Wired* 19, no. 1 (Dec. 2010), www.wired.com, https://www.wired.com/2010/12/ff-angrynerd-geekculture/.

31. For an introduction to (and intellectual history of) the concept of "imaginaries," see Saskia Cousin's "Images, Imaginaries, and Imaginations."

32. Erin Tapken, "50 Facts for 50 Years of San Diego Comic-Con (Part 2)," San Diego Comic-Con Unofficial Blog, 13 June 2019, https://sdccblog.com/2019/06/50-facts-for-50-years-of-san-diego-comic-con-part-2/.

33. Dave Itzkoff quoted in Marleen S. Barr, "Fantastic Language/Political Reporting: The Postcolonial Science Fiction Illocutionary Force Is with Us," in *The Postnational Fantasy: Essays on Postcolonialism, Cosmopolitics and Science Fiction*, edited by Masood Ashraf Raja et al. (Jefferson, NC: McFarland, 2011), 188.

34. Howard Bruce Franklin, *War Stars: The Superweapon and the American Imagination* (Amherst: Univ. of Massachusetts Press, 2008); Thomas M. Disch, *The Dreams Our Stuff Is Made Of: How Science Fiction Conquered the World* (New York: Simon and Schuster, 2000) 78-96; David Meerman Scott and Richard Jurek, *Marketing the Moon: The Selling of the Apollo Lunar Program* (Cambridge, Mass.: MIT Press, 2014), 3–11.

35. Michael Cavna, "Alexandria Ocasio-Cortez Responded to Critics by Quoting Alan Moore. Many Nerds Rejoiced in Response," *Washington Post*, 11 Jan. 2019, https://www.washingtonpost.com/arts-entertainment/2019/01/11/alexandria-ocasio-cortez-responded-critics-by-quoting-alan-moore-many-nerds-rejoiced-response/; Valerie Oved Giovanini, "An Army of Me: Representations of Intersubjective Relations, Ethics, and Political Resistance in The Handmaid's Tale," *Free Associations*, no. 75 (May 2019): 84–101; Kaulingfreks, Ruud, and Femke Kaulingfreks. "In Praise of Anti-Capitalist Consumption: How the V for Vendetta Mask Blows up Hollywood Marketing," *Ephemera* 13, no. 2 (2013): 453-7; Brian Cronin, "A History of The Punisher Logo Being Used By Police, Military & Politicians," *CBR*, July 2019, https://www.cbr.com/punisher-history-logo-used-police-military-politicians/amp/; Arthur Chu, "Your Princess Is in Another Castle: Nerds and Misogyny," *The Daily Beast* (blog), May 27, 2014, https://www.thedailybeast.com/your-princess-is-in-another-castle-misogyny-entitlement-and-nerds-1; John Scalzi, "Straight White Male: The Lowest Difficulty Setting There Is," *Whatever*, 15 May 2012, *whatever.scalzi.com*, https://whatever.scalzi.com/2012/05/15/straight-white-male-the-lowest-difficulty-setting-there-is/; Daniel W. Drezner, "Metaphor of the Living Dead: Or, the Effect of the Zombie Apocalypse on Public Policy Discourse," *Social Research: An International Quarterly* 81, no. 4 (2014): 825-49; John Wagner, "'SANCTIONS ARE COMING': Trump Tweets out 'Game of Thrones'–Inspired Warning to Iran," *Washington Post*, 2 Nov. 2018, https://

www.washingtonpost.com/politics/sanctions-are-coming-trump-tweets-out-game-of-thrones-inspired-warning-to-iran/2018/11/02/86eef714-deba-11e8-85df-7a6b4d25cfbb_story.html; Morten Bay, "Weaponizing the Haters: *The Last Jedi* and the Strategic Politicization of Pop Culture through Social Media Manipulation," *First Monday* 23, no. 11, Nov. 2018.

36. Those perennially excluded from the American Dream could rightly complain, as in the response to the white feminism of *The Handmaid's Tale* and its cosplayers, that "there is no 'normal' to return to," that outside of whiteness' protective dome, apocalypse and dystopia are a continuous present tense (van Veen 64, 69). As N. K. Jemisin reminds us, "an apocalypse is a relative thing": what white people remember as "normal" *is* the end of the world for everyone else (6, 170).

37. Richard Barbrook an Andy Cameron, "The Californian Ideology," Science as Culture 6, no. 1 (1996): 48–50; Tad Friend, "Sam Altman's Manifest Destiny," The New Yorker, 10 Oct. 2016, www.newyorker.com, https://www.newyorker.com/magazine/2016/10/10/sam-altmans-manifest-destiny. And so it is that this afternoon (May 17, 2020), I find the following succinct dialectic unfolding on Twitter:

Elon Musk @elonmusk: Take the red pill 💊
Ivanka Trump (@IvankaTrump): Taken!
Lilly Wachowski @lilly_wachowski: Fuck both of you

38. Daniel W. Drezner, "Metaphor of the Living Dead: Or, the Effect of the Zombie Apocalypse on Public Policy Discourse," *Social Research: An International Quarterly* 81, no. 4 (2014): 825.

39. Barr, "Fantastic Language/Political Reporting," 190.

40. For Jutta Weldes, this reveals a key weakness in Suvin's theorization of SF as an allegory for real-world politics. "SF is not just a 'window' onto an already pre-existing world," she argues. "Rather, SF texts are part of the processes of world politics themselves: they are implicated in *producing and reproducing* the phenomena that . . . [Suvinian critics] assume they merely reflect" (Jutta Weldes, "Popular Culture, Science Fiction and World Politics," in *To Seek out New Worlds: Science Fiction and World Politics*, ed. Jutta Weldes [London: Palgrave, 2003], 12, emphasis mine).

41. Lilly Wachowski and Lana Wachowski, dirs., *The Matrix*, Warner Bros., 1999.

42. Marc Tuters, "LARPing & Liberal Tears: Irony, Belief and Idiocy in the Deep Vernacular Web," in *Post-Digital Cultures of the Far Right: Online Actions and Offline Consequences in Europe and the US*, edited by Maik Fielitz and Nick Thurston (Bielefeld, Germany: Transcript Verlag, 2019), 44; Alexandra Minna Stern, *Proud Boys and the White Ethnostate: How the Alt-Right Is Warping the American Imagination* (Boston: Beacon Press, 2019), 21; Shawn P. Van Valkenburgh, "Digesting the Red Pill: Masculinity and Neoliberalism in the Manosphere," *Men and Masculinities* (Dec. 2018): 4–6.

43. Luke Munn, "Alt-Right Pipeline: Individual Journeys to Extremism Online," *First Monday* 24, no. 6 (June 2019). The gaming metaphor has been literalized, weaponized, most notably in a series of white supremacist mass shootings that have led observers to speak of a "gamification of mass violence": livestreaming massacres with POV video and custom soundtracks, speaking of the body counts as "high scores," redpilled killers turn their victims into NPCs in "a First Person Shooter video game" (Robert Evans, "The El Paso Shooting and the Gamification of Terror," bellingcat, August 4, 2019, https://www.bellingcat.com/news/americas/2019/08/04/the-el-paso-shooting-and-the-gamification-of-terror/).

44. *Shields and Shield Tactics Primer*, 2017, https://www.unicornriot.ninja/wp-content/uploads/2017/09/Shields_and_Shield_Tactics_Primer.pdf, 1.

45. Clio Chang, "The Unlikely Rise of an Alt-Right Hero," *The New Republic*, Mar. 2017, https://newrepublic.com/article/141766/unlikely-rise-alt-right-hero.

46. Justin Wm Moyer and Lindsey Bever. "Vanguard America, a White Supremacist Group, Denies Charlottesville Ramming Suspect Was a Member," *Washington Post*, 15 Aug. 2017, www.washingtonpost.com, https://www.washingtonpost.com/local/vanguard-america-a-white-supremacist-group-denies-charlottesville-attacker-was-a-member/2017/08/15/2ec897c6-810e-11e7-8072-73e1718c524d_story.html.

47. brown, adrienne maree. "Outro," in *Octavia's Brood: Science Fiction Stories from Social Justice Movements*, edited by Walidah Imarisha and Adrienne Maree Brown (Oakland: AK Press, 2015), 279.

48. Walidah Imarisha, "Introduction," in *Octavia's Brood: Science Fiction Stories from Social Justice Movements*, edited by Walidah Imarisha and Adrienne Maree Brown (Oakland: AK Press, 2015), 3.

49. Ethan Gilsdorf, "A Game as Literary Tutorial," *The New York Times*, 13 July 2014, https://www.nytimes.com/2014/07/14/books/dungeons-dragons-has-influenced-a-generation-of-writers.html.

50. Augusto Boal, *Theatre of the Oppressed*, translated by Charles A. McBride and Maria Odilia Leal McBride (New York: Theatre Communications Group, 2013), 122; brown, "Outro," 279; Walidah Imarisha, "Rewriting the Future: Using Science Fiction to Re-Envision Justice," *Bitch Media*, 11 Feb. 2015, https://www.bitchmedia.org/article/rewriting-the-future-prison-abolition-science-fiction.

51. Santos Hugo, Saldanha Lucinda, Pinto Marta, and Ferreira Pedro. "Civic and Political Transgressions in Videogames: The Views and Experiences of the Players," in *DiGRA Nordic '18: Proceedings of 2018 International DiGRA Nordic Conference*, 2018. http://www.digra.org/wp-content/uploads/digital-library/DIGRA-NORDIC-2018-paper-27.pdf, 2; Mark Kaethler, "Failed Feminist Interventions in Wolfenstein II: The New Colossus," in *Feminist War Games?: Mechanisms of War, Feminist Values, and Interventional Games*, edited by Jon Saklofske et al. (Abingdon, UK: Routledge, 2020).

52. Laurie Penny, "Harry Potter and the Conscience of a Liberal," *The Baffler* (2 Sept. 2016), thebaffler.com, https://thebaffler.com/war-of-nerves/harry-potter-laurie-penny.

53. Miriam Rosenberg Roček quoted in Claire Nally, *Steampunk: Gender, Subculture and the Neo-Victorian* (London: Bloomsbury Publishing, 2019), 47.

54. Daniel Colson, "L'Ange de l'histoire," *Le Monde Libertaire* no. 1377 (Nov. 2004), http://www.federation-anarchiste.org/spip.php?article892; Colson, *Lexicon*, 82; Nally, *Steampunk*, 46, 103; Colson, *Lexicon*, 83.

55. Fisher, 4-8; Chantelle Gray, *Anarchism After Deleuze and Guattari: Fabulating Futures* (London: Bloomsbury Publishing, 2022), 14.

56. Bil'in Popular Committee, "Bil'in Weekly Demonstration Reenacts the *Avatar* Film," bilin-village.org, December 2, 2010, https://web.archive.org/web/20100217175438/http://www.bilin-village.org/english/articles/testimonies/Bilin-weekly-demonstration-reenacts-the-Avatar-film.

57. Along similar lines, we might think of the use of the three-finger salute from *The Hunger Games* as a gesture of defiance by dissidents in Thailand, the Philippines, Myanmar, and Hong Kong (Arvyn Cerézo, "THE HUNGER GAMES' Three-Finger Salute: A Symbol of Resistance to Tyranny in Asia," *BOOK RIOT* (blog), August 27, 2021. https://bookriot.com/the-hunger-games-three-finger-salute/).

58. Activestills, "Hundreds Mark 13 Years of Protests against the Wall in Bil'in," *+972 Magazine*, 2 Mar. 2018, https://www.972mag.com/hundreds-mark-13-years-of-protests-against-the-wall-in-bilin/.

59. In an interesting counterpoint to this deployment of fantastic imaginary and fannish practices, on May 7, 2020, Trump campaign manager Brad Parscale boasted on Twitter that the

"juggernaut campaign" he had prepared was a "Death Star," and threatened to "start pressing FIRE for the first time." Mark Hamill, the iconic actor portraying Luke Skywalker in the *Star Wars* films, fired back: "Your overconfidence is your weakness." Here is a case of what Kenneth Burke called "the 'stealing back and forth' of symbols" (Burke, *Attitudes Toward History*, 103): from the entertainment megatext, a political operative draws an image of irresistible power, and an entertainer scores a political point by reminding the audience of the canonical narrative outcome to wide applause. But here it is the more hegemonic actor (in the political sense) who makes a more canonically fannish use of the megatext by reading it anti-canonically, inverting the values attributed to the narrative's actants and actions, whereas the "resistance" to this "textual poaching" comes in the form of an attempt to stitch the stolen imagery back into its fixed, proprietary narrative, the one that everybody knows. Ironically, it is Parscale rather than Hamill who enacts a "resistant" reading in the cultural-studies sense—albeit in the service of the rich and powerful.

60. Gilles Deleuze, *Nietzsche and Philosophy*, translated by Hugh Tomlinson (New York: Continuum, 2006), 3, 6.

61. Yann Moulier Boutang, "Antagonism Under Cognitive Capitalism: Class Composition, Class Consciousness and Beyond," *Immaterial Labour, Multitudes and New Social Subjects: Class Composition in Cognitive Capitalism*, 28 April, 2006, King's College, University of Cambridge, Seminar Presentation.

62. Heather Mendick and Becky Francis, "Boffin and Geek Identities: Abject or Privileged?" *Gender and Education* 24, no. 1 (Jan. 2012): 16.

63. Nicole B. Ellison and Danah M. Boyd, "Sociality Through Social Network Sites," in *The Oxford Handbook of Internet Studies*, edited by William H. Dutton (Oxford: Oxford University Press, 2013), 163.

64. Timothy Morton, *Hyperobjects: Philosophy and Ecology after the End of the World* (Minneapolis: University of Minnesota Press, 2014); Chu, *Do Metaphors Dream of Literal Sleep*, 9, 80–81.

65. Katherine Hayles quoted in Joshua Clover, "Retcon: Value and Temporality in Poetics." *Representations* 126, no. 1 (2014): 17; Clover, "Retcon," 14–15, 10.

66. Joseph William Singer, "The Indian States of America: Parallel Universes & Overlapping Sovereignty," *American Indian Law Review* 38, no. 1 (2013): 6.

67. Junot Díaz, *The Brief Wondrous Life of Oscar Wao* (New York: Penguin, 2008), 6.

68. Nalo Hopkinson, "A Reluctant Ambassador from the Planet of Midnight," *Journal of the Fantastic in the Arts* 21, no. 3 (2010): 346.

69. Silvio O. Funtowicz and Jerome R. Ravetz, "Science for the Post-Normal Age," *Futures* 25, no. 7 (1993): 739–55.

70. Stephen Oghenemuro Okpadah, "The Postnormal Condition and the Politics of Migration in Biyi Bandele's *Half of a Yellow Sun*," *VTU Review: Studies in the Humanities and Social Sciences* 3, no. 1 (2019): 42.

71. Kenneth Burke, *Attitudes Toward History*, 3rd ed. (Berkeley: University of California Press, 1984), 3, 21–22.

12. Acting Supernaturally (with Notes on the Monster of Anarchy)

1. Murray Bookchin, *To Remember Spain: The Anarchist and Syndicalist Revolution of 1936* (Oakland: AK Press, 1994), 17.

2. Thomas M. Disch, "The Embarrassments of Science Fiction," On SF (Ann Arbor: University of Michigan Press, 2005), 10.

3. Colson, *Lexicon*, 27.

4. Mikhail Aleksandrovich Bakunin, *Bakunin on Anarchism*, edited by Sam Dolgoff (Montréal: Black Rose Books, 1980), 327; Adorno and Horkheimer, *Dialectic of Enlightenment*, 9.

5. Paul-François Tremlett, "On the Formation and Function of the Category 'Religion' in Anarchist Writing," *Culture and Religion* 5, no. 3 (November 2004): 369; Paul Breines, "Marxism, Romanticism, and the Case of Georg Lukács: Notes on Some Recent Sources and Situations," *Studies in Romanticism* 16, no. 4 (1977): 473.

6. Kingsley Amis, *New Maps of Hell: A Survey of Science Fiction* (New York: Arno Press, 1975), 20.

7. Georg Lukács, *The Theory of the Novel: A Historico-Philosophical Essay on the Forms of Great Epic Literature*, translated by Anna Bostock (Cambridge, Mass.: MIT Press, 1971), 88; Leslie A. Fiedler, *Love and Death in the American Novel* (Normal, Ill.: Dalkey Archive Press, 1997), 501.

8. Suvin, *Metamorphoses*, 12, italics mine.

9. Lester Del Rey, "2001: A Space Odyssey." *Galaxy* 26, no. 6 (July 1968): 194.

10. Peter Nicholls, "Science Fiction: The Monsters and the Critics," in *Science Fiction at Large: A Collection of Essays, by Various Hands, About the Interface Between Science Fiction and Reality* (New York: Harper & Row, 1977), 175–76.

11. Bookchin, *Re-enchanting Humanity*, 174, 176.

12. Bookchin, *Re-enchanting Humanity*, 92, 120.

13. Brian Stableford, "Why There Is (Almost) No Such Thing as Science Fiction: Observations on Rhetoric and Plausibility in Science and Science Fiction," *Narrative Strategies in Science Fiction and Other Essays on Imaginative Fiction* (Rockville, Md.: Borgo Press/Wildside Press, 2010), 62, 68.

14. Chu, *Do Metaphors Dream of Literal Sleep*, 69-70.

15. Stableford, "Why There Is (Almost) No Such Thing as Science Fiction," 70. Stableford's "we" here may be falsely universalizing, however.

16. Stableford, "Why There Is (Almost) No Such Thing as Science Fiction," 70–71.

17. William A. Covino, "Magic and/as Rhetoric: Outlines of a History of Phantasy," *Journal of Advanced Composition* 12, no. 2 (1992): 350; Ioan P. Couliano, *Eros and Magic in the Renaissance* (Chicago: University of Chicago Press, 1987), xvii, 25, 180.

18. Nicholls, "The Monsters and the Critics," 175.

19. Nils Bubandt, "Sorcery, Corruption, and the Dangers of Democracy in Indonesia," *The Journal of the Royal Anthropological Institute* 12, no. 2 (2006): 413.

20. Yvonne P. Chireau, *Black Magic: Religion and the African American Conjuring Tradition* (Berkeley: University of California Press, 2006), 48.

21. Graeber, *Revolutions in Reverse* (London: Minor Compositions, 2011), 94.

22. Deleuze and Guattari, *A Thousand Plateaus*, 93.

23. David Graeber, *Lost People: Magic and the Legacy of Slavery in Madagascar* (Bloomington, In.: Indiana University Press, 2007), 140.

24. Marcio Goldman, "An Afro-Brazilian Theory of the Creative Process: An Essay in Anthropological Symmetrization," *Social Analysis* 53, no. 2 (2009): 110.

25. Graeber, *Lost People*, 430.

26. Alf Hornborg, *Global Magic: Technologies of Appropriation from Ancient Rome to Wall Street* (New York: Springer, 2016), 107.

27. Alf Hornborg, "Artifacts Have Consequences, Not Agency: Toward a Critical Theory of Global Environmental History," *European Journal of Social Theory* 20, no. 1 (1 Feb. 2017): 98.

28. Hornborg, "Artifacts Have Consequences," 103.

29. Hornborg, "Artifacts Have Consequences," 98–99. This, of course, assumes that the only choices are to distinguish them firmly and absolutely or to make no distinction at all, ever. I find more persuasive Nancy Tuana's suggestion of a "viscous porosity" between the natural

and the cultural: "*Viscosity* is neither fluid nor solid, but intermediate between them. Attention to the *porosity* of interactions helps to undermine the notion that distinctions, as important as they might be in particular contexts, signify a natural or unchanging boundary, a natural kind. At the same time, 'viscosity' retains an emphasis on resistance to changing form, thereby a more helpful image than 'fluidity,' which is too likely to promote a notion of open possibilities and to overlook sites of resistance and opposition or attention to the complex ways in which material agency is often involved in interactions, including, but not limited to, human agency" (Nancy Tuana, "Viscous Porosity: Witnessing Katrina," in *Material Feminisms*, edited by Stacy Alaimo and Susan Hekman [Bloomington, IN: Indiana University Press, 2008], 193–94).

30. As Arturo Escobar writes, "the pluriverse does not assume that worlds are completely separate, interacting with and bumping into one another like so many billiard balls. On the contrary, worlds are completely interlinked, though under unequal conditions of power . . . The dominant modern worlds have globalized themselves and today partially occupy all other worlds on the planet. However, the fact that worlds are interlinked through partial connections does not turn them all into the same thing" (Escobar, *Pluriversal Politics*, 27).

31. Howard Richards, *Letters from Quebec: A Philosophy for Peace and Justice* (San Francisco: International Scholars Publications, 1996), 57.

32. Escobar, *Pluriversal Politics*, 26.

33. Escobar, *Pluriversal Politics*, xvii, 26, emphasis mine.

34. Philippe Pignarre and Isabelle Stengers, *Capitalist Sorcery: Breaking the Spell* (Basingstoke: Palgrave Macmillan UK, 2011) 41, 43.

35. Pignarre and Stengers, *Capitalist Sorcery*, 134.

36. Gordin and Gordin, *Sotsiomagiya i Sotsiotekhnika*, 86, 85; Jesse Cohn, *Underground Passages: Anarchist Resistance Culture, 1848-2011* (Oakland: AK Press, 2015).

37. Starhawk, "How We Really Shut Down the WTO," in *From ACT UP to the WTO: Urban Protest and Community Building in the Era of Globalization*, edited by Benjamin Shepard and Ronald Hayduk (New York: Verso, 2002) 55–56.

38. Starhawk, *Dreaming the Dark: Magic, Sex & Politics* (Boston: Beacon Press, 1988), 155–56.

39. Stephen Duncombe, "Imagining No-Place," *Transformative Works and Cultures* 10 (2012), https://journal.transformativeworks.org/index.php/twc/article/download/350/266.

40. Russell Jacoby, *Picture Imperfect: Utopian Thought for an Anti-Utopian Age* (New York: Columbia University Press, 2005), 87–88.

41. Stephen Duncombe, *Dream or Nightmare: Reimagining Politics in an Age of Fantasy* (New York: OR Books, 2019), xvii.

42. Stephen Duncombe, "It Stands on Its Head: Commodity Fetishism, Consumer Activism, and the Strategic Use of Fantasy," *Culture & Organization* 18, no. 5 (Dec. 2012): 370.

43. Cecilia Nowell, "In the Navajo Nation, Anarchism Has Indigenous Roots," *The Nation*, Sept. 2020. www.thenation.com, https://www.thenation.com/article/activism/anarchism-navajo-aid/. Theresa Warburton (in *The Politics of Make/Believe*) and Erica Lagalisse (in "Marginalizing Magdalena") offer important cautions about how these affinities must be recognized if white anarchists are to avoid falling back into old patterns of patronizing (a form of representation that subordinates the patronized party).

44. Warburton, *Other Worlds Here*, 223.

45. Susan Sontag, *Against Interpretation: And Other Essays* (New York: Macmillan, 2001), 215.

46. Nisi Shawl, "Hope and Vengeance in Post-Apocalyptic Sudan: *Who Fears Death* by Nnedi Okorafor," Tor.Com (7 Nov. 2017), https://www.tor.com/2017/11/07/hope-and-vengeance-in-post-apocalyptic-sudan-who-fears-death-by-nnedi-okorafor/.

Bibliography

@MagsVisaggs [Magdalene Visaggio]. "i'm sorry" *Twitter*, 13 Oct. 2019, https://twitter.com/magsvisaggs/status/1183453555233038341.

@MagsVisaggs [Magdalene Visaggio]. "it's insane" *Twitter*, 17 Dec. 2019, https://twitter.com/MagsVisaggs/status/1207053900286746627.

@MagsVisaggs [Magdalene Visaggio]. "stupidest timeline" *Twitter*, 17 Dec. 2019, https://twitter.com/MagsVisaggs/status/1207054207364354059.

@phillosoraptor. "Why Is Lord of the Flies Happening in an American City?" *Twitter*, 31 July 2020. *twitter.com*, https://twitter.com/philllosoraptor/status/1289179367411597313.

@twakeonline. "Sarah Palin Singing Baby Got Back as the World Burns. This Is the Darkest Timeline." *Twitter*, 11 Mar. 2020, https://twitter.com/twakeonline/status/1237936502664916994.

A Bro Tryna Catch His Unicorn. "Unicorn." *Urban Dictionary*, 12 Feb. 2012, https://www.urbandictionary.com/define.php?term=Unicorn.

Activestills. "Hundreds Mark 13 Years of Protests against the Wall in Bil'in." *+972 Magazine*, 2 Mar. 2018, https://www.972mag.com/hundreds-mark-13-years-of-protests-against-the-wall-in-bilin/.

Adorno, Theodor W. *Minima Moralia: Reflections from Damaged Life*. Translated by E. F. N. Jephcott, Verso, 1999.

Adorno, Theodor W., and Max Horkheimer. *Dialectic of Enlightenment*. Translated by John Cumming, Verso, 1997.

Agoyo, Acee. "'The Elizabeth Warren of the Sci-Fi Set': Author Faces Criticism for Repeated Use of Tribal Traditions." *Indianz.com*, 24 June 2020, https://www.indianz.com/News/2020/06/24/the-elizabeth-warren-of-the-scifi-set-au.asp.

Alexander, Christopher, Sara Ishikawa, Murray Silverstein, Max Jacobson, Ingrid F. King, and Shlomo Angel. *A Pattern Language: Towns, Buildings, Construction*. New York: Oxford University Press, 1977.

Alfred, Gerald R., and Taiaiake Alfred. *Peace, Power, Righteousness: An Indigenous Manifesto*. Oxford University Press, 2009.

Ames, Melissa. *Small Screen, Big Feels: Television and Cultural Anxiety in the Twenty-First Century*. University Press of Kentucky, 2020.

Amis, Kingsley. *New Maps of Hell: A Survey of Science Fiction*. New York: Arno Press, 1975.

Anarkata: A Statement. 19 Oct. 2019. *anarkataastatement.wordpress.com*, https://drive.google.com/file/d/1XU9P2PwO2wEChvsRdvu51hQBQnIiTbzG/view.

Anderson, Benedict Richard O'Gorman. *Under Three Flags: Anarchism and the Anti-Colonial Imagination*. Verso, 2005.

Anderson, Ray Lynn. "Persuasive Functions of Science Fiction: A Study in the Rhetoric of Science." Dissertation, University of Minnesota, 1968.

Andreevskii, Georgii Vasil'evich. *Povsednevnaia zhizn' Moskvy v stalinskuiu epokhu (20–30-e gody)* [*Everyday Life In Moscow In the Stalin Era (20s–30s)*]. Molodaja gvardija, 2003.

Andrenko, Leonid. *La Diversité du Principe Vital dans l'Univers: Immutabilité de la Vie Universelle et Eternelle*. 1924. Russian Academy of Sciences, Inventory 5, no. 3.
Andrenko, Leonid. *La Diversité du Principe vital dans l'Univers: La Vie sur les Planètes aux combinaisons chimiques autres que celles de la Terre*. 1923–1924. Russian Academy of Sciences, Inventory 5, no. 1.
Andrenko, Leonid. "Les Habitants Des Astres (Fin)." *L'En dehors*, vol. 16, no. 286, Sept. 1935, p. 34.
Andrenko, Leonid. "Les Habitants Des Astres." *L'En dehors*, vol. 16, no. 284–85, juillet-août 1935, pp. 17–18.
Andreuko, Leónidas [Leonid Andrenko]. "La Diversidad del Principio Vital en el Universo." *Estudios*, vol. 7, no. 65, Enero 1929, pp. 55–59.
Antliff, Allan. "Anarchy, Power, and Poststructuralism." *SubStance*, vol. 36, no. 2, Johns Hopkins University Press, Aug. 2007, pp. 56–66. Project MUSE, doi:10.1353/sub.2007.0026.
Applebaum, Anne. "You're Not Supposed to Understand the Rumors About Biden." *The Atlantic*, Oct. 2020, https://www.theatlantic.com/ideas/archive/2020/10/smears-against-biden-dont-need-make-any-sense/616824/.
Aquinas, Thomas. *Summa Theologiae: Volume 8, Creation, Variety and Evil: 1a. 44–49*. Edited by Thomas Gilby, Cambridge University Press, 2006.
Arendt, Hannah. *Between Past and Future: Eight Exercises in Political Thought*. Penguin, 1993.
Aristotle. *Aristotle: The Art of Rhetoric*. Harvard University Press, 1967.
Arolovich, Amalia Viktorovna. *Anarkhizm-universalizm v kontekste russkoy "kosmicheskoy paradigmy" nachala XX veka* [*Anarchism-Universalism In the Context of the Russian "Cosmic Paradigm" of the Early Twentieth Century*]. Diss., Moscow State University, 2005.
Arolovich, Amalia Viktorovna. "Kontseptsiya slova i yazyka u russkikh anarkhistov-universalistov nachala 20 v. [The Concept of the Word and Language Among Russian Anarchist Universalists of the Early 20th Century.]" *Anarkhizm: pro et contra: sotsial'no-politicheskoye yavleniye glazami yego rossiyskikh storonnikov, kritikov i otechestvennykh uchenykh-issledovateley: antologiya*, edited by P. I. Talerov and D. K. Bogatyrev, Izdatel'stvo Russkoy khristianskoy gumanitarnoy akademii, 2015, pp. 788–97.
Aryanne's Jewish Sex Slave. *Equality Is A Lie. Derpibooru*, 2018, https://derpibooru.org/1697191?q=starlight+glimmer. Accessed 13 Aug. 2019.
Asimov, Isaac. *Foundation; Foundation and Empire; Second Foundation*. Everyman's Library, 2010.
Asimov, Isaac. "Introduction." *More Soviet Science Fiction*, Collier Books, 1962, pp. 7–13.
Asimov, Isaac. "The Caves of Steel (1)." *Galaxy Science Fiction*, vol. 7, no. 1, Oct. 1953, pp. 4–66.
Asimov, Isaac. "The Caves of Steel (2)." *Galaxy Science Fiction*, vol. 7, no. 2, Nov. 1953, pp. 98–159.
Asimov, Isaac. "The Caves of Steel (3)." *Galaxy Science Fiction*, vol. 7, no. 3, Dec. 1953, pp. 108–59.
Attebery, Brian. *Decoding Gender in Science Fiction*. Routledge, 2002.
Attebery, Brian. *Strategies of Fantasy*. Indiana University Press, 1992.
Attebery, Brian. "Fantasy as an Anti-Utopian Mode." *Reflections on the Fantastic: Selected Essays from the Fourth International Conference on the Fantastic in the Arts*. Ed. Michael R. Collings. New York: Greenwood Press, 1986. 3–8.
Auden, W. H. *The Dyer's Hand and Other Essays*. Random House, 1962.
Aune, James Arnt. *Rhetoric and Marxism*. Taylor & Francis Group, 2020.
Austin, John L. *How to Do Things with Words*. Oxford University Press, 1962.
Author Earnings. 2018 SFWA Nebula Conference Presentation. 24 May 2018, https://web.archive.org/web/20180613083320/http://authorearnings.com/sfwa2018/.
Badino, Massimiliano, and Jaume Navarro. "Introduction: Ether: The Multiple Lives of a Resilient Concept." *Ether and Modernity: The Recalcitrance of an Epistemic Object in the Early Twentieth Century*, edited by Jaume Navarro, Oxford University Press, 2018, pp. 1–13.

Bailey, James O. *Pilgrims Through Space and Time: Trends and Patterns in Scientific and Utopian Fiction*. Greenwood Press, 1977.
Bakhtin, M. M. *The Dialogic Imagination: Four Essays*. Translated by Michael Holquist, University of Texas Press, 1994.
Bakunin, Mikhail Aleksandrovich. *Bakunin on Anarchism*. Edited by Sam Dolgoff. Montréal: Black Rose Books, 1980.
Bakunin, Mikhail Aleksandrovich. *God and the State*. Dover, 1970.
Bakunin, Mikhail Aleksandrovich. *Oeuvres*. PV. Stock, 1895.
Balboa, Elizabeth. "The Fabled Starbucks Unicorn Frappuccino: 'It's Like Watered-Down Cotton Candy.'" *Benzinga*, Apr. 2017, https://www.benzinga.com/general/restaurants/17/04/9319107/the-fabled-starbucks-unicorn-frappuccino-its-like-watered-down-cot.
Ballard, J. G. "Build-Up." *New Worlds Science Fiction*, vol. 19, no. 55, Jan. 1957, pp. 52–70.
Bangsund, John, editor. *John W. Campbell: An Australian Tribute*. Ronald E. Graham and John Bangsund, 1972.
Barad, Karen. *Meeting the Universe Halfway: Quantum Physics and the Entanglement of Matter and Meaning*. Duke University Press, 2007.
Barbrook, Richard, and Andy Cameron. "The Californian Ideology." *Science as Culture*, vol. 6, no. 1, 1996, pp. 44–72.
Barel, Sophie. "La symbolique de la licorne et sa place dans la sémiologie du web?" *Medium*, 29 Feb. 2016, https://medium.com/@SophieB./la-symbolique-dela-licorne-et-sa-place-dans-las%C3%A9miologie-du-web-6ec33ef2745f.
Barr, Marleen S. *Afro-Future Females: Black Writers Chart Science Fiction's Newest New-Wave Trajectory*. Ohio State University Press, 2008.
Barr, Marleen S. "Fantastic Language/Political Reporting: The Postcolonial Science Fiction Illocutionary Force Is with Us." *The Postnational Fantasy: Essays on Postcolonialism, Cosmopolitics and Science Fiction*, edited by Masood Ashraf Raja et al., McFarland, 2011, pp. 188–210.
Barth, John. *Lost in the Funhouse: Fiction for Print, Tape, Live Voice*. Anchor Press, 1988.
Barthes, Roland. *Mythologies*. Translated by Annette Lavers, Farrar, Strauss, and Giroux, 1991.
Bateson, Gregory. *Steps to an Ecology of Mind*. Ballantine Books, 1972.
Baudemann, Kristina. "'I HAVE SEEN THE FUTURE AND I WON'T GO': The Comic Vision of Craig Strete's Science Fiction Stories." *Studies in American Indian Literatures*, vol. 29, no. 4, University of Nebraska Press, 2017, pp. 76–101.
Baudrillard, Jean. *Simulacra and Simulation*. Translated by Sheila Faria Glaser, University of Michigan Press, 1994.
Bay, Morten. "Weaponizing the Haters: The Last Jedi and the Strategic Politicization of Pop Culture through Social Media Manipulation." *First Monday*, vol. 23, no. 11, Nov. 2018. *firstmonday.org*, doi:10.5210/fm.v23i11.9388.
Beaska, Sara Marielle Gaup. *Gulahallat Eatnamiin / WE SPEAK EARTH*. 2015. *YouTube*, https://www.youtube.com/watch?v=VzBmDpDAskQ.
Bear, Greg. *Queen of Angels*. Warner Books, 1991.
Beaver, Harold. "Homosexual Signs (In Memory of Roland Barthes)." *Critical Inquiry*, vol. 8, no. 1, Oct. 1981, pp. 99–119.
Belsey, Catherine. *Critical Practice*. Routledge, 2002.
Ben-Moshe, Liat. "Infusing Disability in the Curriculum: The Case of Saramago's 'Blindness.'" *Disability Studies Quarterly*, vol. 26, no. 2, 2, Mar. 2006. *dsq-sds.org*, doi:10.18061/dsq.v26i2.688.
Benjamin, Walter. *Selected Writings*. Harvard University Press, 2004.
Bennett, Jane. *The Enchantment of Modern Life: Attachments, Crossings, and Ethics*. Princeton University Press, 2016.

Bensusan, Hilan, and Tomás Ribeiro Cardoso. "Por Uma Metafísica de Tramas: O Mundo Sem Arché." *Kriterion*, vol. 53, no. 125, June 2012, pp. 281–98, doi:10.1590/S0100-512X2012000100014.
Bensusan, Hilan. "Polemos Doesn't Stop Anywhere Short of the World: On Anarcheology, Ontology, and Politics." *Anarchist Developments in Cultural Studies*, no. 2, 2013. *journals.uvic.ca*, https://journals.uvic.ca/index.php/adcs/article/view/17151.
Beobi [Gordin, Wolf Lvovich]. *AO-russkiy grammaticheskiy slovar'* [*AO-Russian Grammar Dictionary*]. Chelovechestvo (Pananarchisty), 1920.
Beobi [Gordin, Wolf Lvovich]. *Gnoseologiya: Vvedeniye vo Vseizobretatel'stvo* [*Epistemology: An Introduction to All-Invention*]. Izdaniye Vseizobretal'ni, 1921.
Beobi [Gordin, Wolf Lvovich]. *Grammatika Logicheskogo Yazyka AO* [*Grammar of the Logical Language AO*]. Beobi, 1924.
Beobi [Gordin, Wolf Lvovich]. *Grammatika Yazyka Chelovechestva AO, Perevod s AO* [*Grammar of the Language of Mankind, Translated From AO*]. Chelovechestvo, 1920.
Beres, Derek. "Will the Future Be 'Mad Max' or 'Star Trek'? Coronavirus Offers Clues." *Big Think*, 11 Mar. 2020. bigthink.com, https://bigthink.com/politics-current-affairs/jordan-hall-coronavirus.
Berlant, Lauren Gail. *Cruel Optimism*. Duke University Press, 2011.
Bernico, Matt. "Anthropodicy: An Anarchism of Things." *Anarchist Developments in Cultural Studies*, no. 1 & 2, Nov. 2015. *journals.uvic.ca*, https://journals.uvic.ca/index.php/adcs/article/view/17181.
Bester, Alfred. "5,271,009." *Magazine of Fantasy & Science Fiction*, vol. 6, no. 3, Mar. 1954, pp. 3–32.
Bil'in Popular Committee. "Bil'in Weekly Demonstration Reenacts the Avatar Film." *bilin-village.org*, December 2, 2010. https://web.archive.org/web/20100217175438/http://www.bilin-village.org/english/articles/testimonies/Bilin-weekly-demonstration-reenacts-the-Avatar-film.
Blavier, André. *Les fous littéraires*. Editions des Cendres, 2000.
Boal, Augusto. *Theatre of the Oppressed*. Translated by Charles A. McBride and Maria Odilia Leal McBride, Theatre Communications Group, 2013.
Bookchin, Murray. *Reenchanting humanity: defense of the human spirit against antihumanism, misanthropy, mysticism and primitivism*. Cassell, 1995.
Bookchin, Murray. *The Ecology of Freedom: The Emergence and Dissolution of Hierarchy*. Black Rose Books, 1991.
Bookchin, Murray. *To Remember Spain: The Anarchist and Syndicalist Revolution of 1936*. Oakland: AK Press, 1994.
Bookchin, Murray. *Toward An Ecological Society*. Black Rose Books, 1980.
Bookchin, Murray. "Toward an Ecological Solution." *Ramparts*, vol. 8, no. 11, May 1970, pp. 7–15.
Borges, Jorge Luis. *Other Inquisitions, 1937–1952*. Translated by Ruth L. C. Simms, Washington Square Press, 1966.
Bottici, Chiara. *A Philosophy of Political Myth*. Cambridge: Cambridge University Press, 2010.
Boucher, Geoff. "Syfy's 'Vagrant Queen': Adriyan Rae To Star In Live-Action Space Tale." *Deadline*, 17 May 2019, https://deadline.com/2019/05/syfys-vagrant-queen-adriyan-rae-to-star-in-live-action-space-tale-1202617182/.
Bould, Mark, and Sherryl Vint. "There Is No Such Thing as Science Fiction." *Reading Science Fiction*, edited by James Gunn et al., Palgrave Macmillan, 2009, pp. 43–51.
Boutang, Yann Moulier. "Antagonism Under Cognitive Capitalism: Class Composition, Class Consciousness and Beyond." *Immaterial Labour, Multitudes and New Social Subjects: Class Composition In Cognitive Capitalism*, 28 April 2006, King's College, University of Cambridge. Seminar Presentation.

Boyarin, Jonathan. "Europe's Indian, America's Jew: Modiano and Vizenor." *boundary 2*, vol. 19, no. 3, 1992, pp. 197–222.
Bracher, Mark. *Lacan, Discourse, and Social Change: A Psychoanalytic Cultural Criticism*. Cornell University Press, 2018.
Brat'ia Gordiny [A. L. Gordin and V. L. Gordin]. *Anarkhiia v Mechte: Strana Anarkhiia: Utopiia-poema* [*Anarchy in a Dream: The Land of Anarchy: Utopia-Poem*]. Moskva: Izd. Pervogo tsentral'nogo sotsiotekhnikuma. Vremennyi tekhnikum propagandy i agitatsii, 1919.
Br. Gordin [Abba Gordin]. *Pochemu? ili Kak muzhik popal v stranu 'Anarkhiia'* ["Why": Or, How a Peasant Arrived in the Land of Anarchy]. Moscow: Mosk. federatsiya anarkhist. grupp, 1918.
Breines, Paul. "Marxism, Romanticism, and the Case of Georg Lukács: Notes on Some Recent Sources and Situations." *Studies in Romanticism* 16, no. 4 (1977): 473–89. https://doi.org/10.2307/25600099.
Brin, David. "The Dogma of Otherness." *Otherness*, by David Brin, New York: Bantam Books, 1994, pp. 86–100.
Broderick, Damien. *Reading by Starlight: Postmodern Science Fiction*. Routledge, 2005.
Brooke-Rose, Christine. *A Rhetoric of the Unreal: Studies in Narrative and Structure, Especially of the Fantastic*. Cambridge University Press, 1988.
Brooks, Peter. *Reading for the Plot: Design and Intention in Narrative*. Harvard University Press, 1992.
Brouillette, Sarah. "Corporate Publishing and Canonization: Neuromancer and Science-Fiction Publishing in the 1970s and Early 1980s." *Book History*, vol. 5, 2002, pp. 187–208, doi:10.1353/bh.2002.0001.
brown, adrienne maree. "Outro." *Octavia's Brood: Science Fiction Stories from Social Justice Movements*, edited by Walidah Imarisha and Adrienne Maree Brown (Oakland: AK Press, 2015), 279–82.
Bryan, Peter Cullen. "Geeking Out and Hulking Out: Toward an Understanding of Marvel Fan Communities." *Age of the Geek: Depictions of Nerds and Geeks in Popular Media*, edited by Kathryn E. Lane, Springer International Publishing, 2018, pp. 149–65.
Bubandt, Nils. "Sorcery, Corruption, and the Dangers of Democracy in Indonesia." *The Journal of the Royal Anthropological Institute* 12, no. 2 (2006): 413–31.
Buber, Martin. *Paths in Utopia*. Syracuse: Syracuse University Press, 1996.
Bucholtz, Mary. "'Why Be Normal?': Language and Identity Practices in a Community of Nerd Girls." *Language in Society*, vol. 28, no. 2, 1997, pp. 203–23.
Bui, Long. "Asian Roboticism: Connecting Mechanized Labor to the Automation of Work." *Perspectives on Global Development and Technology*, vol. 19, no. 1–2, Mar. 2020, pp. 110–26.
Burke, Kenneth. *Attitudes Toward History*. 3rd ed., University of California Press, 1984.
Burke, Kenneth. *Language as Symbolic Action: Essays on Life, Literature, and Method*. University of California Press, 1966.
Burke, Kenneth. *Permanence and Change: An Anatomy of Purpose*. Bobbs-Merrill Educational Publishing, 1965.
Burke, Kenneth. *The Philosophy of Literary Form: Studies in Symbolic Action*. Louisiana State University Press, 1967.
Busse, Kristina. "Geek Hierarchies, Boundary Policing, and the Gendering of the Good Fan." *Participations* 10, no. 1 (2013): 73–91.
Butler, Isaac. "V For Vile." *The Hooded Utilitarian*, 17 Sept. 2012, https://www.hoodedutilitarian.com/2012/09/v-for-vile/.
Butler, Judith. *The Psychic Life of Power: Theories in Subjection*. Stanford University Press, 1997.
Campbell, John W., Jr. "The Science of Science-Fiction Writing." *Of Worlds Beyond: The Science of Science-Fiction Writing*, edited by Lloyd Arthur Eshbach, Fantasy Press, 1947, pp. 84–96.

Canavan, Gerry. "*Do Metaphors Dream of Literal Sleep? A Science-Fictional Theory of Representation* (review)." Comparative Literature Studies 49, no. 4 (2012): 616–19.
Carra, Jean-Louis. *Essai Sur La Nautique Aérienne, Contenant l'art de Diriger Les Ballons Aérostatiques à Volonté, & d'accélérer Leur Course Dans Les Plaines de l'air, Avec Le Précis de Deux Expériences Particulières de Météorologie à Faire*, Lu à l'Académie Royale Des Sciences de Paris, Le 14 Janvier 1784. 1784.
Carroll, Noel. *The Philosophy of Horror: Or, Paradoxes of the Heart*. Routledge, 2003.
Carter, J. Kameron, and Sarah Jane Cervenak. "Black Ether." *CR: The New Centennial Review*, vol. 16, no. 2, Michigan State University Press, 2016, pp. 203–24. JSTOR, doi:10.14321/crnewcentrevi.16.2.0203.
Castellanos, Israel, and Alfonso L. Herrera. *La plasmogenia*. Imprenta y papelería de Rambla, Bouza y ca., 1921.
Castoriadis, Cornelius. *The Imaginary Institution of Society*. Translated by Kathleen Blamey, MIT Press, 1997.
Cavna, Michael. "Alexandria Ocasio-Cortez Responded to Critics by Quoting Alan Moore. Many Nerds Rejoiced in Response." *Washington Post*, 11 Jan. 2019, https://www.washingtonpost.com/arts-entertainment/2019/01/11/alexandria-ocasio-cortez-responded-critics-by-quoting-alan-moore-many-nerds-rejoiced-response/.
Center for Consciousness Studies. "A Russian Chinese Room Story Antedating Searle's 1980 Discussion." *Center for Consciousness Studies, Philosophy Department of Moscow State University*, 15 June 2018, http://www.hardproblem.ru/en/posts/Events/a-russian-chinese-room-story-antedating-searle-s-1980-discussion/index.php.
Cerézo, Arvyn. "THE HUNGER GAMES' Three-Finger Salute: A Symbol of Resistance to Tyranny in Asia." *BOOK RIOT* (blog), August 27, 2021. https://bookriot.com/the-hunger-games-three-finger-salute/.
Césaire, Aimé. *Discourse on Colonialism*. Translated by Joan Pinkham, Monthly Review Press, 2000.
Chait, Jonathan. "'We Do That All the Time, Get Over It.' Mulvaney Boasts About Ukraine Plot." *New York Magazine*, Oct. 2019, https://nymag.com/intelligencer/2019/10/mulvaney-ukraine-get-over-it.html.
Chalmers, David J. *The Conscious Mind: In Search of a Fundamental Theory*. OUP USA, 1996.
Chang, Clio. "The Unlikely Rise of an Alt-Right Hero." *The New Republic*, Mar. 2017. *The New Republic*, https://newrepublic.com/article/141766/unlikely-rise-alt-right-hero.
Chant, Ian. "Bad News, Everyone—North Korean Unicorn Is Actually Just A Stupid Kirin." *The Mary Sue*, 7 Dec. 2012, https://www.themarysue.com/korean-unicorn-actually-a-kirin/.
Chant, Ian. "North Korea Finds Ancient Unicorn Lair, Because Sure, Why Not." *The Mary Sue*, 30 Nov. 2012, https://www.themarysue.com/north-korea-unicorn/.
Chekhov, Anton. *Sakhalin Island*. Surrey, UK: Alma Books, 2018.
Chen, Mel Y. *Animacies: Biopolitics, Racial Mattering, and Queer Affect*. Durham, NC: Duke University Press, 2012.
Chess, Shira, and Adrienne Shaw. "A Conspiracy of Fishes, or, How We Learned to Stop Worrying About #GamerGate and Embrace Hegemonic Masculinity." *Journal of Broadcasting & Electronic Media*, vol. 59, no. 1, 2015, pp. 208–20. Zotero, doi:0.1080/08838151.2014.999917.
Cheyne, Ria. "'She Was Born a Thing': Disability, the Cyborg and the Posthuman in Anne McCaffrey's The Ship Who Sang." *Journal of Modern Literature*, vol. 36, no. 3, Indiana University Press, July 2013, pp. 138–56.
Chiang, Ted. *Stories of Your Life and Others*. Vintage Books, 2016.
Chickera, Amal de, et al. "The World's Stateless: a new report on why size does and doesn't matter." *European Network on Statelessness*, 15 Dec. 2014. www.statelessness.eu, https://www

.statelessness.eu/blog/world%E2%80%99s-stateless-new-report-why-size-does-and-doesn%E2%80%99t-matter.

Christian, Barbara. "The Race for Theory." *Feminist Studies* 14, no. 1 (1988): 67–79. https://doi.org/10.2307/3177999.

Chu, Arthur. "Your Princess Is in Another Castle: Nerds and Misogyny." *The Daily Beast* (blog), May 27, 2014. https://www.thedailybeast.com/your-princess-is-in-another-castle-misogyny-entitlement-and-nerds-1.

Chu, Seo-Young. *Do Metaphors Dream of Literal Sleep?: A Science-Fictional Theory of Representation*. Cambridge, Mass: Harvard University Press, 2010.

Chumley, Cheryl K. "AOC's Unicorn Flight of 'Green New Deal' Fancy." *The Washington Times*, 9 Feb. 2019, https://www.washingtontimes.com/news/2019/feb/9/alexandria-ocasio-cortezs-unicorn-flight-green-new/.

Cioffi, Frank L. *Formula Fiction? An Anatomy of American Science Fiction, 1930–1940*. Westport, Conn.: Greenwood Press, 1982.

Clark, P. Djèli. *Ring Shout, or, Hunting Ku Kluxes In the End Times*. New York: Tom Doherty Associates, 2020.

Clarke, Arthur C. *Profiles of the Future*. Harper & Row, 1973.

Cleminson, Richard. *Anarchism and Eugenics: An Unlikely Convergence, 1890–1940*. Manchester University Press, 2019.

Clover, Joshua. *The Matrix*. BFI, 2004.

Clover, Joshua. "Retcon: Value and Temporality in Poetics." *Representations*, vol. 126, no. 1, 2014, pp. 9–30. JSTOR, *JSTOR*, doi:10.1525/rep.2014.126.1.9.

Clute, John. *Pardon This Intrusion*. London: Orion Publishing Group, 2016.

Cohn, Jesse. *Underground Passages: Anarchist Resistance Culture, 1848–2011*. Oakland: AK Press, 2015.

Cohen, Roger. "Robert Mueller in the Age of the Unicorn." *The New York Times*, 19 Apr. 2019, https://www.nytimes.com/2019/04/19/opinion/mueller-trump-russia-report.html.

Coleman, Daniel B. "Countering Afropessimist Ontological Nihilism: The Afrofuturism of Lovecraft Country's 'Magic' Perceived through Afrodiasporic Spiritual Cosmology." Forthcoming.

Coleridge, Samuel Taylor. "Frost at Midnight." *The Broadview Anthology of Literature of the Revolutionary Period 1770–1832*, edited by D. L. Macdonald and Anne McWhir, Broadview Press, 2010, pp. 754–55.

Collard, Rebecca. "Why Syrian 'Safe Zones' Could Be Dangerous for Civilians and U.S. Policy." *Time*, July 2015, https://time.com/3979906/syria-safe-zones/.

Colombo, Eduardo. *L'Espace politique de l'anarchie: Esquisses pour une philosophie politique de l'anarchisme*. Atelier de création libertaire, 2008.

Colombo, Eduardo. "À propos du *Petit lexique philosophique de l'anarchisme* de Daniel Colson." *Réfractions*, no. 8, Printemps-été 2002, pp. 127–41.

Colson, Daniel. *A Little Philosophical Lexicon of Anarchism from Proudhon to Deleuze*. Translated by Jesse Cohn, Minor Compositions, 2019.

Colson, Daniel. *Trois essais de philosophie anarchiste: Islam, histoire, monadologie*. Léo Scheer, 2004.

Colson, Daniel. "L'Ange de l'histoire." *Le Monde Libertaire*, no. 1377, Nov. 2004, http://www.federation-anarchiste.org/spip.php?article892.

Colson, Daniel. "Réponse de Daniel Colson à Eduardo Colombo." *Réfractions*, no. 8, Printemps-été 2002, pp. 143–53.

Comisión Nacional para el Conocimiento y Uso de la Biodiversidad. "Alfonso Luis Herrera, 1868–1942." *Curiosos y Comprometidos*, https://www.biodiversidad.gob.mx/biodiversidad/curiosos/sXIX/AlfonsoLHerrera.php. Accessed 9 Oct. 2020.

Copjec, Joan. *Supposing the Subject*. Verso, 1994.
Couliano, Ioan P. *Eros and Magic in the Renaissance*. University of Chicago Press, 1987.
Cousin, Saskia. "Images, Imaginaries, and Imaginations: French Notes." *Tourism Imaginaries at the Disciplinary Crossroads: Place, Practice, Media*, edited by Maria Gravari-Barbas and Nelson Graburn, Routledge, 2016, pp. 35–47.
Covino, William A. "Magic and/as Rhetoric: Outlines of a History of Phantasy." *Journal of Advanced Composition* 12, no. 2 (1992): 349–58.
Cramer, Florian. "Rosa Einhörner [Pink Unicorns]." *pleintekst.nl*, http://cramer.pleintekst.nl/essays/rosa_einhoerner/. Accessed 9 Aug. 2019.
"Crisis in Syria: Decision Point." *CNN.Com*, 12 Sept. 2013, http://transcripts.cnn.com/TRANSCRIPTS/1309/12/se.01.html.
Cronin, Brian. "A History of The Punisher Logo Being Used By Police, Military & Politicians." *CBR*, July 2019. www.cbr.com, https://www.cbr.com/punisher-history-logo-used-police-military-politicians/amp/.
Csicsery-Ronay, Istvan. *The Seven Beauties of Science Fiction*. Middletown: Wesleyan University Press, 2012.
Culler, Jonathan. *Theory of the Lyric*. Harvard University Press, 2015.
Culler, Jonathan. "Apostrophe." *The Pursuit of Signs: Semiotics, Literature, Deconstruction*, Cornell University Press, 2002, pp. 135–54.
Daly, Mary. *Gyn/Ecology: The Metaethics of Radical Feminism*. Boston: Beacon Press, 1978.
Damien, Robert. "Transport ferroviaire et ordre politique: Proudhon, une pensée philosophique des réseaux?" *Penser les réseaux*, edited by Daniel Parrochia, Champ Vallon, 2001, pp. 218–32.
Dankosky, John. "Writing The Fantastic In 2017." *Science Friday*, National Public Radio, 13 Oct. 2017, https://www.sciencefriday.com/segments/writing-the-fantastic-in-2017/.
DaSilva, Brianna, and Jeannette Ng. *Othering in Fantasy with Jeannette Ng*. 2019. *soundcloud.com*, https://soundcloud.com/femalesinfantasy/othering-in-fantasy-with-jeannette-ng.
Davis, Arianna. "What's Really Behind Unicorn Fever." *Refinery29*, https://www.refinery29.com/en-us/2017/05/152423/unicorn-trend-explanation-history. Accessed 9 Aug. 2019.
Davis, Lennard J. *Enforcing Normalcy: Disability, Deafness, and the Body*. Verso, 1995.
Davis, Peter. *Hearts and Minds*. Warner Bros., 1974.
Dawkins, Richard. *The God Delusion*. Boston: Houghton Mifflin, 2006.
Day, Richard J. F., and Adam Lewis. "Radical Subjectivity and the N-Row Wampum: A General Model for Autonomous Relations Against and Beyond the Dominant Global Order?" *Subjectivity in the Twenty-First Century: Psychological, Sociological, and Political Perspectives*, edited by Romin W. Tafarodi, Cambridge University Press, 2013, pp. 169–89.
de Acosta, Alejandro. "Two Undecidable Questions for Thinking in Which Anything Goes." *Contemporary Anarchist Studies: An Introductory Anthology of Anarchy in the Academy*, edited by Randall Amster et al., Routledge, 2009, pp. 26–34.
De la Mettrie, Julien Offray. *Man a Machine; And, Man a Plant*. Hackett Publishing, 1994.
Deckard, Sharae. "Trains, Stone, and Energetics: African Resource Culture and the Neoliberal World-Ecology." *World Literature, Neoliberalism, and the Culture of Discontent*, edited by Sharae Deckard and Stephen Shapiro, Springer, 2019, pp. 239–62.
deFord, Miriam Allen. "Frustrated Frankenstein: Alfonso Herrera and His Colpoids." *The Magazine of Fantasy and Science Fiction*, vol. 7, no. 2, Aug. 1954, pp. 55–60.
Degener, Theresia. "Disabled Persons and Human Rights: The Legal Framework." *Human Rights and Disabled Persons: Essays and Relevant Human Rights Instruments*, edited by Theresia Degener and Yolan Koster-Dreese, Martinus Nijhoff Publishers, 1995, pp. 9–39.
del Rey, Lester. "2001: A Space Odyssey." *Galaxy Science Fiction*, vol. 26, no. 6, July 1968, pp. 193–94.

del Valle, Adrián. "El Alma Estructural (1)." *La Revista Blanca*, vol. 4, no. 78, Aug. 1926, pp. 171–74.
del Valle, Adrián. "El Alma Estructural (1)." *Revista Bimestre Cubana*, vol. 21, no. 5, Oct. 1926, pp. 656–68.
del Valle, Adrián. "El Alma Estructural (2)." *Revista Bimestre Cubana*, vol. 21, no. 6, Dec. 1926, pp. 821–38.
del Valle, Adrián. "El Movimiento Creador." *Revista Bimestre Cubana*, vol. 47, 1941, pp. 336–47.
del Valle, Adrián. "Nuestro Universo. Ensayo de Una Concepción Sintética." *Cuba Contemporanea*, vol. 37, no. 145, Jan. 1925, pp. 33–39.
del Valle, Adrián. "Social Environment as a Psychological Factor." *Inter-América*, vol. 7, no. 8, Aug. 1924.
del Valle, Adrián. "The World as a Plurality." *Inter-América*, vol. 8, no. 2, Dec. 1924, pp. 99–113.
del Valle, Adrián. "Vida Universal." *Cuba y América*, vol. 21, no. 22, Aug. 1906, p. 357.
Delany, Samuel R. *Silent Interviews: On Language, Race, Sex, Science Fiction, and Some Comics—A Collection of Written Interviews*. Hanover, NH: Wesleyan University Press, 2011.
Delany, Samuel R. *The American Shore: Meditations on a Tale of Science Fiction by Thomas M. Disch—"Angouleme."* Wesleyan University Press, 2014.
Delany, Samuel R. *The Jewel-Hinged Jaw: Notes on the Language of Science Fiction*. Wesleyan University Press, 2011.
Deleuze, Gilles, and Félix Guattari. *A Thousand Plateaus: Capitalism and Schizophrenia*. Translated by Brian Massumi, University of Minnesota Press, 1987.
Deleuze, Gilles, and Félix Guattari. *Anti-Oedipus: Capitalism and Schizophrenia*. Translated by Robert Hurley et al., University of Minnesota Press, 1983.
Deleuze, Gilles, and Félix Guattari. *L'Anti-Oedipe: Capitalisme et Schizophrénie*. Minuit, 1972.
Deleuze, Gilles, and Félix Guattari. *Kafka: Toward a Minor Literature*. Translated by Dana Polan. Minneapolis, MN: University of Minnesota Press, 1986.
Deleuze, Gilles, and Guattari Félix. *A Thousand Plateaus: Capitalism and Schizophrenia*. Translated by Brian Massumi, University of Minnesota Press, 1987.
Deleuze, Gilles. *Nietzsche and Philosophy*. Translated by Hugh Tomlinson, Continuum, 2006.
Deleuze, Gilles. *Spinoza: Practical Philosophy*. Translated by Robert Hurley, City Lights Books, 1988.
Deloria, Philip Joseph. *Playing Indian*. Yale University Press, 1998.
Denevan, William M. "The Pristine Myth: The Landscape of the Americas in 1492." *Annals of the Association of American Geographers*, vol. 82, no. 3, 1992, pp. 369–85.
Dennett, Daniel C. *Elbow Room: The Varieties of Free Will Worth Wanting*. Clarendon Press, 1984.
Dennett, Daniel Clement. *Consciousness Explained*. Little, Brown and Co., 1991.
Derrida, Jacques. *Writing and Difference*. Translated by Bass Alan, University of Chicago Press, 1978.
Descartes, René. *Discourse on Method and Meditations on First Philosophy*. Translated by Donald A. Cress, Fourth ed., Hackett Publishing, 1999.
Díaz, Junot. *The Brief Wondrous Life of Oscar Wao*. Penguin, 2008.
Dick, Philip K. "The Electric Ant." *The Magazine of Fantasy and Science Fiction*, vol. 37, no. 4, Oct. 1969, pp. 100–115.
Dick, Philip K. *A Scanner Darkly*. New York: Vintage Books, 1991.
Dillon, Grace L. "Imagining Indigenous Futurisms." *Walking the Clouds: An Anthology of Indigenous Science Fiction*, edited by Grace L Dillon, Tucson: University of Arizona Press, 2012, pp. 1–12.
Disch, Thomas M. *The Dreams Our Stuff Is Made Of: How Science Fiction Conquered the World*. Simon and Schuster, 2000.

Disch, Thomas M. "The Embarrassments of Science Fiction." *On SF*, University of Michigan Press, 2005, pp. 3–15.
Disch, Thomas M. "The Embarrassments of Science Fiction." *Science Fiction at Large: A Collection of Essays, by Various Hands, About the Interface Between Science Fiction and Reality*, Harper & Row, 1977, pp. 139–56.
Dneprov, Anatoliy. "Igra." *Znanie-Sila*, no. 5, 1961, pp. 39–42.
Doctorow, Cory. "Cold Equations and Moral Hazard." *Locus Online*, 2 Mar. 2014, https://locusmag.com/2014/03/cory-doctorow-cold-equations-and-moral-hazard/.
Doctorow, Cory. "Cold Equations and Moral Hazard." *Locus Online*, 2 Mar. 2014, https://locusmag.com/2014/03/cory-doctorow-cold-equations-and-moral-hazard/.
Drezner, Daniel W. "Metaphor of the Living Dead: Or, the Effect of the Zombie Apocalypse on Public Policy Discourse." *Social Research: An International Quarterly*, vol. 81, no. 4, 2014, pp. 825–49.
Dubey, Madhu. "Becoming Animal in Black Women's Science Fiction." *Afro-Future Females: Black Writers Chart Science Fiction's Newest New-Wave Trajectory*, edited by Marleen S. Barr, Ohio State University Press, 2008, pp. 31–51.
Dubois, Raphaël. "Problème de la création artificielle de l'être vivant." *Les Cahiers de l'Université populaire : revue mensuelle* (10 Jan. 1906): 625–45.
Duchesne Winter, Juan. "Literary Communism: A Manifesto of the Rearguard." *Journal of Latin American Cultural Studies: Travesia*, vol. 19, no. 3, 2010, pp. 225–36, doi:10.1080/13569325.2010.528889.
Duncombe, Stephen. Dream or Nightmare: Reimagining Politics in an Age of Fantasy. OR Books, 2019.
Duncombe, Stephen. "Does It Work?: The Æffect of Activist Art." *Social Research: An International Quarterly*, vol. 83, no. 1, Spring 2016, pp. 115–34.
Duncombe, Stephen. "Imagining No-Place." *Transformative Works and Cultures*, vol. 10, 2012, doi:10.3983/twc.2012.0350. https://journal.transformativeworks.org/index.php/twc/article/download/350/266.
Duncombe, Stephen. "It Stands on Its Head: Commodity Fetishism, Consumer Activism, and the Strategic Use of Fantasy." *Culture & Organization*, vol. 18, no. 5, Routledge, Dec. 2012, pp. 359–75. *EBSCOhost*, doi:10.1080/14759551.2012.733856.
Eagleton, Terry. *Criticism and Ideology: A Study in Marxist Literary Theory*. London, New York: Verso, 2006.
Eco, Umberto. *Foucault's Pendulum*. Translated by William Weaver, Harcourt Brace Jovanovich, 1989.
Eco, Umberto. *The Role of the Reader: Explorations in the Semiotics of Texts*. Indiana University Press, 1984.
"Ecstasy, n." *OED Online*, Oxford University Press, June 2019, www.oed.com/view/Entry/59423. Accessed 13 August 2019.
Egan, Jennifer. "Black Box." *The New Yorker*, May 2012. www.newyorker.com, https://www.newyorker.com/magazine/2012/06/04/black-box-2.
Eliade, Mircea. *The Sacred and the Profane: The Nature of Religion*. Houghton Mifflin Harcourt, 1959.
Eliot, George. *Middlemarch: An Authoritative Text, Backgrounds, Criticism*. Edited by Bert G. Hornback, W. W. Norton & Company, 2000.
Ellison, Nicole B., and Danah M. Boyd. "Sociality Through Social Network Sites." *The Oxford Handbook of Internet Studies*, edited by William H. Dutton, Oxford University Press, 2013, pp. 151–72.

Enns, Anthony. "Pseudoscience." *The Oxford Handbook of Science Fiction*, edited by Rob Latham, Oxford University Press, 2014, pp. 498–511.
Escobar, Arturo. *Pluriversal Politics: The Real and the Possible*. Translated by David Frye. Durham, NC: Duke University Press, 2020.
Estour, J. "Plasmogénie." *Encyclopédie anarchiste*, edited by Sébastien Faure, vol. 4, La Librairie internationale, 1934, pp. 2060–62.
Evans, Robert. "The El Paso Shooting and the Gamification of Terror." *bellingcat*, August 4, 2019. https://www.bellingcat.com/news/americas/2019/08/04/the-el-paso-shooting-and-the-gamification-of-terror/.
Farsayt, A. "The Worker's Movement." *Smorgonie, District Vilna; Memorial Book and Testimony (Smarhon, Belarus)*, edited by Marc D. Hodies, translated by Janie Respitz, pp. 237–72, https://www.jewishgen.org/yizkor/smorgon/smo237.html. Accessed 22 May 2020.
Faust, Wolfgang Max, and R. Baird Shuman. "Comics and How to Read Them." *The Journal of Popular Culture*, vol. 5, no. 1, 1971, pp. 195–202.
Featherstone, Liza. "Radical Academics for the Status Quo." *Jacobin*, Dec. 2019, https://jacobinmag.com/2019/12/radical-academics-judith-butler-kamala-harris-donation.
Feenberg, Andrew. *Lukács, Marx, and the Sources of Critical Theory*. New York: Oxford University Press, 1986.
Feynman, Richard P. *QED: The Strange Theory of Light and Matter*. Princeton University Press, 2014.
Fiedler, Leslie A. *Love and Death in the American Novel*. Dalkey Archive Press, 1997.
Fiedler, Leslie A. "Introduction." *In Dreams Awake : A Historical-Critical Anthology of Science Fiction*, Dell, 1975, pp. 11–24.
Fiks, Yevgeny. *Alphabet of the Language AO (after Volf Gordin's Display at the 1st International Exhibition of Interplanetary Machines and Mechanisms in Moscow in 1927)*. 2017.
Fink, Bruce. *Lacan to the Letter: Reading Ecrits Closely*. U of Minnesota Press, 2004.
Fiori de Lima, Nabylla. *Maria Lacerda de Moura Na Revista Estudios (1930–1936): Anarquismo Individualista e Filosofia da Natureza*. Universidade Tecnológica Federal do Paraná, 2016.
Fisher, Mark. *Capitalist Realism: Is There No Alternative?* Zero Books, 2010.
Flammarion, Camille. *Astronomy for Amateurs*. Translated by Frances A. Welby, D. Appleton and Company, 1910.
Folch, Christine. "Why the West Loves Sci-Fi and Fantasy: A Cultural Explanation." *The Atlantic*, 13 June 2013, https://www.theatlantic.com/entertainment/archive/2013/06/why-the-west-loves-sci-fi-and-fantasy-a-cultural-explanation/276816/.
Foucault, Michel. *Discipline and Punish: The Birth of the Prison*. Translated by Alan Sheridan, Vintage Books, 1977.
Foucault, Michel. *The Order of Things: An Archaeology of the Human Sciences*. London: Routledge, 2002.
Fox, Sidney W., and Nicols Fox. *Emergence Of Life*. Basic Books, 1988.
Frank, Tom. "Alternative to What?" *The Baffler*, no. 5, 1993, pp. 5–14.
Franklin, Howard Bruce. *War Stars: The Superweapon and the American Imagination*. Univ of Massachusetts Press, 2008.
Frazer, Sir James George. *The Golden Bough, Part 1: The Magic Art and the Evolution of Kings*. Macmillan and Co., Ltd., 1921.
Freire, Paulo. *Pedagogy of the Oppressed*. Translated by Myra Bergman Ramos, Continuum, 1970.
Freud, Sigmund. *The Interpretation of Dreams*. Translated by James Strachey, Basic Books, 2010.
Freud, Sigmund. *The Standard Edition of the Complete Psychological Works of Sigmund Freud*. Translated by James Strachey and Anna Freud, vol. 17, Hogarth Press, 1971.

Friedenthal, Andrew J. *The World of DC Comics*. Routledge, Taylor & Francis Group, 2019.
Friend, Tad. "Sam Altman's Manifest Destiny." *The New Yorker*, 10 Oct. 2016. www.newyorker.com, https://www.newyorker.com/magazine/2016/10/10/sam-altmans-manifest-destiny.
Funtowicz, Silvio O., and Jerome R. Ravetz. "Science for the Post-Normal Age." *Futures*, vol. 25, no. 7, 1993, pp. 739–55.
Fussell, Paul. *Class: A Guide through the American Status System*. New York: Summit Books, 1983.
Galilei, Galileo. *Discoveries and Opinions of Galileo*. Translated by Stillman Drake, Doubleday & Co., 1957.
Garjavin, A. N., and A. V. Pashkin. "Problemy Vospitaniya, Obrazovaniya i Prosveshcheniya v Trudakh A. L. i V. L. Gordinykh [Problems of Upbringing, Education and Enlightenment in the Works of A. L. and V. L. Gordin]." *Prosveshcheniye Na Rusi, v Rossii: Istoricheskiy Opyt: Materialy Devyatnadtsatoy Vserossiyskoy Zaochnyy Nauchnoy Konferentsii [Enlightenment in Russia, in Russia: Historical Experience: Materials of the Nineteenth All-Russian Absentee Scientific Conference]*, edited by Sergey Nikolaevich Poltorak, "Nestor," 2000, pp. 98–105.
Garroutte, Eva Marie, and Kathleen Delores Westcott. "The Story Is a Living Being: Companionship with Stories in Anishinaabe Studies." *Centering Anishinaabeg Studies: Understanding the World through Stories*, edited by Jill Doerfler et al., MSU Press, 2013, pp. 61–80.
Gernsback, Hugo. "A New Sort of Magazine." *Amazing Stories*, vol. 1, no. 1, Apr. 1926, p. 3.
Gibbs, Alan. *Contemporary American Trauma Narratives*. Edinburgh University Press, 2014.
Gibson, William, and Bruce Sterling. *The Difference Engine*. Random House Publishing Group, 2011.
Gifford, James. "Place, Personalism, Anarchism, & Fantasy: Recasting Late Modernism." *Literature Compass*, vol. 12, no. 7, 2015, pp. 322–32.
Gilsdorf, Ethan. "A Game as Literary Tutorial." *The New York Times*, 13 July 2014. *NYTimes.com*, https://www.nytimes.com/2014/07/14/books/dungeons-dragons-has-influenced-a-generation-of-writers.html.
Gilsdorf, Ethan. "A Game as Literary Tutorial." *The New York Times*, 13 July 2014. *NYTimes.com*, https://www.nytimes.com/2014/07/14/books/dungeons-dragons-has-influenced-a-generation-of-writers.html.
Giovanini, Valerie Oved. "An Army of Me: Representations of Intersubjective Relations, Ethics, and Political Resistance in *The Handmaid's Tale*." *Free Associations*, no. 75, May 2019, pp. 84–101. *freeassociations.org.uk*, doi:10.1234/fa.v0i75.254.
Gittlitz, A. M. "Let Them Drink Blood." *The New Inquiry*, 27 Dec. 2016. *thenewinquiry.com*, https://thenewinquiry.com/let-them-drink-blood/.
Godwin, Tom. "The Cold Equations." *Astounding Science Fiction*, vol. 53, no. 6, Aug. 1954, pp. 59–84.
Gold, H. L. "That's Life, Etc." *Galaxy Science Fiction*, vol. 7, no. 2, Nov. 1953, pp. 3–4.
Goldman, Marcio. "An Afro-Brazilian Theory of the Creative Process: An Essay in Anthropological Symmetrization." *Social Analysis*, vol. 53, no. 2, 2009, pp. 108–29.
Goncharok, Moshe. *Pepel Nashikh Kostrov: Ocherki Istorii Yevreyskogo Anarkhistskogo Dvizheniya (Idish-anarkhizm) [Ashes From Our Fires: Essays on the History of the Jewish Anarchist Movement (Yiddish-Anarchism)]*. Problemen, 2002.
Goodman, Paul. *Speaking and Language: Defence of Poetry*. Random House, 1971.
Google Trends, https://trends.google.com/trends/. Accessed 17 Mar. 2020.
GoogleTrends. "Search: Unicorn, Zombies, 2009–2019." Accessed August 9, 2019. https://trends.google.com/trends/explore?date=2009-01-01%202019-08-09&q=unicorn,zombies.
Google Trends. "Search: 'the darkest timeline,' 2010–2022." Accessed October 26, 2022. https://trends.google.com/trends/explore?date=2010-01-01%202022-10-26&geo=US&q=%22the%20darkest%20timeline%22.

Gopnik, Adam. "Did the Oscars Just Prove That We Are Living in a Computer Simulation?" *The New Yorker*, Feb. 2017. www.newyorker.com, https://www.newyorker.com/culture/cultural-comment/did-the-oscars-just-prove-that-we-are-living-in-a-computer-simulation.

Gordin, Abba. *Draysik Yor in Lite un Poyln: Oytabyagrafye*. Buenos Aires: Bukhgemaynshaft bay der Yidisher Ratsyonalistisher gezelshaft, 1958.

Gordin, A[bba]. L[vovich]., and V[olf]. L[vovich]. Gordin. "Deklaratsiya: Pervyy Tsentral'nyy Sotsiotekhnikum [Declaration of the First Central Sociotechnicum]." In *Anarkhisty. Dokumenty i Materialy. 1883–1935 Gg.*, vol. 2, ROSSPEN, 1999, pp. 199–205.

Gordin, Abba, and Wolf Gordin. *Podrazhatel'no-Ponimatel'nyy Metod' dlya Obucheniya Gramote*. Vilna: Novaya Pedagogika, 1909.

Gordin, Abba, and Wolf Gordin. *Sotsiomagiya i sotsiotekhnika, ili Obshcheznakharstvo i obshchestroitel'stvo [Sociomagic and Sociotechnics, or Generalized Quackery Versus Global Construction]*. Moscow: Pervyy tsentral'nyy sotsiotekhnikum, 1918.

Gordin, Abba, and Wolf Gordin. *Strana Anarkhiya (Utopii)*. Edited by Evgeniy Kuchinov. Moscow: Common place, 2019.

Gordin, Abba, and Wolf Gordin. "Deklaratsiya Moskovskoy Organizatsii Anarkho-Universalistov (K Vos'momu s"yezdu Sovetov)." In *Anarkhisty. Dokumenty i Materialy. 1883–1935 Gg.*, vol. 2, ROSSPEN, 1999, pp. 406–14.

Gordin, Abba, and Wolf Gordin. "Manifest Pananarkhistov." In *Strana Anarkhiia (Utopii)*, edited by Evgeniy Kuchinov. Moscow: Common place, 2019, pp. 214–300.

Gordin, Abba, and Wolf Gordin. "Strana Anarkhiia: Utopiia-poema." *Strana Anarkhiia (utopii)*, edited by Evgeniy Kuchinov, Common place, 2019, pp. 74–213.

Gordin, Abba. "Our Works That Were Published in Smorgon." *Smorgonie, District Vilna: Memorial Book and Testimony (Smarhon, Belarus)*, edited by Marc D. Hodies et al., translated by Jerrold Landau, JewishGen, 2019, pp. 209–36, https://www.jewishgen.org/yizkor/smorgon/smo209.html.

Gordin, Abba. "Social Mythology." *The Clarion*, vol. 1, no. 4, pp. 3–6.

Gordin, V[olf]. L[vovich]. *Inventism or Eurologism, Being the Teaching of Invention*. The All-Invention House, 1925.

Gordin, Wolf et al. "Ni Za Kogo [No One]!" *Burevestnik*, 14 Nov. 1917, p. 1.

Gould, Stephen Jay. *Wonderful Life: The Burgess Shale and the Nature of History*. W. W. Norton & Company, 1990.

Graeber, David. *Debt: The First 5,000 Years*. Melville House, 2014.

Graeber, David. *Fragments of an Anarchist Anthropology*. Prickly Paradigm Press, 2004.

Graeber, David. *Lost People: Magic and the Legacy of Slavery in Madagascar*. Indiana University Press, 2007.

Graeber, David. *Revolutions in Reverse*. London: Minor Compositions, 2011.

Graeber, David. *The Utopia of Rules: On Technology, Stupidity, and the Secret Joys of Bureaucracy*. Melville House, 2015.

Graeber, David. "Consumption." *Current Anthropology*, vol. 52, no. 4, The University of Chicago Press, Aug. 2011, pp. 489–511, doi:10.1086/660166.

Graeber, David. "Fetishism as Social Creativity: Or, Fetishes Are Gods in the Process of Construction." *Anthropological Theory*, vol. 5, no. 4, Dec. 2005, pp. 407–38. *SAGE Journals*, doi:10.1177/1463499605059230.

Graeber, David. "Radical Alterity Is Just Another Way of Saying 'Reality': A Reply to Eduardo Viveiros de Castro." *HAU: Journal of Ethnographic Theory*, vol. 5, no. 2, The University of Chicago Press, Sept. 2015, pp. 1–41. *journals.uchicago.edu (Atypon)*, doi:10.14318/hau5.2.003.

Grant, Caesar. "All God's Chillen Had Wings." *The Book of Negro Folklore*, edited by Langston Hughes and Arna Wendell Bontemps, New York, Dodd, Mead, 1958, pp. 62–65.

Gray, Chantelle. *Anarchism After Deleuze and Guattari: Fabulating Futures*. London: Bloomsbury Publishing, 2022.
Graziosi, Graig. "Herschel Walker Moans about Obama Mocking His Vampire vs Werewolf Debate." *The Independent*, December 5, 2022. https://www.independent.co.uk/news/world/americas/us-politics/herschel-walker-obama-werewolf-debate-b2239441.html.
Grierson, Tim. "'Unicorn Store': Toronto Review." *Screen Daily*, 13 Sept. 2017, https://www.screendaily.com/reviews/unicorn-store-toronto-review/5122310.article.
Groopman, Jerome. "Can Brain Science Help Us Break Bad Habits?" *The New Yorker*, Oct. 2019, https://www.newyorker.com/magazine/2019/10/28/can-brain-science-help-us-break-bad-habits.
Gross, Lawrence William. *Anishinaabe Ways of Knowing and Being*. Ashgate, 2014.
Gunn, James. *Isaac Asimov: The Foundations of Science Fiction*. Scarecrow Press, 1996.
Gunn, James. "A Touch of Stone." *The Road to Science Fiction*, vol. 3, Penguin Group (Canada), 1981, pp. 244–46.
Haas, Angela M. "Wampum as Hypertext: An American Indian Intellectual Tradition of Multimedia Theory and Practice." *Studies in American Indian Literatures*, vol. 19, no. 4, University of Nebraska Press, 2007, pp. 77–100. *Project MUSE*, doi:10.1353/ail.2008.0005.
Haddish, Tiffany. *The Last Black Unicorn*. Simon and Schuster, 2019.
Halberstam, Judith. "Automating Gender: Postmodern Feminism in the Age of the Intelligent Machine." *Feminist Studies*, vol. 17, no. 3, Feminist Studies, Inc., 1991, pp. 439–60. JSTOR, *JSTOR*, doi:10.2307/3178281.
Hall, Melinda. "Horrible Heroes: Liberating Alternative Visions of Disability in Horror." *Disability Studies Quarterly*, vol. 36, no. 1, Mar. 2016. dsq-sds.org, doi:10.18061/dsq.v36i1.3258.
Hamill, Mark. "'Your overconfidence is your weakness'-'Once you start down the dark path, forever will it dominate your destiny'-'Who's the more foolish . . . the fool or the fool who follows him?' (& many MANY more) #Red5StandingBy." *Twitter*, 7 May 2020. twitter.com, https://twitter.com/hamillhimself/status/1258509463658434560.
Hantke, Steffen. *Monsters in the Machine: Science Fiction Film and the Militarization of America after World War II*. Univ. Press of Mississippi, 2016.
Haraway, Donna J. *Staying with the Trouble: Making Kin in the Chthulucene*. Duke University Press, 2016.
Haraway, Donna J. *When Species Meet*. U of Minnesota Press, 2013.
Haraway, Donna J. "Teddy Bear Patriarchy: Taxidermy in the Garden of Eden, New York City, 1908–1936." *Social Text*, no. 11, Duke University Press, 1984, pp. 20–64.
Hardt, Michael. "Anti-Oedipus (1)," *Reading Notes on Deleuze and Guattari*, Capitalism & Schizophrenia, Duke University, December 18, 2012, https://people.duke.edu/~hardt/a01.htm.
Hardt, Michael, and Antonio Negri. *Empire*. Harvard University Press, 2001.
Harman, Graham. *Bruno Latour: Reassembling the Political*. Pluto Press, 2014.
Harman, Graham. *Weird Realism: Lovecraft and Philosophy*. Zero Books, 2012.
Hartman, Chester W., and Gregory D. Squires, editors. *There Is No Such Thing as a Natural Disaster: Race, Class, and Hurricane Katrina*. Taylor & Francis, 2006.
Harvey, Graham. *Animism: Respecting the Living World*. Columbia University Press, 2006.
Heidegger, Martin. *The Fundamental Concepts of Metaphysics: World, Finitude, Solitude*. Translated by William McNeill and Nicholas Walker, Indiana University Press, 1995.
Heller, Leonid. "Brat'ya Gordiny, Anarkhizm i Russkii Avangard." *The Russian Word in the Land of Israel, the Jewish Word in Russia*, edited by V. Khazan and W. Moskovich, Hebrew University, Center for Slavic Languages and Literatures, 2006, pp. 129–47, http://hylaea.ru/uploads/files/page_3993_1557303179.pdf.

Heller, Leonid. "Voyage au pays de l'anarchie: Un itinéraire: l'utopie." *Cahiers Du Monde Russe*, vol. 37, no. 3, 1996, pp. 249–75, doi:10.3406/cmr.1996.2460.

Herbert, Frank. *Dune*. Penguin, 2005.

"Herd of Unicorns." *Servant Financial News*, 18 Aug. 2016, https://www.servantfinancialnews.com/2016/08/herd-of-unicorns.html.

Heron, Kai. "[Review] *Lacan and Deleuze: A Disjunctive Synthesis*." *Psychoanalysis, Culture & Society*, vol. 24, no. 2, June 2019, pp. 230–33. *Springer Link*, doi:10.1057/s41282-019-00125-9.

Herrera, Alfonso L. *Herrera's 'Plasmogenia' and Other Collected Works: Early Writings on the Experimental Study of the Origin of Life*. Edited by Henderson James Cleaves et al., Translated by Henderson James Cleaves and Fabiola Barraclough, Springer, 2014.

Herrera, Alfonso L. *La vie sur les hauts plateaux: Influence de la pression barométrique sur la constitution et le développement des êtres organisés. Traitement climatérique de la tuberculose*. I. Escalante, 1899.

Herrera, Alfonso L. "El origen del pensamiento." *Estudios*, vol. 10, no. 106, June 1932, pp. 23–27.

Herrera, Alfonso L. "Filosofía Comparada: El Animal y El Salvaje." *Memorias de La Sociedad Científica Antonio Alzate*, no. 9, 1896 1895, pp. 77–96.

Herrera, Alfonso L. "La aspiración gigantesca de la plasmogenia." *Nervio*, vol. 1, no. 3, July 1931, pp. 17–19.

Herrera, Alfonso L. "La sordera mística de la ciencia." *Estudios* vol. 12, no. 130, June 1934, pp. 10–11.

Herrera, Alfonso L. "Mi labor revolucionaria en la enseñanza." *Crisol*, no. 73, Jan. 1935, pp. 55–58.

Hignett, Katherine. "AOC Is in 'Bananaland': Fox News Contributor Slams Green New Deal in Unicorns-and-Cannibalism Tirade." *Newsweek*, 3 Mar. 2019, https://www.newsweek.com/alexandria-ocasio-cortez-cannibalism-green-new-deal-fox-news-katherine-timpf-1349999.

Hilgartner, Stephen. "Overflow and Containment in the Aftermath of Disaster." *Social Studies of Science*, vol. 37, no. 1, Feb. 2007, pp. 153–58.

Hill, Andrew. "Fresh Blood: Why Everyone Fell for Theranos." *Financial Times*, 19 Mar. 2018, https://www.ft.com/content/a45ffdf0-2850-11e8-b27e-cc62a39d57a0.

Hobbes, Michael. "What Is The Internet Doing To Boomers' Brains?" *Huffington Post*, 29 Oct. 2020, https://www.huffpost.com/entry/internet-baby-boomers-misinformation-social-media_n_5f998039c5b6a4a2dc813d3d.

Hodies, Marc D., editor. *Smorgonie, District Vilna; Memorial Book and Testimony*. Translated by Jerrold Landau and Sara Mages, JewishGen, 2019, https://www.jewishgen.org/yizkor/smorgon/Smorgon.html.

Hofstadter, Richard. *Social Darwinism in American Thought: 1860–1915*. University of Pennsylvania Press, 1944.

Holmes à Court, Simon. "Just How Bullshit Is the Fantasy of 'Clean Coal' and CCS?" *Crikey*, 16 June 2017, https://www.crikey.com.au/2017/06/16/just-how-bullshit-is-the-fantasy-of-clean-coal-and-ccs/.

Hopkinson, Nalo. "A Reluctant Ambassador from the Planet of Midnight." *Journal of the Fantastic in the Arts*, vol. 21, no. 3, 2010, pp. 339–50.

Hornborg, Alf. *Global Magic: Technologies of Appropriation from Ancient Rome to Wall Street*. Springer, 2016.

Hornborg, Alf. "Artifacts Have Consequences, Not Agency: Toward a Critical Theory of Global Environmental History." *European Journal of Social Theory* 20, no. 1 (February 1, 2017): 95–110. https://doi.org/10.1177/1368431016640536.

Huizinga, Johan. *Homo Ludens: A Study of the Play-Element in Culture*. London: Routledge & Kegan Paul, 1980.

Hume, David. *An Enquiry Concerning Human Understanding: With Hume's Abstract of A Treatise of Human Nature and A Letter from a Gentleman to His Friend in Edinburgh*. Indianapolis: Hackett Publishing, 1993.
Hume, Kathryn. *Fantasy and Mimesis: Responses to Reality in Western Literature*. Routledge, 2014.
Hurley, Jessica, and N. K. Jemisin. "An Apocalypse Is a Relative Thing: An Interview with N. K. Jemisin." *ASAP/Journal*, vol. 3, no. 3, Dec. 2018, pp. 467–77. *Project MUSE*, doi:10.1353/asa.2018.0035.
Hutchins, Edwin. *Cognition in the Wild*. MIT Press, 1995.
Hynes, Gerard. "Locations and Borders." *The Routledge Companion to Imaginary Worlds*, edited by Mark J. P. Wolf, Routledge, 2017, pp. 3–10.
Imarisha, Walidah. "Introduction." *Octavia's Brood: Science Fiction Stories from Social Justice Movements*, edited by Walidah Imarisha and Adrienne Maree Brown, 2015, pp. 3–6.
Imarisha, Walidah. "Rewriting the Future: Using Science Fiction to Re-Envision Justice." *Bitch Media*, 11 Feb. 2015, https://www.bitchmedia.org/article/rewriting-the-future-prison-abolition-science-fiction.
Imarisha, Walidah. "Walidah Imarisha on Visionary Sci Fi." Presented at the Are the Gods Afraid of Black Sexuality? Conference, Columbia University, March 3, 2015. https://www.youtube.com/watch?v=iYnR9_2UgLE.
ImperialAce. *Frau Oberst Twilight Sparkle*. DeviantArt. 8 Oct. 2013, https://www.deviantart.com/imperialace/art/Frau-Oberst-Twilight-Sparkle-406046273. Accessed 13 Aug. 2019.
Ingebretsen, Edward. *At Stake: Monsters and the Rhetoric of Fear in Public Culture*. Chicago: University of Chicago Press, 2001.
Ingold, Tim. "Being Alive to a World without Objects." *The Handbook of Contemporary Animism*, edited by Graham Harvey, Acumen Pub., 2013, pp. 213–25.
Irwin, W. R. *The Game of the Impossible: A Rhetoric of Fantasy*. University of Illinois Press, 1976.
it-wasnt-me. "Madmaxing." Urban Dictionary, February 10, 2016. https://www.urbandictionary.com/define.php?term=madmaxing.
Jablon, Madelyn. *Black Metafiction: Self-Consciousness in African American Literature*. Iowa City: University of Iowa Press, 1997.
Jackson, Rosemary. *Fantasy: The Literature of Subversion*. Routledge, 2013.
Jackson, Steve, and John M. Ford. *GURPS Infinite Worlds*. Steve Jackson Games, 2004.
Jackson, Zakiyyah Iman. *Becoming Human: Matter and Meaning in an Antiblack World*. NYU Press, 2020.
Jacoby, Russell. *Picture Imperfect: Utopian Thought for an Anti-Utopian Age*. New York: Columbia University Press, 2005.
James, Edward. *Science Fiction in the Twentieth Century*. Oxford University Press, 1994.
James, William. *Pragmatism, a New Name for Some Old Ways of Thinking: Popular Lectures on Philosophy*. Longmans, Green, and Company, 1907.
Janicka, Iwona. "Who Can Speak? Rancière, Latour and the Question of Articulation." *Humanities*, vol. 9, no. 4, 4, Multidisciplinary Digital Publishing Institute, Dec. 2020, p. 123. www.mdpi.com, doi:10.3390/h9040123.
Jayadev, Arjun, and Samuel Bowles. "Guard Labor." *Journal of Development Economics*, vol. 79, no. 2, Apr. 2006, pp. 328–48. *ScienceDirect*, doi:10.1016/j.jdeveco.2006.01.009.
Jemisin, N. K. *The City We Became*. New York: Orbit, 2020.
Jemisin, N. K. *The Fifth Season*. New York: Orbit, 2015.
Jemisin, N. K. *The Obelisk Gate*. New York: Orbit, 2016.
Jemisin, N. K. *The Stone Sky*. New York: Orbit, 2017.
Jemisin, N. K. "Once again, since I keep getting asked: the Broken Earth series' setting was not intended to be our Earth in the future. I take it that's a commonality of the 'dying Earth'

subgenre? But I've never read any of those, sorry, so IDK." *Twitter*, 20 Oct. 2020, https://twitter.com/nkjemisin/status/1318717990528057345.

Jemisin, N. K., and Elizabeth Flock. "How the Protests in Ferguson Helped Inspire This Fantasy Novel." *PBS NewsHour*, 12 June 2019. *www.pbs.org*, https://www.pbs.org/newshour/arts/how-the-protests-in-ferguson-helped-inspire-this-fantasy-novel.

Jenkins, Henry, Sam Ford, and Joshua Green. *Spreadable Media: Creating Value and Meaning in a Networked Culture*. NYU Press, 2018.

Johnson, Charles Richard. *Being & Race: Black Writing Since 1970*. Bloomington, Ind.: Indiana University Press, 1988.

Jones, Adam Garnet. "History of the New World." *Love After the End: An Anthology of Two-Spirit and Indigiqueer Speculative Fiction*, edited by Joshua Whitehead, Arsenal Pulp Press, 2020. 35–60.

Joreth. "Unicorn Hunter." *Urban Dictionary*, 8 Apr. 2010, https://www.urbandictionary.com/define.php?term=Unicorn%20Hunter.

Judkis, Maura. "Starbucks' Unicorn Frappuccino Tastes like Sour Birthday Cake and Shame." *Washington Post*, 19 Apr. 2017, https://www.washingtonpost.com/news/food/wp/2017/04/19/starbucks-unicorn-frappuccino-tastes-like-sour-birthday-cake-and-shame/?noredirect=on.

Justice, Daniel Heath. "Indigenous Wonderworks and the Settler-Colonial Imaginary." *Apex Magazine*, 10 Aug. 2017, https://apex-magazine.com/indigenous-wonderworks-and-the-settler-colonial-imaginary/.

Kaethler, Mark. "Failed Feminist Interventions in Wolfenstein II: The New Colossus." *Feminist War Games?: Mechanisms of War, Feminist Values, and Interventional Games*, edited by Jon Saklofske et al., Routledge, 2020.

Kafka, Franz. *The Complete Stories*. Edited by Nahum N. Glatzer, Schocken Books, 1995.

Kamelamela, Katie L. "Tweet: 'For the Record . . .'" *Twitter*, 14 July 2019, https://twitter.com/kteabam/status/1150561729421516800?s=20.

Kane, Vivian. "We Are in Love with the New She-Ra Design (And the Awful Men Hating on It Is Just a Bonus)." *The Mary Sue*, 16 July 2018, https://www.themarysue.com/new-she-ra-trolls-gtfo/.

Kant, Immanuel. *Correspondence*. Translated by Arnulf Zweig, Cambridge University Press, 1999.

Kant, Immanuel. *Groundwork of the Metaphysics of Morals*. Translated by Mary Gregor, Cambridge University Press, 1998.

Kant, Immanuel. *Lectures on Ethics*. Edited by Peter Heath and J. B. Schneewind, Translated by Peter Heath, Cambridge University Press, 2001.

Kant, Immanuel. *Political Writings*. Edited by Hans Reiss, Translated by H. B. Nisbet, Cambridge University Press, 1991.

Kant, Immanuel. "Observations on the Feeling of the Beautiful and Sublime." *Anthropology, History, and Education*, translated by Paul Guyer, Cambridge University Press, 2007, pp. 18–62.

"Karén:na." *Kanienkeha: An Open Source Endangered Language Initiative (Mohawk Dictionary)*, 17 Sept. 2015, https://kanienkeha.net/body-parts/karenna/.

Karl, Frederick Robert. *Franz Kafka, Representative Man*. Fromm International Publishing Corporation, 1993.

Karlamangla, Soumya. "I've Seen 'Contagion' Four Times. No, the Coronavirus Outbreak Isn't the Same." *Los Angeles Times*, 11 Mar. 2020. *www.latimes.com*, https://www.latimes.com/california/story/2020-03-11/coronavirus-contagion-outbreak-movie-comparison.

Kaulingfreks, Ruud, and Femke Kaulingfreks. "In Praise of Anti-Capitalist Consumption: How the V for Vendetta Mask Blows up Hollywood Marketing." *Ephemera*, vol. 13, no. 2, Nick Butler (On Behalf of the Editorial Collective of Ephemera), 2013, pp. 453–7.

Kehe, Jason. "WIRED Book Club: Fantasy Writer N. K. Jemisin on the Weird Dreams That Fuel Her Stories." *Wired*, June 2016. www.wired.com, https://www.wired.com/2016/06/wired-book-club-nk-jemisin/.

Kelsey, Penelope Myrtle. *Reading the Wampum: Essays on Hodinöhsö:ni' Visual Code and Epistemological Recovery*. Syracuse University Press, 2014.

Kerr, Clark. *The Great Transformation in Higher Education*, 1960–1980. SUNY Press, 1991.

King, Noel. "I Agree With DHS Overcrowding Report 100%, Border Agents' President Says." *Morning Edition*, National Public Radio, 3 July 2019. www.npr.org, https://www.npr.org/2019/07/03/738401304/i-agree-with-dhs-overcrowding-report-100-border-agents-president-says.

Kinna, Ruth, et al. "Occupy and the Constitution of Anarchy." Global Constitutionalism, vol. 8, no. 2, 2019, pp. 357–90, doi:10.1017/S204538171900008X.

Kirkman, Robert, and Charles Adlard. *The Walking Dead Omnibus, Vol. 1*. Image, 2013.

Klingensmith-Parnell, Elsa Ruth. "Cultural Appropriation or Much-Needed Representation? On Rebecca Roanhorse's *Trail of Lightning*." *SFRA Review*, no. 330, Fall 2019, pp. 90–98.

Klock, Geoff. *How to Read Superhero Comics and Why*. London: A&C Black, 2002.

Koelb, Clayton. "The Language of Presence in Varley's 'The Persistence of Vision.'" *Science Fiction Studies*, vol. 11, no. 2, 1984, pp. 154–65.

Kolodnyi, Lev Efimovich. *Moskva v Ulitsakh i Litsakh: Zamoskvoreche: Avtorskii Putevoditel.'* Golos-press, 2002.

Kottke, Jason. "Decrease in Economic Activity Due to COVID-19 Reduced Air Pollution and Saved Lives." *Kottke.Org: Home of Fine Hypertext Products*, 11 Mar. 2020. kottke.org, https://kottke.org/20/03/covid-19-reduced-air-pollution.

Krivenkiy, V. V., ed. *Anarkhisty: dokumenty i materialy, 1883–1935 gg. v. 2: 1917–1935 gg.* (Moscow: ROSSPĖN, 1999).

Kropotkin, Peter. *Mutual Aid: A Factor of Evolution*. McClure, Philips & Co., 1902.

Kropotkin, Peter. "Anarchist Morality (Continued)." *Mother Earth*, vol. 11, no. 10, Dec. 1916, pp. 710–18.

Kropotkin, Peter. "The Morality of Nature." *The Nineteenth Century*, vol. 57, Mar. 1905, pp. 407–26.

Kuchinov, Eugene. "Anarchy as Access to Space: Language, Imagination, Technics." *Science Fiction and Communism Conference, Sofia, 25–27 May*. www.academia.edu, https://www.academia.edu/41807554/Anarchy_as_Access_to_Space_Language_Imagination_Technics_eng_. Accessed 26 July 2020.

Kuchinov, Eugene. "Fictio Audaciae: Po Tu Storonu Samosokhraneniya Prosveshcheniya." *Logos*, vol. 29, no. 4, 2019, pp. 107–26, doi:10.22394/0869-5377-2019-4-109-25.

Kuchinov, Eugene. "Ot pananarkhizma k AOizmu i AIIZu: ocherk istorii i mifologii odnogo inoplanetnogo plemeni." *Etnograficheskoye obozreniye*, no. 6, Dec. 2019, pp. 34–48. ras.jes.su, doi:10.31857/S086954150007767-3.

Kuchinov, Eugene. "Russkiy Mars [Russian Mars]." *KROT*, no. 4, 2018, https://krot.me/articles/mars-is-ours.

Kuchinov, Evgeniy. *Anarchy as Access to Space: Language, Imagination, Technics*. https://www.youtube.com/watch?v=KXplBjuWBCg&feature=youtu.be&t=1296. Science Fiction and Communism Conference, American University in Bulgaria, Jun 28, 2018.

Kuznecov, Sergei N. "Interlingvistiko en 'Kosma Dimensio': Vojaĝo Inter Kosmoglotiko Kaj Kosmolingvistiko." *Interlinguistische Informationen*, no. 39, 2001, http://www.lingviko.net/db/10_kuznecov.htm.

Kuznecov, Sergei N. "Linguistica Cosmica: Rozhdeniye 'Kosmicheskoy Paradigmy.'" *Sovremennaya Nauka*, no. 2, 2014, pp. 39–65.

Lacan, Jacques. *Ecrits: The First Complete Edition in English*. Translated by Bruce Fink, W. W. Norton & Company, 2006.

Lacan, Jacques. *The Seminar of Jacques Lacan, Book II: The Ego in Freud's Theory and in the Technique of Psychoanalysis, 1954–1955*. Edited by Jacques-Alain Miller, Translated by Sylvana Tomaselli, Cambridge University Press, 1988.

Lacan, Jacques. *The Seminar of Jacques Lacan, Book III: The Psychoses, 1955–1956*. Edited by Jacques-Alain Miller, Translated by Russell Grigg, Routledge, 1993.

Lacan, Jacques. *The Seminar of Jacques Lacan, Book XXIII: The Sinthome*. Edited by Jacques-Alain Miller, Translated by A. R. Price, Polity Press, 2016.

Lagalisse, Erica Michelle. "'Marginalizing Magdalena': Intersections of Gender and the Secular in Anarchoindigenist Solidarity Activism." *Signs: Journal of Women in Culture and Society*, vol. 36, no. 3, Spring 2011, pp. 653–78. Semantic Scholar, doi:10.1086/657526.

Lagalisse, Erica. "Occult Features of Anarchism." *Essays in Anarchism and Religion*, edited by Alexandre Christoyannopoulos and Matthew S. Adams, vol. 2, Stockholm University Press, 2018, pp. 278–332.

Landon, Brooks. *Science Fiction After 1900: From the Steam Man to the Stars*. Routledge, 2014.

Landon, Brooks. "Synthespians, Virtual Humans, and Hypermedia: Emerging Contours of Post-SF Film." *Edging into the Future : Science Fiction and Contemporary Cultural Transformation*, edited by Veronica Hollinger and Joan Gordon, University of Pennsylvania Press, 2002, pp. 57–74.

Lane, Sabra. "Carbon Capture a 'Unicorn' for Power Generation." Interview with Simon Holmes à Court. *AM*, Australian Broadcasting Corporation. https://www.abc.net.au/am/content/2016/s4677884.htm.

Larbalestier, Justine. *The Battle of the Sexes in Science Fiction*. Wesleyan University Press, 2002.

Latour, Bruno, and T. Hugh Crawford. "An Interview with Bruno Latour." *Configurations*, vol. 1, no. 2, Johns Hopkins University Press, May 1993, pp. 247–68. Project MUSE, doi:10.1353/con.1993.0012.

Latour, Bruno. *Reassembling the Social: An Introduction to Actor-Network-Theory*. OUP Oxford, 2005.

Latour, Bruno. *Science in Action: How to Follow Scientists and Engineers Through Society*. Harvard University Press, 1987.

Latour, Bruno. *War of the Worlds: What about Peace?* Prickly Paradigm Press, 2002.

Latour, Bruno. *We Have Never Been Modern*. Translated by Catherine Porter, Harvard University Press, 1993.

Latour, Bruno. "Agency at the Time of the Anthropocene." *New Literary History*, vol. 45, no. 1, Johns Hopkins University Press, Apr. 2014, pp. 1–18. Project MUSE, doi:10.1353/nlh.2014.0003.

Latour, Bruno. "On Interobjectivity." *Mind, Culture, and Activity*, vol. 3, no. 4, Routledge, Oct. 1996, pp. 228–45. Taylor and Francis+NEJM, doi:10.1207/s15327884mca0304_2.

Latour, Bruno. "Why Has Critique Run Out of Steam?: From Matters of Fact to Matters of Concern." *Critical Inquiry*, vol. 30, no. 2, 2004, pp. 225–48.

Le Guin, Ursula K. *The Dispossessed*. Perennial Classics : Harper, 2003.

Le Guin, Ursula K. "Introduction." *The Norton Book of Science Fiction: North American Science Fiction, 1960–1990*, edited by Ursula K. Le Guin and Brian Attebery, W. W. Norton & Co., 1993, pp. 15–42.

Le Guin, Ursula K. "The Direction of the Road." *Orbit 12*, edited by Damon Knight, Putnam, 1973, pp. 31–38.

Le Guin, Ursula K. "'The Author of the Acacia Seeds' and Other Extracts from the *Journal of the Association of Therolinguistics*." *Fellowship of the Stars: Nine Science Fiction Stories*, edited by Terry Carr, Simon and Schuster, 1974, pp. 170–78.

Lebadeva, Ekaterina. "Svyatogor about Volcanism as a True Art." *Constructivists, Biocosmists, and the New Human: Technology and Philosophy in Petrograd and Leningrad*, 28 May 2020. *techpeterburg.wixsite.com*, https://techpeterburg.wixsite.com/mysite/blog/categories/cosmism/.

Ledesma Prietto, Nadia. "Apuntes sobre la eugenesia y la libertad sexual en el discurso de dos médicos anarquistas. Argentina, 1930–1940." *Nomadías*, no. 16, 2012, pp. 75–97.

Leidner, D., et al. "Antecedents to Stigma: Factors That Diminish IT Value." *2013 46th Hawaii International Conference on System Sciences*, 2013, pp. 4697–708. IEEE Xplore, doi:10.1109/HICSS.2013.634.

Leighten, Patricia. "The White Peril and *L'Art Nègre*: Picasso, Primitivism, and Anticolonialism." *The Art Bulletin*, vol. 72, no. 4, 1990, pp. 609–30, doi:10.1080/00043079.1990.10786458.

Lem, Stanislaw. "On the Structural Analysis of Science Fiction." *Microworlds*, edited by Franz Rottensteiner, translated by Franz Rottensteiner et al., Harcourt Brace Jovanovich, 1986, pp. 31–44.

"A Letter of Solidarity from the Year 3017." *Puget Sound Anarchists*, 30 Nov. 2017, https://pugetsoundanarchists.org/a-letter-of-solidarity-from-the-year-3017/.

Lévi-Strauss, Claude. "The Structural Study of Myth." *The Journal of American Folklore*, vol. 68, no. 270, Dec. 1955, pp. 428–44.

Lewis, C. S. *The Silver Chair: Book 4 in the Chronicles of Narnia*. New York: Collier Books, 1976.

Ley, Willy. "Homemade Pseudo-Cells." *Galaxy Science Fiction*, vol. 7, no. 12, Dec. 1953, pp. 52–54.

Ley, Willy. "Is Artificial Life Possible?" *Galaxy Science Fiction*, vol. 7, no. 11, Nov. 1953, pp. 35–40.

Linnemann, Travis. "Bad Cops and True Detectives: The Horror of Police and the Unthinkable World." *Theoretical Criminology* 23, no. 3 (2019): 355–74.

Linnemann, Travis. *The Horror of Police*. Minneapolis: University of Minnesota Press, 2022.

Little Badger, Darcie. "Nkásht íí." Strange Horizons, December 15, 2014. http://strangehorizons.com/fiction/nksht/.

Livesay, Nora, and John D. Nichols, editors. "Migizi (Na)." *The Ojibwe People's Dictionary*, https://ojibwe.lib.umn.edu/main-entry/migizi-na. Accessed 5 June 2021.

Lolcano, Abbey. "I Teach Social Movements at My Job . . . " *Facebook*, 2 June 2020, https://www.facebook.com/abbeyvolcano/posts/10221550293817910.

Long, Marion. "Paradise Tossed: Personal Utopias of Elie Wiesel, Tammy Faye Bakker, Max Headroom, Hans Kung, Coretta Scott King, Stephen Jay Gould, Jesse Jackson, Grace Slick, David Rockefeller, Kurt Vonnegut, and Others." *Omni*, vol. 10, no. 7, Apr. 1988, pp. 36–42, 96–108.

Lovecraft, Howard P. *The Annotated H. P. Lovecraft*. Edited by Sunand T Joshi, Dell, 1997.

Lowe, Frank. "Editor's Note." *Raised by Unicorns: Stories from People with LGBTQ+ Parents*, edited by Frank Lowe. Jersey City, NJ: Red Wheel/Weiser, 2018, pp. 1–4.

Lukács, Georg. *The Theory of the Novel: A Historico-Philosophical Essay on the Forms of Great Epic Literature*. Translated by Anna Bostock, MIT Press, 1971.

Lundberg, Christian. *Lacan in Public: Psychoanalysis and the Science of Rhetoric*. Tuscaloosa, Ala.: University of Alabama Press, 2012.

Lupton, Deborah. *Digital Sociology*. Routledge, 2014.

Lyons, Scott Richard. *X-Marks: Native Signatures of Assent*. U of Minnesota Press, 2010.

Lyons, Scott Richard. "Rhetorical Sovereignty: What Do American Indians Want from Writing?" *College Composition and Communication*, vol. 51, no. 3, Feb. 2000, pp. 447–68. *DOI.org (Crossref)*, doi:10.2307/358744.

Machado Sáez, Elena. "Dictating Desire, Dictating Diaspora: Junot Díaz's 'The Brief Wondrous Life of Oscar Wao' as Foundational Romance." *Contemporary Literature*, vol. 52, no. 3, 2011, pp. 522–55.

Malm, Andreas. *The Progress of This Storm: Nature and Society in a Warming World.* New York: Verso Books, 2018.
Malmgren, Carl D. "Self and Other in SF: Alien Encounters." *Science Fiction Studies*, vol. 20, no. 1, SFTH Inc, 1993, pp. 15–33. JSTOR.
Maltby, Richard. "Film Noir: The Politics of the Maladjusted Text." *Journal of American Studies*, vol. 18, no. 1, Apr. 1984, pp. 49–71, doi:10.1017/S0021875800018235.
Manjoo, Farhad. "Fitting Label, Yet Unicorns Are No Myth." *The New York Times*, 6 July 2015, p. B1.
Martel, James. *Textual Conspiracies: Walter Benjamin, Idolatry, and Political Theory.* University of Michigan Press, 2011.
Martin, Karin A. "Becoming a Gendered Body: Practices of Preschools." *American Sociological Review*, vol. 63, no. 4, 1998, pp. 494–511.
Martynov, Mikhail. "Âzyk i vlast'. Anarhičeskie praktiki russkogo hudožestvennogo avangarda." *Zbornik Matice srpske za slavistiku*, vol. 84, Matica srpska, 2013, pp. 159–73.
Marut, Ret. "Die Zerstörung unseres Welt-Systems durch die Markurve." *Der Ziegelbrenner*, vol. 4, no. 20–22, Jan. 1920, pp. 1–48
Marx, Karl, and Friedrich Engels. *Marx Engels Collected Works.* Lawrence & Wishart, 1975.
Marx, Karl. *Grundrisse: Foundations of the Critique of Political Economy.* Vintage Books, 1973.
McGee, Ryan. "How to Approach Peak TV in a Post-Trump Era." *Pajiba*, 3 Jan. 2017. www.pajiba.com, https://www.pajiba.com/think_pieces/how-to-approach-peak-tv-in-a-posttrump-era.php.
McGrath, Matt. "Caution Urged over 'Carbon Unicorns.'" *BBC News*, 5 Oct. 2018. www.bbc.com, https://www.bbc.com/news/science-environment-45742191.
McHale, Brian. *Constructing Postmodernism.* Routledge, 2012.
McLuhan, Marshall. *The Gutenberg Galaxy: The Making of Typographic Man.* Toronto: University of Toronto Press, 1962.
Mellamphy, Deborah. "Dead Eye: The Spectacle of Torture Porn in Dead Rising." In *Game On, Hollywood!: Essays on the Intersection of Video Games and Cinema*, edited by Gretchen Papazian and Joseph Michael Sommers, 35–46. Jefferson, NC: McFarland, 2013.
Mendick, Heather, and Becky Francis. "Boffin and Geek Identities: Abject or Privileged?" *Gender and Education*, vol. 24, no. 1, Jan. 2012, pp. 15–24. Taylor and Francis+NEJM, doi:10.1080/09540253.2011.564575.
Mendlesohn, Farah. *Rhetorics of Fantasy.* Wesleyan University Press, 2008.
Michael, Mike. "The Power-Persuasion-Identity Nexus: Anarchism and Actor Networks." *Anarchist Studies*, vol. 2, no. 1, Spring 1994.
Miéville, China. *October: The Story of the Russian Revolution.* Verso, 2018.
Miéville, China. *Perdido Street Station.* Del Rey Books, 2000.
Miéville, China. "Marxism and Fantasy: Editorial Introduction." *Historical Materialism*, vol. 10, no. 4, 2002, pp. 39–49.
Milner, Andrew. *Locating Science Fiction.* Oxford University Press, 2012.
Mitchell, David T., and Sharon L. Snyder. *Narrative Prosthesis: Disability and the Dependencies of Discourse.* University of Michigan Press, 2000.
Moore, Alan, and David Lloyd. *V For Vendetta.* DC Comics, 1989.
Moore, Jasmine A. *Sankofa: Framing Afrofuturistic Dialectical Utopias in N. K. Jemisin's The Fifth Season, Nisi Shawl's Everfair, and Nnedi Okorafor's Binti.* The University of Alabama in Huntsville, 2018. ProQuest, http://search.proquest.com/docview/2103322493/abstract/BC7C7CC23DDE4DEEPQ/1.
Morais, Renata Lemos, and Massimo Canevacci. "Digital Ubiquity in the Anthropocene: The Non-Anthropocentric Anthropology of Massimo Canevacci." *Antrocom: Online Journal of Anthropology*, vol. 12, no. 1, 2016, pp. 5–12.

Moreton-Robinson, Aileen. *The White Possessive: Property, Power, and Indigenous Sovereignty.* University of Minnesota Press, 2015.

Morse, Brandon. "My Little Pony To Children: Marxism Is Not Magic." *The Federalist*, 8 Apr. 2015, https://thefederalist.com/2015/04/08/my-little-pony-to-children-marxism-is-not-magic/.

Morton, Timothy. *Hyperobjects: Philosophy and Ecology after the End of the World.* University of Minnesota Press, 2014.

Moten, Fred, and Stefano Harney. *The Undercommons: Fugitive Planning & Black Study.* Minor Compositions, 2013.

Moyer, Justin Wm, and Lindsey Bever. "Vanguard America, a White Supremacist Group, Denies Charlottesville Ramming Suspect Was a Member." *Washington Post*, 15 Aug. 2017. *www.washingtonpost.com*, https://www.washingtonpost.com/local/vanguard-america-a-white-supremacist-group-denies-charlottesville-attacker-was-a-member/2017/08/15/2ec897c6-810e-11e7-8072-73e1718c524d_story.html.

Munn, Luke. "Alt-Right Pipeline: Individual Journeys to Extremism Online." *First Monday*, vol. 24, no. 6, June 2019. *journals.uic.edu*, doi:10.5210/fm.v24i6.10108.

Murphy, Bernice M. *Key Concepts in Contemporary Popular Fiction.* Edinburgh: Edinburgh University Press, 2017.

Musk, Elon. "Take the Red Pill 💊." *Twitter*, 17 May 2020. *twitter.com*, https://twitter.com/elonmusk/status/1262076474565242880.

Nagel, Thomas. "What Is It Like to Be a Bat?" *The Philosophical Review*, vol. 83, no. 4, 1974, pp. 435–50. *JSTOR*, doi:10.2307/2183914.

Nagle, Angela. *Kill All Normies: Online Culture Wars From 4Chan and Tumblr to Trump and the Alt-Right.* Winchester, UK: Zero Books, 2017.

Nally, Claire. *Steampunk: Gender, Subculture and the Neo-Victorian.* Bloomsbury Publishing, 2019.

Nash Information Services, LLC. "Leading Creative Types 1995–2019." *The Numbers: Where Data and the Movie Business Meet*, 15 June 2019, https://www.the-numbers.com/market/creative-types.

Natsional'na Akademiya Nauk Ukrayini, Komisiya z Naukovoyi Spadshchyny Akademika V. I. Vernads'koho, Natsional'na Biblioteka Ukrayiny Imeni V. I. Vernads'koho, Instytut Arkhivoznavstva. *V. I. Vernads'kyy I: Ukrayina z Lystuvannya.* 2018. V. I. Vernads'kyy Archive.

Ndalianis, Angela. *Science Fiction Experiences.* New Academia Publishing, LLC, 2011.

Nedava, Joseph. "Abba Gordin: A Portrait of a Jewish Anarchist." *Soviet Jewish Affairs* vol. 4, no. 2 (1974), pp. 73–79.

Neruda, Pablo. "Oda a La Piedra/Ode to the Stone." *Fifty Odes*, by Pablo Neruda, translated by George Schade, Host Publications, Inc., 1996, pp. 162–69.

Newshub, "Kanye West Claims Elon Musk Is a Clone in Latest Bizarre Rant," sec. Entertainment. Accessed December 7, 2022, https://www.newshub.co.nz/home/entertainment/2022/12/kanye-west-claims-elon-musk-is-a-clone-in-latest-bizarre-rant.html.

Nicholls, Peter. "Science Fiction: The Monsters and the Critics." *Science Fiction at Large: A Collection of Essays, by Various Hands, About the Interface Between Science Fiction and Reality*, New York: Harper & Row, 1977, pp. 157–84.

Nietzsche, Friedrich. *On the Genealogy of Morals and Ecce Homo.* Edited by Walter Kaufmann, Translated by Walter Kaufmann and R. J. Hollingdale, Vintage Books, 1989.

Nietzsche, Friedrich. "Twilight of the Idols." *The Anti-Christ and Other Writings*, edited by Aaron Ridley, translated by Judith Norman, Cambridge University Press, 2005, pp. 153–229.

Nolte, John. "Benjamin Kunkel: The Lena Dunham of Literature." *Breitbart*, 12 Mar. 2014, https://www.breitbart.com/entertainment/2014/03/12/benjamin-kunkel-the-lena-dunham-of-literature/.

Nowell, Cecilia. "In the Navajo Nation, Anarchism Has Indigenous Roots." *The Nation*, Sept. 2020. *www.thenation.com*, https://www.thenation.com/article/activism/anarchism-navajo-aid/.

Noys, Benjamin. *The Lovecraft Event*. https://www.academia.edu/548596/The_Lovecraft_Event. Accessed 13 Dec. 2020.

O'Brien, Jack. "Classics of Science Fiction: 'The Cold Equations.'" *Asimov's Science Fiction*, vol. 32, no. 3, Mar. 2008, p. 107.

OECD. *States of Fragility 2018*. OECD, 2018. *DOI.org (Crossref)*, doi:10.1787/9789264302075-en.

Okorafor, Nnedi, et al. "The NATIVE Exclusive: Nnedi Okorafor on Africanfuturism and the Challenges of Pioneering." *The Native*, 5 Nov. 2018, https://thenativemag.com/interview/native-exclusive-nnedi-okorafor-africanfuturism-challenges-pioneering/.

Okorafor, Nnedi. *The Book of Phoenix*. New York: DAW Books, 2017.

Okpadah, Stephen Ogheneruro. "The Postnormal Condition and the Politics of Migration in Biyi Bandele's *Half of a Yellow Sun*." *VTU Review: Studies in the Humanities and Social Sciences*, vol. 3, no. 1, 2019, pp. 42–50.

Oparin, A. I. *The Origin of Life on the Earth*. Trans. Ann Synge. Academic Press, 1957.

Oswalt, Patton. "Wake Up, Geek Culture. Time to Die." *Wired*, vol. 19, no. 1, Dec. 2010. *www.wired.com*, https://www.wired.com/2010/12/ff-angrynerd-geekculture/.

Otten, Sam. "DC v. Marvel Comic Sales Before Rebirth." *Comic and Screen*, 23 Jan. 2016, http://comicandscreen.blogspot.com/2016/01/dcv-marvel-comic-sales-before-rebirth.html.

Padilla, Javier. "Science Fiction as Theory Fiction." *Modernism/Modernity Print Plus*, May 16, 2022. https://modernismmodernity.org/forums/posts/padilla-science-fiction-theory-fiction.

Palin, Sarah. "Statement on the Current Health Care Debate." *Facebook*, 7 Aug. 2009, https://www.facebook.com/note.php?note_id=113851103434&ref=mf.

Pan, David. "Kafka as a Populist: Re-Reading 'In the Penal Colony.'" *Telos*, vol. 1994, no. 101, Telos Press, Sept. 1994, pp. 3–40. *journal.telospress.com*, doi:10.3817/0994101003.

Pan, Landyn, and Anna Moore. *The Gender Unicorn*. TSER: Trans Student Educational Resources, 2014, http://www.transstudent.org/gender.

Pantozzi, Jill. "Coming to New York Comic Con Next Week? Party with the Mary Sue & Her Universe!" *The Mary Sue*, 4 Oct. 2012, https://www.themarysue.com/nycc-party-2012/.

paperson, la [K. Wayne Yang]. *A Third University Is Possible*. U of Minnesota Press, 2017.

Park, Katrin. "COVID-19: Life in Seoul Amid the Epidemic." *The Diplomat*, 11 Mar. 2020. *thediplomat.com*, https://thediplomat.com/2020/03/covid-19-life-in-seoul-amid-the-epidemic/.

Parker, Arthur Caswell. *Seneca Myths and Folk Tales*. Buffalo Historical Society, 1923.

Parker, Ian. *Psychology After Lacan: Connecting the Clinic and Research*. Routledge, 2014.

Parloff, Roger. "This CEO's out for Blood." *Fortune*, vol. 169, no. 9, June 2014, pp. 64–72.

Parrinder, Patrick. "The Alien Encounter: Or, Ms Brown and Mrs Le Guin." *Science Fiction Studies*, vol. 6, no. 1, SFTH Inc, 1979, pp. 46–58.

Parscale, Brad. "For nearly three years we have been building a juggernaut campaign (Death Star). It is firing on all cylinders. Data, Digital, TV, Political, Surrogates, Coalitions, Etc. In a few days we start pressing FIRE for the first time." *Twitter*, 7 May 2020. *twitter.com*, https://twitter.com/parscale/status/1258388669544759296.

Penley, Constance. "Time Travel, Primal Scene, and the Critical Dystopia." *Camera Obscura: Feminism, Culture, and Media Studies*, vol. 5, no. 3, 1986, pp. 66–85. *DOI.org (Crossref)*, doi:10.1215/02705346-5-3_15-66.

Penny, Laurie. "Harry Potter and the Conscience of a Liberal." *The Baffler*, 2 Sept. 2016. *thebaffler.com*, https://thebaffler.com/war-of-nerves/harry-potter-laurie-penny.

Penny, Laurie. "On Nerd Entitlement." *New Statesman America*, Dec. 2014, https://www.newstatesman.com/laurie-penny/on-nerd-entitlement-rebel-alliance-empire.

Pervushin, Anton Ivanovich. *Kosmonavty Stalina. Mezhplanetnyy Proryv Sovetskoy Imperii*. Eksmo, 2005.

Pescovitz, David. "Unicorn Chaser Beverage from ThinkGeek!" *Boing Boing*, 1 Apr. 2009, https://boingboing.net/2009/04/01/unicorn-chaser-bever.html.

Peters, Mark. "The Linguistic Sex Appeal of the Unicorn." *The Boston Globe*, 21 Feb. 2016, https://www.bostonglobe.com/ideas/2016/02/21/see-any-unicorns-lately/oCEhFoIgea Hw71hjbKSWVO/story.html.

"Peters: Obama Is 'Chasing a Unicorn' in Iraq and Syria." *Fox News Insider*, 2 Mar. 2015, https://insider.foxnews.com/2015/03/02/peters-sounds-after-us-backed-syrian-group-disbands.

Pietz, William. "The Problem of the Fetish, I." *RES: Journal of Anthropology and Aesthetics*, no. 9, Mar. 1985, pp. 5–17. *journals.uchicago.edu (Atypon)*, doi:10.1086/RESv9n1ms20166719.

Pietz, William. "The Problem of the Fetish, II: The Origin of the Fetish." *RES: Journal of Anthropology and Aesthetics*, no. 13, Mar. 1987, pp. 23–45. *journals.uchicago.edu (Atypon)*, doi:10.1086/RESv13n1ms20166762.

Pietz, William. "The Problem of the Fetish, IIIa: Bosman's Guinea and the Enlightenment Theory of Fetishism." *RES: Journal of Anthropology and Aesthetics*, no. 16, Sept. 1988, pp. 105–24. *journals.uchicago.edu (Atypon)*, doi:10.1086/RESv16n1ms20166805.

Pignarre, Philippe, and Isabelle Stengers. *Capitalist Sorcery: Breaking the Spell*. Basingstoke: Palgrave Macmillan UK, 2011.

Planey, Arrianna. "Thoughts on Margaret Atwood's 'A Handmaid's Tale': Part 2." *Arrianna Planey's Blog*, 30 Dec. 2013, https://arriannaplaney.wordpress.com/2013/12/30/thoughts-on-margaret-atwoods-a-handmaids-tale-part-2/.

Porter, Tom. "Soldiers of Odin: Name of Far-Right Group Patented for Use by \'Glittery Unicorn\' Clothing Range." *International Business Times UK*, 18 May 2016, https://www.ibtimes.co.uk/soldiers-odin-name-far-right-group-patented-use-by-glittery-unicorn-clothing-range-1560781.

Pottermore News Team, The. "500 Million Harry Potter Books Sold Worldwide." *Pottermore*, Feb. 1, 2018, https://www.pottermore.com/news/500-million-harry-potter-books-have-now-been-sold-worldwide.

Poulakos, John. "Toward a Sophistic Definition of Rhetoric." *Philosophy & Rhetoric*, vol. 16, no. 1, JSTOR, 1983, pp. 35–48.

Powell, John A., et al. "Toward a Transformative View of Race: The Crisis and Opportunity of Katrina." *There Is No Such Thing as a Natural Disaster: Race, Class, and Hurricane Katrina*, edited by Gregory D. Squires and Chester W. Hartmann, Routledge, 2006, pp. 59–84.

Pratt, Scott L. *Native Pragmatism: Rethinking the Roots of American Philosophy*. Indiana University Press, 2002.

Prescod-Weinstein, Chanda. "Making Black Women Scientists under White Empiricism: The Racialization of Epistemology in Physics." *Signs: Journal of Women in Culture and Society*, vol. 45, no. 2, The University of Chicago Press, Jan. 2020, pp. 421–47. *journals.uchicago.edu (Atypon)*, doi:10.1086/704991.

Primack, Dan. "Theranos Controversy Has Little to Do with 'Unicorns.'" *Fortune.com*, Oct. 22, 2015.

Probyn, Clive. "Swift and Linguistics: The Context Behind Lagado and Around the Fourth Voyage." *Neophilologus*, vol. 58, no. 4, 1974, pp. 425–39.

Proudhon, Pierre-Joseph. *Oeuvres complètes*. Edited by Célestin Charles Alfred Bouglé and Henri Moysset, Slatkine, 1982.

Proudhon, Pierre-Joseph. *System of Economical Contradictions: Or, The Philosophy of Misery*. Translated by Benjamin R. Tucker, Benjamin R. Tucker, 1888.

Pyle, Kai Minosh. "How to Survive the Apocalypse for Native Girls." *Love After the End: An Anthology of Two-Spirit and Indigiqueer Speculative Fiction*, edited by Joshua Whitehead, Vancouver: Arsenal Pulp Press, 2020. 77–94

Pynchon, Thomas. *The Crying of Lot 49*. Harper & Row, 1986.

Quintana-Navarrete, Jorge. "Biopolítica y vida inorgánica: La plasmogenia de Alfonso Herrera." *Revista Hispánica Moderna*, vol. 72, no. 1, 2019, pp. 79–95.

Rachel. "Presidential Unicorn Rides: Dan Lacey Creates Creepy Barack Obama Drawings." *TrendHunter.com*, 31 July 2009, https://www.trendhunter.com/trends/barack-obama-dan-lacey-art.

Rampell, Palmer. "The Science Fiction of Roe v. Wade." *ELH*, vol. 85, no. 1, Johns Hopkins University Press, Mar. 2018, pp. 221–52. *Project MUSE*, doi:10.1353/elh.2018.0008.

Rancière, Jacques. *The Politics of Aesthetics*. Edited and translated by Gabriel Rockhill, London: A&C Black, 2013.

Rappeport, Alan. "On Budget, No Clarity, and So Far No 'Magic Unicorn.'" *New York Times*, 25 May 2017, p. A15.

Reay, Diane. "Universities and the Reproduction of Inequalities." *A Manifesto for the Public University*, edited by John Holmwood, Bloomsbury, 2011, pp. 112–26.

Reynolds, Renee H. *The Rise of Geek Chic: An Analysis of Nerd Identity in a Post-Cult Market*. The University of Arizona, 2019.

Rhee, Margaret. "Racial Recalibration: Nam June Paik's *K-456*." *Asian Diasporic Visual Cultures and the Americas*, no. 1, 2015, pp. 285–309.

Richards, Howard. *Letters from Quebec: A Philosophy for Peace and Justice*. San Francisco: International Scholars Publications, 1996.

Richter, Kathleen. "My Little Homophobic, Racist, Smart-Shaming Pony." *Ms. Magazine*, 9 Dec. 2010, https://msmagazine.com/2010/12/09/my-little-homophobic-racist-smarts-shaming-pony/.

Rieder, John. *Colonialism and the Emergence of Science Fiction*. Wesleyan University Press, 2007.

Rieder, John. "Spectacle, Technology and Colonialism in SF Cinema: The Case of Wim Wenders' *Until the End of the World*." *Red Planets: Marxism and Science Fiction*, edited by Mark Bould and China Miéville, Wesleyan University Press, 2009, pp. 83–102.

Rippetoe, Rita Elizabeth. *Booze and the Private Eye: Alcohol in the Hard-Boiled Novel*. Jefferson, North Carolina: McFarland, 2004.

Roanhorse, Rebecca. "Welcome to Your Authentic Indian Experience™." *Apex Magazine*, 8 Aug. 2017, https://www.apex-magazine.com/welcome-to-your-authentic-indian-experience/.

Roberts, Adam. *The History of Science Fiction*. Basingstoke: Palgrave Macmillan, 2006.

Robinson, Andrew, and Simon Tormey. "A Ticklish Subject? Žižek and the Future of Left Radicalism." *Thesis Eleven*, vol. 80, no. 1, SAGE Publications Ltd, Feb. 2005, pp. 94–107. *SAGE Journals*, doi:10.1177/0725513605049126.

Robinson, Andrew. "The Political Theory of Constitutive Lack: A Critique." *Theory & Event*, vol. 8, no. 1, 2005. *Project MUSE*, doi:10.1353/tae.2005.0016.

Robinson, Christopher L. "Teratonymy: The Weird and Monstrous Names of H. P. Lovecraft." *Names*, vol. 58, no. 3, 2010, pp. 127–38.

Robinson, Lillian S. *Wonder Women: Feminisms and Superheroes*. New York: Routledge, 2004.

Rosiek, Jerry Lee, and Jimmy Snyder. "Narrative Inquiry and New Materialism: Stories as (Not Necessarily Benign) Agents." *Qualitative Inquiry*, vol. 0, no. 0, Aug. 2018, pp. 1–12, doi:10.1177/1077800418784326.

Rosiek, Jerry Lee, Jimmy Snyder, and Scott L. Pratt. "The New Materialisms and Indigenous Theories of Non-Human Agency: Making the Case for Respectful Anti-Colonial Engagement." *Qualitative Inquiry*, vol. 26, no. 3–4, SAGE Publications Inc, Mar. 2020, pp. 331–46. *SAGE Journals*, doi:10.1177/1077800419830135.

Roth, David. "The Man Who Was Upset: Making Sense of Donald Trump's Petulant Reign." *The New Republic*, June 2019. *The New Republic*, https://newrepublic.com/article/154100/making-sense-donald-trump-petulant-presidency.

Rucker, Rudy. "Alien Contact." *Alien Tongue*, by Stephen Leigh, Bantam Spectra, 1991, pp. 301–27.

Rumberger, Russell W. "Education and the Reproduction of Economic Inequality in the United States: An Empirical Investigation." *Economics of Education Review*, vol. 29, no. 2, Apr. 2010, pp. 246–54. *ScienceDirect*, doi:10.1016/j.econedurev.2009.07.006.

Russ, Joanna. *To Write Like a Woman: Essays in Feminism and Science Fiction*. Indiana University Press, 1995.

Russell, John G. "Darkies Never Dream: Race, Racism, and the Black Imagination in Science Fiction." *CR: The New Centennial Review*, vol. 18, no. 3, Winter 2018, pp. 255–77.

Ruthrof, Horst. *Language and Imaginability*. Cambridge Scholars Publishing, 2014.

Ruthrof, Horst. *Pandora and Occam*. Taylor & Francis Limited, 2019.

Ruthrof, Horst. *Semantics and the Body: Meaning from Frege to the Postmodern*. University of Toronto Press, 1997.

Ruthrof, Horst. *The Body in Language*. Bloomsbury Publishing, 2015.

Ryner, Han. *The Human Ant*. Translated by Brian Stableford, Black Coat Press, 2014.

Rynin, N. A. *Interplanetary Flight and Communication Vol. II, No.4: Rockets*. National Aeronautics and Space Administration and the National Science Foundation, 1971.

Sagan, Carl. *Communication with Extraterrestrial Intelligence (CETI)*. MIT Press, 1973.

Saldívar, Ramón. "Historical Fantasy, Speculative Realism, and Postrace Aesthetics in Contemporary American Fiction." *American Literary History*, vol. 23, no. 3, Sept. 2011, pp. 574–99. *DOI.org (Crossref)*, doi:10.1093/alh/ajro26.

Samman, Khaldoun. *Clash of Modernities: The Making and Unmaking of the New Jew, Turk, and Arab and the Islamist Challenge*. Routledge, 2015.

Sanchez-Taylor, Joy. *Diverse Futures: Science Fiction and Authors of Color*. Columbus, Oh.: Ohio State University Press, 2021.

"San Diego Comic-Con—Frequently Asked Questions." *San Diego Comic-Con Unofficial Blog*, 16 Dec. 2019, https://sdccblog.com/san-diego-comic-con-frequently-asked-questions/.

Santos, Hugo, Saldanha Lucinda, Pinto Marta, and Ferreira Pedro. "Civic and Political Transgressions in Videogames: The Views and Experiences of the Players." In *DiGRA Nordic '18: Proceedings of 2018 International DiGRA Nordic Conference*, 2018. http://www.digra.org/wp-content/uploads/digital-library/DIGRANORDIC-2018-paper-27.pdf.

"Save Me from This Epidemic of Unicorns." *Daily Telegraph (London)*, Jan. 2019, p. 21.

Saxton, Josephine. "The Wall." *Science Fantasy*, vol. 24, no. 78, Nov. 1965, pp. 72–78.

Scalzi, John. "Straight White Male: The Lowest Difficulty Setting There Is." *Whatever*, 15 May 2012. *whatever.scalzi.com*, https://whatever.scalzi.com/2012/05/15/straight-white-male-the-lowest-difficulty-setting-there-is/.

Scammell, Margaret. *Consumer Democracy: The Marketing of Politics*. Cambridge University Press, 2014.

Schakel, Peter J. *The Way into Narnia: A Reader's Guide*. William B. Eerdmans Pub. Company, 2005.

Schmidt, Vivien A. "Britain-out and Trump-in: A Discursive Institutionalist Analysis of the British Referendum on the EU and the US Presidential Election." *Review of International Political Economy*, vol. 24, no. 2, Routledge, Mar. 2017, pp. 248–69. *Taylor and Francis+NEJM*, doi:10.1080/09692290.2017.1304974.

Scholes, Robert E. *Textual Power: Literary Theory and the Teaching of English*. Yale University Press, 1985.

Scott, Damion Kareem. "Afrofuturism and Black Futurism: Some Ontological and Semantic Considerations." *Critical Black Futures: Speculative Theories and Explorations*, edited by Philip Butler, Springer, 2021, pp. 139–63. *Springer Link*, doi:10.1007/978-81-15-7880-9_8.

Scott, David Meerman, and Richard Jurek. *Marketing the Moon: The Selling of the Apollo Lunar Program*. MIT Press, 2014.
Scott, James C. *Seeing Like a State: How Certain Schemes to Improve the Human Condition Have Failed*. Yale University Press, 1998.
Search Term "Unicorn" in All Text, 1855–2019, ABI/INFORM Collection (ProQuest). https://search-proquest-com.pnw.idm.oclc.org/abicomplete/results/E48DB4E2B0CC409CPQ/1?accountid=13361. Accessed 8 Aug. 2019.
Searle, John R. "Minds, Brains, and Programs." *The Behavioral and Brain Sciences*, vol. 3, 1980, pp. 417–57.
Searle, John R. "The Case for a Traditional Liberal Education." *The Journal of Blacks in Higher Education*, no. 13, Autumn 1996, pp. 91–98.
Searle, John R. "The Failures of Computationalism." *Think*, no. 2 June 1993, pp. 68–71.
Sedgwick, Eve Kosofsky. *Epistemology of the Closet*. University of California Press, 1990.
Sedgwick, Eve Kosofsky. "Paranoid Reading and Reparative Reading, or, You're So Paranoid, You Probably Think This Essay Is About You." *Touching Feeling: Affect, Performativity, Pedagogy*, Duke University Press, 2003, pp. 123–51.
Serge, Victor. "New Tendencies in Russian Anarchism." *Anarchists Never Surrender: Essays, Polemics, and Correspondence on Anarchism, 1908–1938*, edited and translated by Mitchell Abidor, PM Press, 2015, pp. 177–93.
Serpell, C. Namwali. "A Heap of Cliché." *Critique and Postcritique*, edited by Elizabeth S. Anker and Rita Felski, Duke University Press, 2017, pp. 153–82.
Serres, Michel. *The Natural Contract*. University of Michigan Press, 1995.
Shapiro, Ben. "AOC's Green New Deal Proposal Is One of the Stupidest Documents Ever Written." *The Daily Wire*, 7 Feb. 2019, https://dailywire.com/news/43194/aocs-green-new-deal-proposal-one-stupidest-ben-shapiro.
Shaviro, Steven. *The Universe of Things: On Speculative Realism*. University of Minnesota Press, 2014.
Shawl, Nisi. *Everfair*. New York: Tor Books, 2016.
Shawl, Nisi. "Hope and Vengeance in Post-Apocalyptic Sudan: *Who Fears Death* by Nnedi Okorafor." *Tor.Com*, 7 Nov. 2017, https://www.tor.com/2017/11/07/hope-and-vengeance-in-postapocalyptic-sudan-who-fears-death-by-nnedi-okorafor/.
Sheldon, Rebekah. "Reading for Transgression: Queering Genres." *After Queer Studies*, edited by Tyler Bradway and E. L. McCallum, Cambridge University Press, 2019, pp. 171–87.
Shields and Shield Tactics Primer. 2017, https://www.unicornriot.ninja/wp-content/uploads/2017/09/Shields_and_Shield_Tactics_Primer.pdf.
Shipley, Maynard. "Growing Living Chemicals." *Science and Invention*, vol. 15, no. 9, Jan. 1928, pp. 805, 855.
Siddiqi, Asif A. *The Red Rockets' Glare: Spaceflight and the Soviet Imagination, 1857–1957*. New York: Cambridge University Press, 2013.
Siddiqi, Asif A. "Imagining the Cosmos: Utopians, Mystics, and the Popular Culture of Spaceflight in Revolutionary Russia." *Osiris*, vol. 23, 2008, pp. 260–88.
Silverberg, Robert. "What We Learned from This Morning's Newspaper." *Infinity Four*, edited by Robert Hoskins, Lancer Books, 1972, pp. 17–35.
Silverman, G. G., and Nisi Shawl. "Women in Speculative Fiction: Interview with Nisi Shawl, Author of EVERFAIR." *G. G. Silverman*, 1 Feb. 2017, http://www.ggsilverman.com/women-in-speculative-fiction-interview-with-nisi-shawl-author-of-everfair/.
Singer, Joseph William. "The Indian States of America: Parallel Universes & Overlapping Sovereignty." *American Indian Law Review*, vol. 38, no. 1, 2013, pp. 1–33.
Sirotkina, Irina, and Marina Kokorina. "The Dialectics of Labour in a Psychiatric Ward: Work Therapy in the Kaschenko Hospital." *Psychiatry in Communist Europe*, edited by Sarah Marks and Matt Savelli, Palgrave Macmillan, 2015, pp. 27–49.

Sloterdijk, Peter. *The Critique of Cynical Reason*. Translated by Michael Eldred, University of Minnesota Press, 1987.

Smith, Michael G. *Rockets and Revolution: A Cultural History of Early Spaceflight*. University of Nebraska Press, 2014.

Smith, Kathleen. *The Fangirl Life: A Guide to Feeling All the Feels and Learning How to Deal*. New York: TarcherPerigee, 2016.

Sokal, Alan D., and Jean Bricmont. *Fashionable Nonsense: Postmodern Intellectuals' Abuse of Science*. New York: Picador, 1999.

Sollenberger, Roger. "Laura Ingraham Is Working from Home. She Says Dems Are Pushing 'Panic Porn,' Want 'Purple Unicorns.'" *Salon*, 17 May 2020. www.salon.com, https://www.salon.com/2020/05/17/laura-ingraham-working-from-home-democrats-pushing-panic-porn-because-they-dont-like-you/.

Solstein, Eric, and Gregory Moosnick. *John W. Campbell's Golden Age of Science Fiction: Text Supplement to the DVD*. Digital Media Zone, 2002, http://www.sfcenter.ku.edu/JWC_Study_Supplement.pdf.

Sontag, Susan. *Against Interpretation: And Other Essays*. Macmillan, 2001.

Spade, Dean. *Normal Life: Administrative Violence, Critical Trans Politics, and the Limits of Law*. Duke University Press, 2015.

Spector, Shmuel, and Geoffrey Wigoder. *The Encyclopedia of Jewish Life Before and During the Holocaust*. Yad Vashem, 2001.

Speyer, Lea. "Israel Advocates at Columbia University to Hold 'Indigenous People Unite' Event to 'Reclaim Narrative' about Jewish Right to Native Land." *Algemeiner.com*. Accessed May 23, 2020. https://www.algemeiner.com/2016/12/04/columbia-university-students-reclaiming-israels-narrative-event-aimed-countering-false-palestinian-claims-jewish-rights-land/.

Spiers, Miriam C. Brown. "Reimagining Resistance: Achieving Sovereignty in Indigenous Science Fiction." *Transmotion*, vol. 2, no. 1 & 2, 1 & 2, Nov. 2016, pp. 52–52. *journals.kent.ac.uk*, doi:10.22024/UniKent/03/tm.224.

Spinoza, Baruch. *Ethics: With The Treatise on the Emendation of the Intellect and Selected Letters*. Edited by Seymour Feldman, Translated by Samuel Shirley, Hackett Publishing, 1992.

Stableford, Brian. "Why There Is (Almost) No Such Thing as Science Fiction: Observations on Rhetoric and Plausibility in Science and Science Fiction." *Narrative Strategies in Science Fiction and Other Essays on Imaginative Fiction*, Borgo Press / Wildside Press, 2010, pp. 61–73.

Starhawk. *Dreaming the Dark: Magic, Sex & Politics*. Beacon Press, 1988.

Starhawk. "How We Really Shut Down the WTO." *From ACT UP to the WTO: Urban Protest and Community Building in the Era of Globalization*, edited by Benjamin Shepard and Ronald Hayduk, Verso, 2002, pp. 52–56.

Sterling, Bruce. "Digital Dolphins in the Dance of Biz." *Science Fiction Eye*, no. 9, Nov. 1991, pp. 13–16.

Sterling, Bruce. "My Rihla." *Science Fiction EYE*, vol. 2, no. 2, August 1990, pp. 12–16.

Sterling, Bruce. "Preface." *Burning Chrome*, Ace Books, 1987, pp. ix–xii.

Stern, Alexandra Minna. *Proud Boys and the White Ethnostate: How the Alt-Right Is Warping the American Imagination*. Beacon Press, 2019.

Stetson, Charlotte Perkins. "The Yellow Wall-Paper." *The New England Magazine*, vol. 5, no. 5, Jan. 1892, pp. 647–56.

Stirner, Max. *The Ego and Its Own*. Edited by David Leopold, Translated by Steven T. Byington, Cambridge University Press, 2000.

Strete, Craig. "A Horse of a Different Technicolor." *Galaxy*, vol. 36, no. 1, Jan. 1975, pp. 76–82.

Stryker, Susan. "Performing Transgender Rage: My Words to Victor Frankenstein Above the Village of Chamounix." *GLQ* 1, no. 3 (1994): 237–54.

Suler, John. "The Online Disinhibition Effect." *Cyberpsychology & Behavior: The Impact of the Internet, Multimedia and Virtual Reality on Behavior and Society*, vol. 7, no. 3, July 2004, pp. 321–26.

Suvin, Darko. *Metamorphoses of Science Fiction: On the Poetics and History of a Literary Genre*. Yale University Press, 1979.

Suvin, Darko. "Preface." *Other Worlds, Other Seas: Science-Fiction Stories from Socialist Countries*. New York: Random House, 1970, pp. xi–xxxiii.

Swann, Thomas. *Anarchist Cybernetics: Control and Communication in Radical Left Social Movements*. University of Leicester, 2016. ethos.bl.uk, https://figshare.com/articles/Anarchist _Cybernetics_Control_and_Communication_in_Radical_Left_Social_Movements/10164140.

Swann, Thomas. "Are Postanarchists Right to Call Classical Anarchisms 'Humanist'?" *Anarchism and Moral Philosophy*, edited by Benjamin Franks and Matthew Wilson, Palgrave Macmillan UK, 2010, pp. 226–42. *Springer Link*, doi:10.1057/9780230289680_12.

Swann, Thomas. "Towards an Anarchist Cybernetics: Stafford Beer, Self-Organisation and Radical Social Movements." *Ephemera: Theory & Politics in Organization*, vol. 18, no. 3, Loughborough University, Jan. 2018, pp. 427–56.

Tallis, Raymond. *Not Saussure: A Critique of Post-Saussurean Literary Theory*. Palgrave, 1995.

Tamaki, Mariko. *Unicorn Power!* ABRAMS, 2017. *Open WorldCat*, http://link.overdrive.com /?websiteID=220&titleID=3591351.

Tapken, Erin. "50 Facts for 50 Years of San Diego Comic-Con (Part 2)." *San Diego Comic-Con Unofficial Blog*, 13 June 2019, https://sdccblog.com/2019/06/50-facts-for-50-years-of-san -diego-comic-con-part-2/.

Tárrida del Mármól, Fernando. *Problemas Trascendentales. Estudios de Sociología y Ciencia Moderna*. Paris: Sociedad de Ediciones Literarias y Artísticas, 1908.

Taylor, Drew Hayden. *Take Us to Your Chief: And Other Stories: Classic Science-Fiction with a Contemporary First Nations Outlook*. D & M Publishers, 2016.

Team Unicorn. "About." *Team Unicorn FTW*, 17 July 2011, https://web.archive.org/web/2011 0717163826/http://teamunicornftw.com/about/.

Thomas, Ebony Elizabeth. *The Dark Fantastic: Race and the Imagination from Harry Potter to the Hunger Games*. New York: NYU Press, 2020.

Thomas, Kelly Lynn, and Nisi Shawl. "Spitting in the Face of Empire: The Millions Interviews Nisi Shawl." *The Millions*, 2 Oct. 2017, https://themillions.com/2017/10/spitting-in-the-face -of-empire-the-millions-interviews-nisi-shawl.html.

Thorpe, Devin. "Successful African-American Silicon Valley Entrepreneur Feels 'Like a Black Unicorn.'" *Forbes*, Jan. 2018, https://www.forbes.com/sites/devinthorpe/2018/01/15 /successful-african-american-silicon-valley-entrepreneur-feels-like-a-black-unicorn /#58220a6e29ab.

Times Editorial Board. "Editorial: Panic Is Making Coronavirus Worse for Everyone." *Los Angeles Times*, 11 Mar. 2020, https://www.latimes.com/opinion/story/2020-03-11/panic -covid-19-worse-for-everyone.

Tolstoy, Leo. "Yasnaya Polyana School." *The Novels and Other Works of Lyof N. Tolstoi: The Long Exile and Other Stories*, translated by Nathan H. Dole, New York: C. Scribner's Sons, 1911, pp. 164–300.

Tomorrow, Tom [Dan Perkins]. *To Serve Man*. 22 Sept. 2020, https://thenib.com/author/tom -tomorrow/.

Topash-Caldwell, Blaire. "Sovereign Futures in Neshnabé Speculative Fiction." *Borderlands Journal*, vol. 19, no. 2, Mar. 2021. www.exeley.com, doi:10.21307/borderlands-2020-009.

"Transcript: Republican Presidential Debate." *The New York Times*, 15 Dec. 2015, https://www .nytimes.com/2015/12/16/us/politics/transcript-main-republican-presidential-debate.html.

Tremlett, Paul-François. "On the Formation and Function of the Category 'Religion' in Anarchist Writing." *Culture and Religion*, vol. 5, no. 3, Nov. 2004, pp. 367–81. *DOI.org (Crossref)*, doi:10.1080/0143830042000294424.

Trow, George W. S. *Within the Context of No Context*. Atlantic Monthly Press, 1997.

Trump, Ivanka. "Taken!" *Twitter*, 17 May 2020. *twitter.com*, https://twitter.com/IvankaTrump/status/1262095075963736064.

Tuana, Nancy. "Viscous Porosity: Witnessing Katrina." *Material Feminisms*, edited by Stacy Alaimo and Susan Hekman, Indiana University Press, 2008, pp. 188–213.

Turing, Alan M. "Computing Machinery and Intelligence." *Mind*, vol. 59, no. 236, Oct. 1950, pp. 433–60.

Tuters, Marc. "LARPing & Liberal Tears: Irony, Belief and Idiocy in the Deep Vernacular Web." In *Post-Digital Cultures of the Far Right: Online Actions and Offline Consequences in Europe and the US*, edited by Maik Fielitz and Thurston Nick, 71:37–48. Political Science. Transcript Verlag, 2019.

Tzinovitz, Moshe. "Rabbis of the Community of Smorgon." *Smorgonie, District Vilna; Memorial Book and Testimony (Smarhon, Belarus)*, edited by Marc D. Hodies, translated by Jerrold Landau, pp. 78–104, https://www.jewishgen.org/yizkor/smorgon/smo078.html. Accessed 22 May 2020.

Un Medico Rural [Isaac Puente]. "Ideales Redentores." *Estudios*, vol. 10, no. 101, 1932, pp. 3–6.

Unicorn Strike. Occupy May 1st/General Strike! 18 April 2012, https://web.archive.org/web/20120501105533/http://www.occupymay1st.org/2012/04/18/image-unicorn-strike/.

Van Camp, Richard. "Aliens." *Love beyond Body, Space, and Time*, edited by Hope Nicholson, Bedside Press, 2016, pp. 20–30.

Van Valkenburgh, Shawn P. "Digesting the Red Pill: Masculinity and Neoliberalism in the Manosphere." *Men and Masculinities*, Dec. 2018, pp. 1–20. *Google Scholar*, doi:10.1177/1097184X18816118.

van Veen, Tobias. "The Armageddon Effect: Afrofuturism and the Chronopolitics of Alien Nation." *Afrofuturism 2.0: The Rise of Astro-Blackness*, edited by Reynaldo Anderson and Charles E. Jones, Lexington Books, 2015, pp. 63–90.

Varley, John. "Blue Champagne." *Blue Champagne*, by John Varley, Berkeley Books, 1986, pp. 17–79.

Varley, John. "The Persistence of Vision." *The Magazine of Fantasy and Science Fiction*, vol. 54, no. 3, Mar. 1978, pp. 6–50.

Vary, Adam. "The Year of the Geek." *Entertainment Weekly*, 21 Dec. 2007, https://ew.com/article/2007/12/21/year-geek/.

Vint, Sherryl. *Science Fiction: A Guide for the Perplexed*. A&C Black, 2014.

Viveiros de Castro, Eduardo. "Who Is Afraid of the Ontological Wolf? Some Comments on an Ongoing Anthropological Debate." *The Cambridge Journal of Anthropology*, vol. 33, no. 1, Jan. 2015. *DOI.org (Crossref)*, doi:10.3167/ca.2015.330102.

Vizenor, Gerald Robert. *Fugitive Poses: Native American Indian Scenes of Absence and Presence*. U of Nebraska Press, 2000.

Volcano, Abbey. "Solidarity Unicornism and the Future of the Left: WSA's May Day Poster Explained." *Ideas and Action*, 25 Apr. 2012, http://ideasandaction.info/2012/04/solidarity-unicornism-and-the-future-of-the-left-wsas-may-day-poster-explained/.

Vowel, Chelsea May. *Where No Michif Has Gone Before: The Form and Function of Métis Futurisms*. University of Alberta, 2020.

Wachowski, Lilly. "@IvankaTrump Fuck Both Of You." *Twitter*, 17 May 2020. *twitter.com*, https://twitter.com/lilly_wachowski/status/1262104754496339968.

Wachowski, Lilly, and Lana Wachowski, dirs. *The Matrix*. Warner Bros., 1999.

Wagner, John. "'SANCTIONS ARE COMING': Trump Tweets out *Game of Thrones*–Inspired Warning to Iran." *Washington Post*, 2 Nov. 2018, https://www.washingtonpost.com/politics/sanctions-are-coming-trump-tweets-out-game-of-thrones-inspired-warning-to-iran/2018/11/02/86eef714-deba-11e8-85df-7a6b4d25cfbb_story.html.

Walter, K[urt]. "Astronomy in Poland During the Second World War: Memories of a Participating Astronomer." *Journal of the British Astronomical Association*, vol. 97, no. 5, 1987, pp. 270–73.

Warburton, Theresa Anne. "The Politics of Make/Believe: Locating the Intersections of Contemporary Anarchism, Indigenous Decolonization and Women of Color Feminisms." *Dissertations & Theses @ SUNY Buffalo, ProQuest Dissertations & Theses Global*, 2014. ubir.buffalo.edu, http://ubir.buffalo.edu/xmlui/handle/10477/51238.

Warburton, Theresa. *Other Worlds Here: Honoring Native Women's Writing in Contemporary Anarchist Movements*. Evanston, Ill.: Northwestern University Press, 2021.

Wark, McKenzie. *The Spectacle of Disintegration: Situationist Passages Out of the Twentieth Century*. Verso Books, 2013.

Warner, Michael. *Publics and Counterpublics*. Zone Books, 2005.

Warren, Calvin L. *Ontological Terror: Blackness, Nihilism, and Emancipation*. Duke University Press, 2018.

Weber, Leanne. *Policing Non-Citizens*. Routledge, 2013.

Weber, Max. *The Vocation Lectures*. Edited by David S. Owen and Tracy B. Strong, Translated by Rodney Livingstone, Hackett Publishing, 2004.

Weldes, Jutta. "Popular Culture, Science Fiction and World Politics." In *To Seek out New Worlds: Science Fiction and World Politics*, edited by Jutta Weldes. London: Palgrave, 2003. 1–27.

Welles, Abigail. "Recovery." *Economic Inequality, Neoliberalism, and the American Community College*, by Patrick Sullivan, 2017, pp. 101–10.

Wells, Charlie. "Unicorns, Unicorns Everywhere." *The Wall Street Journal*, 11 Mar. 2015, p. D1.

Wells, H. G. *The Complete Science Fiction Treasury of H. G. Wells*. Crown Publishers, 1978.

Wells, H. G. (Herbert George). *The Sleeper Awakes*. London: Collins, 1921.

Westbrook, Amy Deen, and David A. Westbrook. "Unicorns, Guardians, and the Concentration of the U. S. Equity Markets." *Nebraska Law Review*, vol. 96, no. 3, 2018, pp. 688–741.

Wetherbee, Ben. "Dystopoi of Memory and Invention: The Rhetorical 'Places' of Postmodern Dystopian Film." *Journal of Multimodal Rhetorics*, vol. 2, no. 2, Fall 2018, pp. 116–34.

Whitman, Walt. "Song of Myself." *Leaves of Grass: Authoritative Texts, Prefaces, Whitman on His Art, Criticism*, edited by Sculley Bradley and Harold William Blodgett, New York, Norton, 1973, pp. 28–89.

Wilde, Fran. "An Op-Ed From the Future: Please, Stop Printing Unicorns." *The New York Times*, 26 Aug. 2019. *NYTimes.com*, https://www.nytimes.com/2019/08/26/opinion/3d-printed-unicorns.html.

Wilderson III, Frank. *Afropessimism*. W. W. Norton & Co., 2020.

Wilkins, John. *An Essay Towards a Real Character: And a Philosophical Language*. Printed for Sa. Gellibrand, and for John Martyn, 1668.

Winn, Philip, editor. *Dictionary of Biological Psychology*. Routledge, 2003.

Winter, Frank. *Prelude to the Space Age: The Rocket Societies, 1924–1940*. Smithsonian Institution Press, 1983.

Wittkower, Dylan E. "On The Origins of the Cute as a Dominant Aesthetic Category in Digital Culture." *Putting Knowledge to Work and Letting Information Play*, edited by Timothy W. Luke and Jeremy W. Hunsinger, Springer, 2012, pp. 212–21.

Wolfe, Gary K. *Evaporating Genres: Essays on Fantastic Literature*. Wesleyan University Press, 2011.

Womack, Ytasha. *Afrofuturism: The World of Black Sci-Fi and Fantasy Culture*. Chicago Review Press, 2013.

Woods, Baynard. "How Unicorn Riot Covers the Alt-Right without Giving Them a Platform." *Columbia Journalism Review*, 1 Nov. 2017, https://www.cjr.org/united_states_project/charlottesville-alt-right-unicorn-riot.php.

Wordsworth, William. *The Prelude, 1799, 1805, 1850 : Authoritative Texts, Context and Reception, Recent Critical Essays*. Edited by Jonathan Wordsworth et al., Norton, 1979.

Wordsworth, William. "The World Is Too Much with Us; Late and Soon." *The Poems, Volume One*, Penguin, 1990, pp. 568–69.

Worth, Aaron. "Introduction." *The Great God Pan and Other Horror Stories*, by Arthur Machen, Oxford University Press, 2018, pp. ix–xxx.

Wynter, Sylvia. "A Utopia from the Semi-Periphery: Spain, Modernization, and the Enlightenment." *Science Fiction Studies*, vol. 6, no. 1, 1979, pp. 100–7.

Wynter, Sylvia. "Unsettling the Coloniality of Being/Power/Truth/Freedom: Towards the Human, after Man, Its Overrepresentation—An Argument." *cr: The New Centennial Review* 3, no. 3 (2003): 257–337.

Yaszek, Lisa. "The Domestic SF Parabola." *Parabolas of Science Fiction*, edited by Brian Attebery and Veronica Hollinger, Wesleyan University Press, 2013, pp. 106–21, 267–68.

Yelensky, Boris. *In sotsyaln shturem: zikhroynes fun der Rusisher Revolutsye*. Bukhgemeynshaft bay der Yidisher natsyonalistisher gezelshaft, 1967.

Young, Helen. *Race and Popular Fantasy Literature: Habits of Whiteness*. Routledge, 2016.

Young, Kevin. *Bunk: The Rise of Hoaxes, Humbug, Plagiarists, Phonies, Post-Facts, and Fake News*. Graywolf Press, 2018.

Young, Kevin. *The Grey Album: On the Blackness of Blackness*. Graywolf Press, 2012.

Zalce y Rodríguez, Luis J. *Apuntes para la historia de la masoneria en México (de mis lecturas y mis recuerdos)*. Talleres Tipográficos de la Penitenciaría del Distrito Federal, 1950.

Zijderveld, Anton C. *On Clichés : The Supersedure of Meaning by Function in Modernity*. London ; Boston : Routledge and Kegan Paul, 1979.

Žižek, Slavoj. *Absolute Recoil: Towards A New Foundation Of Dialectical Materialism*. Verso Books, 2014.

Žižek, Slavoj. "Cyberspace, or the Unbearable Closure of Being." *Endless Night: Cinema and Psychoanalysis, Parallel Histories*, edited by Janet Bergstrom, University of California Press, 1999, pp. 96–125.

Žižek, Slavoj. *Iraq: The Borrowed Kettle*. Verso, 2005.

Žižek, Slavoj. "The Matrix; Or, The Two Sides of Perversion." In William Irwin, ed., *The Matrix and Philosophy: Welcome to the Desert of the Real* Chicago: Open Court Publishing, 2002), pp. 240–66.

Žižek, Slavoj. *The Sublime Object of Ideology*. Verso, 1989.

Žižek, Slavoj. *Tarrying with the Negative: Kant, Hegel, and the Critique of Ideology*. Duke University Press, 1993.

Zoline, P. A. "The Heat-Death of the Universe." *New Worlds Speculative Fiction*, July 1967, pp. 33–39.

Zotov, Vladimir. "Na Poroge Zvezdnogo Veka." *Ural'skiy Sledopyt*, no. 11, 1977, pp. 52–54.

Index

Academy of Outrageous Books, 159–60
Actor-Network Theory, 5, 64–69
Adams, Douglas, 114
Adorno, Theodor W., 66, 184–85, 250
affinity, 6, 96–97, 142, 180–81, 188–89, 259, 263, 310n43
Afrofuturism, 136, 271n108. *See also* New Black Fantastic
Afropessimism, 136–38
agency, 21–22, 64–65, 67, 80–82, 91, 186, 255–56, 310n29
alchemy, 46, 52, 253, 274n79. *See also* magic
Aldiss, Brian, 251
algorithms, 10–11, 35
aliens, 20, 29, 43, 83, 85–88, 90, 92, 94–96, 102, 109, 121, 126, 134, 148–49, 151–53, 189, 238, 244, 251, 258
"Aliens" (Van Camp), 151–53
allegory, 29, 86, 130, 134–35, 187, 195, 202, 266n14, 306n40
Allegory of the Cave, 40, 43, 127
alt-right, 17, 230–31, 239, 241–42
Amis, Kingsley, 236
analogy, 22, 36, 39, 42, 46–47, 87, 109, 134–35, 137, 141, 171, 189, 223, 250, 268n57, 274n65
"Analytical Language of John Wilkins, The" (Borges), 191, 192
anarchism, nonmodern, 6, 261–63
anarcho-biocosmism, 297n127. *See also* Association Internationale Biocosmique
Anarchy In a Dream: The Land of Anarchy: Utopia-Poem, 182, 184, 189–90, 192–96, 198–203, 209, 295n57, 297n108
Anarkata Statement, 137
ancestrality, 62
Anderson, Ray Lynn, 113, 116
Andrenko, Leonid Leonidovich, 53, 95–96, 164, 175, 281n51

animism, 5, 21, 44, 51, 57, 191
Anishinaabe people, 154. *See also* Ojibwe people; Potawatomi people
antiquum, 102–3, 295n57
antisemitism, 231
anti-trans discourse, 19, 105
AO language, 165, 169–71, 176, 190–93, 209, 211–13, 216, 292n29
Apache people, 145, 153–54
apocalypse, 3, 9, 13–14, 18, 22, 29, 63, 102, 110–11, 122, 126–27, 146, 154–55, 233, 238, 241, 259, 306n36
Appadurai, Arjun, 62
Arntz, Gerd, 170
Asimov, Isaac, 19, 31, 33–35, 110, 116, 128, 175, 271n4
Association Internationale Biocosmique, 53, 96
Association of Inventives-Inventists (AIIZ), 212–13, 216
atheism, 54, 161–62, 163, 177, 250, 263
Atwood, Margaret, 102–3, 237
Auden, W. H., 125
Austin, John L., 209, 272n34
"'Author of the Acacia Seeds' and Other Extracts from the *Journal of the Association of Therolinguistics*, The" (Le Guin), 88–92
Avatar, 244–45
Avatar: The Last Airbender, 116
Avengers: Infinity War, 233
Avengers, The, 140, 237
Avrich, Paul, 218

Bad Writing Contest, 160
Bakunin, Mikhail, 3, 40, 61, 66, 68, 91–92, 188, 204, 250
Ballad of Black Tom, The, 93
Ballard, J. G., 123–25, 251

Barbrook, Richard, 241
Barred Other, 83, 88, 92, 94
Barry, John M., 80–81
Barthelme, Donald, 150
Barthes, Roland, 21, 104, 109, 164
Batman, 237
Baudemann, Kristina, 148, 150
Bauhaus architecture, 193
Bear, Greg, 172
Beaska, Sara Marielle Gaup, 91
Belsey, Catherine, 236
Benallie, Brandon, 259
Benjamin, Walter, 135
Ben-Moshe, Liat, 130
Bensusan, Hilan, 132–33
Berlant, Lauren, 126, 131
Bester, Alfred, 18
Bhaba, Homi K., 160
Big Bang Theory, The, 239
Big Other, 16, 172–73
biocosmism, 52–55, 96, 164, 297n127
Black, Holly, 222
"Black Box" (Egan), 119
Black Mirror, 9
Black Panther, 113, 271n108
Black people, 4, 12, 13, 15, 19, 22–23, 29, 75, 93, 102, 111, 122, 134–37, 141–44, 150, 227, 262, 274n71
blackboxing, 21–23, 106, 113, 117, 163, 172
Blade Runner, 35
Bloch, Ernst, 102
"Blue Champagne" (Varley), 129–30
Book of Phoenix, 107
Bookchin, Murray, 61, 68, 91, 162, 177, 184–85, 249, 251, 261
Borges, Jorge Luis, 150, 191–92
Bould, Mark, 13
Brave New World, 10
Bricmont, Jean, 172
Brin, David, 92, 148–49
Broderick, Damien, 104, 116, 152, 234. See also megatext
Broken Earth Trilogy, 114, 116, 135–36, 139
Brooke-Rose, Christine, 104, 108, 281n3
brown, adrienne maree, ix, 29
Bubandt, Nils, 253
"Build-Up" (Ballard), 123–24
Burke, Kenneth, 33, 39, 134, 176, 195, 247, 308n59
Bush v. Gore, 254

Butler, Isaac, 127
Butler, Judith, 131, 160
Butler, Octavia, 29, 110, 271n108
Butler, Samuel, 111

Callon, Michel, 67–68
Cameron, Andy, 241
Cameron, James, 244
camp, 102, 226, 230, 258
Campbell, John W., Jr., 19–20, 27–28, 108, 110, 129, 175, 251
Campbell Award, 19
Campbell Era, 19, 236
Canavan, Gerry, ix, 12, 266n14
Canevacci, Massimo, 62
Canticle for Leibowitz, A, 63
Čapek, Karel, 116
Capital, 73, 165–66, 176, 291n19
Captain America: Civil War, 242, 245
Carey, M. R., 129
Carpenter, John, 109
Carroll, Lewis, 6
Carroll, Noel, 15, 105–6, 126
Carter, J. Kameron, 274n71
Castellanos, Israel, 54
Castoriadis, Cornelius, 169, 268n57
Cervenak, Sarah Jane, 274n71
Chalmers, David, 42–43
Charlottesville, Virginia, 242
Chen, Mel Y., 272–73n34
Chernyi, Lev, 214
Cherokee people, 145, 148–50
Chiang, Ted, 90
Chinese Room Experiment, 37–44, 58–59, 83
Chireau, Yvonne P., 253
Chomsky, Noam, 67, 267n43
Chronicles of Narnia, 113, 237
Chu, Seo-Young, ix, 11–12, 106–7, 135, 182, 246, 252, 265n7, 266n14, 304n8
Cioffi, Frank, 103–4
City We Became, The, 136
Clark, P. Djèli, 93
Clarke, Arthur C., 92, 113, 116, 152
Clarke's Third Law, 113, 133, 187, 194–95, 260
Cleminson, Richard, 55
clichés, 22, 32–34, 36, 58, 148, 151, 162, 245, 271n6
climate change, 3, 81, 101, 163, 221, 229, 234, 240, 246, 253, 266n9

Clover, Joshua, 65, 114, 246
Clute, John, 29, 233
cognitively estranging referents, 11, 107, 246, 252
"Cold Equations, The" (Godwin), 19–22, 24–28, 119, 251, 269n68, 270n98
Coleman, Daniel B., 93
Coleridge, Samuel Taylor, 56
Colombo, Eduardo, 70–71, 93, 255
colonialism, 89, 102, 135, 142, 144, 146–50, 155, 180, 188–89, 259, 261, 263, 293n7
Colson, Daniel, 62, 70–71, 78–142, 180–81, 183, 185–89
Comicsgate, 17
commodity fetish, 72–73, 76–77, 184
Community, 234, 240
Comte, Auguste, 74
constructed languages, 169. *See also* AO language
Contagion, 233
Coover, Robert, 150
cosmism, ix, 189–90, 193, 195–96. *See also* anarcho-biocosmism; biocosmism
cosplay, 240, 242–44, 306n36
Couliano, Ioan, 252–53
Coulthard, Glen, 261
counterpower, 195–97
counterpublics, 29, 235–36, 241, 245
Covino, William, 252
Creagh, Ronald, 70
Cree people, 147, 153
Crying of Lot 49, The, 133
Csicsery-Ronay, Istvan, 234
Customs and Border Patrol (CPB), 23–24, 28, 94
cyberpunk, 10, 175, 212
cyberspace, 175, 212
cyborgs, 143
Cyrus, Miley, 226

Daly, Mary, 105
Damien, Robert, 69–70
Damnation Alley, 63
Dark Forest, The, 87
Dark Forest Theory, 87
Day, Richard J. F., 14, 75
de Acosta, Alejandro, 63
De la Terre à la Lune, 202
de Saussure, Fernand, 173, 176

deanimation, 62
death drive, 124, 298n128
Deckard, Sharae, 142
deficit model, 149, 153
deFord, Miriam Allen, 47, 50–51
Del Rey, Lester, 251
del Valle, Adrián, 92
Delany, Samuel R., 122, 235
Deleuze, Gilles, 36, 70–71, 161, 190, 198, 213, 265n3
Deloria, Philip Joseph, 150
Dene people, 145, 151–53
Dennett, Daniel, 43–44
Department of Homeland Security (DHS), 23–24
Derrida, Jacques, 61
Descartes, René, 84, 173, 252, 286n49
Díaz, Junot, 235, 243, 246
Dick, Philip K., 33–35, 117, 188, 240
Difference Engine, The, 172
Dillon, Grace L., ix, 151
"Direction of the Road" (Le Guin), 90
disability, 89, 122, 125–26, 129–31, 142, 233, 262
Disch, Thomas M., 26, 249–50
disenchantment, 22, 31, 36, 39, 43, 50–51, 88, 102, 187, 191, 225, 250, 271n6
Dispossessed, The, 140, 190
Divergent, 237
DNA, 46
Doctorow, Cory, 11, 19, 22, 145
Dot-Com Bubble, 245
Drake Equation, 94–95
Driskell, Jay, 80
Dubey, Madhu, ix, 93–94
Due, Tananarive, 29
Duncombe, Stephen, 174, 257–58
Dune, 118, 176, 239
Dungeons & Dragons, 63, 243, 245
Dupuis-Déri, Francis, 70
dystopia, 9, 13, 102–3, 109–11, 126, 129–30, 154–55, 237, 282–83n28, 306n36

Eagleton, Terry, 144
Eco, Umberto, 121, 253
Egan, Jennifer, 119
Eighteenth Brumaire of Louis Napoleon, The, 166
Electric Monk, ix, 114
Eliot, George, 35–36, 236

Ellis, Trey, ix, 29, 270n107
Emerson, Ralph Waldo, 63
empiricism, 68, 160–63, 176–77; white, 135, 145–46
Emrys, Ruthanna, 93
Enlightenment, 46, 57, 79, 111, 147, 191, 199, 250–51, 261
equations, 22–23, 27–28, 128, 164–68, 171–75, 292n46. *See also* Drake Equation
Escauriza, Bettina, ix, 262
Escobar, Arturo, 255–56, 266n9, 310n30
Esperanto, 169–70, 191, 213
Essay towards a Real Character and a Philosophical Language, 191
estrangement, 18–19, 29, 34, 108, 118, 131, 193, 234, 241–42, 246, 303n5; cognitive, 234, 246, 250, 266n14; double, 12. *See also* cognitively estranging referents
ether, 47, 183, 195, 274n71
ethical spectacle, 258
eugenics, 54–55, 144, 200, 261
Everfair, 141–43
exclusion, 4, 28–29, 92, 135, 137, 139, 143, 145, 242, 246, 262, 270n107, 306n36
exobiology, 47, 95–97, 175
extrapolation, 27, 90, 110–11, 202
extro-science fiction, 182–83

Fabian Society, 141–43
family, 53, 75, 125, 140, 142, 146, 155, 190
fandoms, 4, 17, 27–28, 93, 114, 234–39, 256–58
fantastic, post-normal, 246
fantastic religions, 175–76
fascism, 3, 6, 19–20, 78, 101, 185, 213, 231–32, 240–43, 259, 262
fetish, 5, 21, 56, 59, 62–63, 74–75, 80, 82–83, 93, 130, 155, 176, 184, 190–91, 194, 202, 254–55, 293n52. *See also* commodity fetish; metafetishism
Feyerabend, Paul, 66
Fiallo Cabral, Luis Aristedes, 53
fictive languages, 169, 190
Fiedler, Leslie, 26, 235–36
Fields, James Alex, Jr., 243
Fifth Season, The, 13–16, 103, 121, 136
Fink, Bruce, 172
First Central Sociotechnicum, 215
First Men in the Moon, The, 202
Fisher, Mark, 5, 109–10, 155
Folch, Christine, 237–38

Forbidden Planet, 109
Foucault, Michel, 166, 192
fragile states, 14–15
Frank, Thomas, 236
Frankenstein (Shelley), 5, 46, 50, 51–52, 55–56, 77, 137
Freud, Sigmund, 35, 56, 103, 124, 126–27, 161, 184, 298n128
Funtowicz, Silvio O., 246
Fussell, Paul, 225
Futurological Congress, The, 117

Gaiman, Neil, 118
Galeano, Eduardo, 262
Galilei, Galileo, 163, 192
Galton, Francis, 54
Gamergate, 17
García Márquez, Gabriel, 186
genitive, the, 190
genre boundaries, 17, 26–29, 116, 145, 147, 151, 182, 234, 237, 246, 251, 259–60, 304n8
Gernsback, Hugo, 26
Gibbons, Dave, 127
Gibson, William, 104, 129–30, 172, 175, 212
Gilman, Charlotte Perkins, 57–58, 133, 285n14
Girl with All the Gifts, The, 129
Godwin, Tom, 19–21, 27–28, 119
Gogol, Nikolai, 148
Gold, Herbert L., 33
Golding, William, 164
Goldman, Emma, 244
Goldman, Marcio, 254
Goodman, Paul, 209
Gopnik, Adam, 241
Gordin, Abba, 6, 92, 133, 177, 179–89, 193–204, 206–10, 213–15, 217–18, 221, 249, 253, 255–56, 269n84, 285n14, 294n19, 294n21, 295n57, 296n84. *See also* First Central Sociotechnicum
Gordin, Wolf, 6, 92, 133, 165, 169–71, 175–77, 179–92, 193–204, 208–18, 221, 253, 255–56, 269n84, 285n14, 290n7, 294n19, 294n21, 295n57, 296n84, 298n139. *See also* AO language; First Central Sociotechnicum
Gordin, Yehuda Leib, 204
Gothic, 27, 29, 55–59
Gould, Stephen Jay, 96, 162–63
Graeber, David, 74, 122, 146, 196, 209, 253–55, 291n19
Grande, Ariana, 226

Grant, Caesar, 106–7
gravity, 21, 108, 122–23, 154, 162, 187, 189, 202–4
Gray, Chantelle, ix, 5, 109
Great Cities Duology, 93, 136
Great Sorting Machine, 6, 17–18, 55, 59, 62, 79, 144, 160, 180, 187, 189, 217, 259, 268n57, 293n7. *See also* eugenics; hybridity; Kerr, Clark; purification
Green New Deal, 116, 229
Gross, Paul R., 162

Haddish, Tiffany, 227
Halberstam, Judith, 37
Haldane, J. B. S., 44, 52
Handmaid's Tale, The, 102, 237, 240
Haraway, Donna, 143, 165, 173–75
Harman, Graham, 94, 221–22
Harry Potter series, 237, 243, 256
Harvey, Graham, 184, 186
Hayles, Katherine, 41
Hegel, G. W. F., 43, 74, 92, 128, 137, 189
Heinlein, Robert A., 18, 108, 175
Herbert, Frank, 118, 176, 239
Herrera, Alfonso L., 5, 44–55, 59, 164, 167, 175, 218
heteronormativity, 129, 137, 142
Heyer, Heather, 243
HGK477, 118
His Master's Voice, 86, 171
Hobbes, Michael, 108
Hobbes, Thomas, 87, 163–64
homoeroticism, 200–202
Horkheimer, Max, 184–85, 250
Hornborg, Alf, 255–56
Houllebecq, Michel, 93
"How to Survive the Apocalypse for Native Girls" (Pyle), 146, 154–55
Hubbard, L. Ron, 175, 212
Hugo Awards, 17
Huizinga, Johan, 118, 231
humanism, liberal, 64, 148, 187, 192
Humanities, 159–62
Hume, David, 160, 164
Hume, Kathryn, 6, 39, 122–23
Hundred Thousand Kingdoms, The, 134, 136
Hunger Games, 11, 134, 233, 237
Hurricane Katrina, 22–24
hybridity, 16, 18, 26–31, 73, 117, 119, 143, 188, 234, 236, 246, 259, 268n57, 270n98, 304n8

Imarisha, Walidah, 29, 103, 112, 243
imperatives (grammar), 209
"Imperial Messenger, The" (Kafka), 123
impossibility, 4, 6, 89, 102, 108, 110–11, 116, 121–42, 194, 199, 201–4, 243, 247, 252
indian (as phantasm), 148, 150
Indigenous Futurism, 151, 155
Indigenous futurity, 259
Ingold, Timothy, 11
Ingraham, Laura, 116
innocence, 20, 25, 224–26, 230–31
intersectionality, 188
intuition pumps, 43–44
Intuitionist, The, 135
irrationality, 21, 74, 114, 251, 253
Irwin, W. R., 115, 131, 153

Jackson, Rosemary, 123, 126
Jackson, Zakiyyah Iman, ix, 74
James, William, 176
Jameson, Fredric, 160, 266n14
January 6, 2021, 3
Jemisin, N. K., 4, 9, 13–18, 28–29, 91–94, 97, 103, 114, 121, 133–41, 143–44, 260, 266n20, 271n108, 306n36
Jemisin's Corollary, 133, 260. *See also* Clarke's Third Law
Jenkins, Barry, 271n108
Jenkins, Henry, 238
Jesuit Relations, 259
Jones, Adam Garnet, 147
Jung, Carl, 236
Justice, Daniel Heath, ix, 149, 152

Kabbalism, 188, 192
Kafka, Franz, 39, 115, 123, 134, 148, 180, 189, 290n7
Kamelamela, Katie L., 227
Kanien'kehá:ka Gaswënta' (Two-Row Wampum), 75
Kant, Immanuel, 31, 56–57, 59, 74, 79, 81, 163, 183, 185, 199, 215
K'é Infoshop, 259
Keats, John, 35, 118
Kepler, Johannes, 253
Kerr, Clark, 17
Kickapoo people, 63
Kinna, Ruth, 75
Koelb, Clayton, 130
Kristeva, Julia, 126

Kropotkin, Peter, 96, 167, 180
Kubrick, Stanley, 152, 235, 251
Kuchinov, Eugene, ix, 166, 182, 189, 199, 208
Kuckuck, Martin, 44, 53
Kunkel, Benjamin, 228–29

La Mettrie, 36
la paperson, 215. *See also* Yang, K. Wayne
Lacan, Jacques, 16, 36, 79, 83, 86, 91–92, 109, 125, 132, 136, 165, 171–73, 175, 203–4, 267n43, 286n49, 298n128. *See also* Barred Other; lack; University Discourse
Lacaze-Duthiers, Gerard de, 53
Lacerda de Moura, Maria, 53
lack, 92, 132, 172–73, 250
Lagalisse, Erica, 290n7
l'anarchie (journal), 4
Landauer, Gustav, 67, 69
Lane, Homer, 210
Larbalestier, Justine, 222
Larson, Brie, 226
Latour, Bruno, 11–12, 18, 21–22, 40, 50, 62, 66–69, 80–81, 113, 116, 179, 183–84, 255, 277n18
LaValle, Victor, 93
laws, 14, 18, 21–22, 24–27, 58, 64–65, 71, 80, 91, 93, 103, 105, 119, 122, 129, 133, 182–83, 185–87, 194–95, 199, 202–3, 205–8, 231–32, 241, 270n98. *See also* Clarke's Third Law
Le Guin, Ursula, 88, 90–92, 110, 123, 140, 190
League of the Ho-dé-no-sau-nee or Iroquois, The, 259
Lecha-Marzo, Antonio, 54
Lem, Stanislaw, 86–87, 117, 171, 182
Lévi-Strauss, Claude, 134
Levitt, Norman, 162
Lewis, Adam, 14, 75
Lewis, C. S., 210
Ley, Willy, 44–45, 48, 51
liberalism, 75–76, 79, 81, 115, 126, 148, 228, 240–43, 262
Lichtenstein, Roy, 235
Lilly, John C., 92
Linneman, Travis, 12, 105, 108–9, 286n49
Little Badger, Darcie, 105, 153
Liu, Cixin, 87–88, 94
live-action roleplaying games (LARPs), 234, 241–43
Lovecraft, H. P., 93–96, 101, 126, 169, 253, 280n46

Lovecraft Country, 93
Lukács, Georg, 250
Lyons, Scott, 145, 147

Mad Max, 63, 146, 233, 241
magic, 4, 14, 24, 39, 46, 52, 74, 83, 89, 91, 93, 105, 107, 113, 114, 118, 121, 132–33, 152, 164, 176, 186–87, 199, 202, 208–9, 223–29, 231, 249–57, 259–60, 271n6, 274n79. *See also* Clarke's Third Law; disenchantment; fetish; reenchantment; sociomagic
Malagasy people, 197. *See also* Merina people
Mallarmé, Stéphane, 36
Malm, Andreas, 221–22, 255
Man in the High Castle, The, 240, 245
Manifesto of the Communist Party, 166
Mann, Thomas, 53
Mannoni, Octave, 176
March 4 Trump, 242
Marches for Science, 81
Markey, Ed, 116, 229
Martí Ibáñez, Félix, 93
Marx, Karl, 41, 72–74, 161, 165–66, 175, 184–85, 197, 244, 294n41
Marxism, 53, 65, 73, 102, 128, 182, 185, 194, 213–14, 216, 229, 231, 249–50, 254, 259
Masked Singer, The, 233
master signifier, 79
materialism, 3, 41, 54, 91–92, 147, 175, 250, 255, 295n57
Matrix, The, 9, 65, 109, 127, 203, 241, 245
May, Todd, 70
Maze Runner, The, 237
McCaffrey, Anne, 129
McHale, Brian, 11
McLuhan, Marshall, 198, 297n108
McTeigue, James, 127
megatext, 116–17, 152, 154, 234, 304n8, 308n59
Meillassoux, Quentin, 62, 199
Mendlesohn, Farah, 113, 147, 247, 281n3
Merina people, 254–56. *See also* Malagasy people
Merlin, 135–36
metabolic rift, 135
metafetishism, 6, 62–63, 176, 218, 256–57, 259. *See also* fetish
mētis, 186, 256
Métis people, 145–47
Miami people, 63
Michael, Mike, 66–69

Miéville, China, 73, 77, 93, 213
Miller, Frank, 242
Miller, Stanley, 44, 50, 52
Ministry for the Future, The, 64–65
Mitchell, David T., 129–30
Modern Constitution, 12, 30, 62, 81, 136, 163, 183, 187, 198, 255, 260
modern *epistème*, 192, 253
modernism, high, 186, 190, 218
modernity, 4–7, 10, 12, 15–17, 22, 29–30, 39, 46, 62, 73–74, 91, 117, 126, 130, 146, 154, 159, 179–88, 190, 192–93, 197, 249–50, 255, 260–63, 266n9, 268n57, 271n6, 293n4, 310n30. *See also* nonmodernity; postmodernism
monsters, 15, 16, 18, 28, 30, 46, 50, 52, 101, 105, 126, 141, 143, 146, 197, 251, 253
Moore, Alan, 15, 93, 127
Morrison, Grant, 93
Morse, Brandon, 231
Morton, Timothy, 11, 62
Mulvaney, Mick, 115
Musk, Elon, 241, 303n5, 306n37
My Little Pony: Friendship Is Magic, 225, 230–31
myth, 21, 71, 104, 115, 126, 128, 134, 164, 180, 182, 190, 194, 222, 225–28, 245, 250–51, 253

Nagel, Thomas, 90
Nagle, Andrea, 231
nature, 5, 11–12, 14–16, 21–23, 25, 30–31, 36, 47, 54, 56–57, 65, 71, 73, 80–81, 88, 91–92, 104–5, 108, 114, 128–29, 131, 133, 141, 163–64, 171, 174, 179–80, 183–87, 197–98, 200–202, 208, 252–53, 255, 261, 263, 269n84, 274n65, 274n79, 309–10n29; human, 20; laws of, 21–22, 71, 183, 185, 203, 208; machinification or technicization of, 197–98, 200, 202; state of, 88, 164, 197. *See also* science
Nazis, 15, 185, 231, 240, 243, 281n51
Nebula Awards, 17
Neill, A. S., 210
neoliberalism, 5, 17, 160, 164, 228–29, 241
networks, 21, 53–54, 64–72, 75, 87, 142, 155, 173, 193, 254, 256
Neurath, Otto, 170
Neuromancer, 104
New Atheism, 161–62, 177
New Black Fantastic, 13

New Materialism, 5, 91–92. *See also* Actor-Network Theory
Newton, Isaac, 163, 183, 185, 253, 270n98
Ng, Jeannette, 19, 83–86, 125, 134, 197
Nicholls, Peter, 251
Nietzsche, Friedrich, 70–71, 191
nihilism, 138, 143, 221, 251
Nishnaabeg people, 145–46
"Nkásht Íí" (Little Badger) 105, 153–54
nonbinary people, 262
nonmodernity, 6, 16, 91, 184, 218, 261–63
novum, 11, 102–4, 108, 111–12, 115–16, 129, 148, 152–53, 252, 295n57
Noys, Benjamin, 280n46

objects, 5, 11–12, 16, 20–21, 31, 38, 40, 55, 58, 61–62, 67, 73–75, 78–83, 111, 136–37, 171, 176, 182, 184, 186, 191, 230, 245–46, 252–53, 254–56, 265n5. *See also* cognitively estranging referents; fetish
Öcalan, Abdullah, 261
Ocasio Cortez, Alexandria, 116, 229
Occupation, Israeli, 244
Očeti Šakówiŋ (Sioux) people, 63
OGPU (Ob"yedinonnoye Gosudarstvennoye Politicheskoye Upravleniye), 169, 216–17, 281n51
Ohkay-Owingeh people, 145, 150
Ojibwe people, 145, 148–49, 182, 186, 188
Okorafor, Nnedi, 4, 29, 107, 259, 271n108
"Ones Who Walk Away from Omelas, The" (Le Guin), 109, 186
Onödowá'ga:' people, 182, 188
Oparin, Alexander, 44, 46, 52–53
order-words, 208–9, 254
Orwell, George, 6, 10, 169
Oswalt, Patton, 238–39

paganism, 175, 251, 256–57
Palin, Sarah, 233–34
panpsychism, 44
paranoia, 36, 94, 128, 213, 276–77n16, 277n18, 280n46
Pataud, Émile, 112
Pattern Language, A, 193
Paz, Octavio, 106
Peele, Jordan, 93
Pelletier, Madeleine, 53
Penny, Laurie, 243

Peoria people, 63
Perdido Street Station, 77–78
Perry, Katy, 226
"Persistence of Vision, The" (Varley), 91, 129–30
Piaroa people, 197
Planey, Arrianna, ix, 102
plasmogeny, 5, 44–55, 59, 175
plausibility: psychological, 202–3, 251–52; rational, 251–52
playing Indian, 150–51
pluriversality, 75, 159, 174–75, 189, 255–56, 263, 310n30
Poe, Edgar Allan, 26, 159
"Poseidon" (Kafka), 134
postmodernism, 69, 265n3
postnormal times, 6, 246–47
Potawatomi people (Neshnabék), 63, 145
Pouget, Émile, 112
Pravic language, 190
Pritchard, Alex, 75
Proudhon, Pierre-Joseph, 41, 69–74, 92, 167–68, 204, 250
pseudoscience, 45, 53, 117, 159–60, 162, 281n51
publics, 6, 25, 27–28, 29, 116, 125, 193, 216, 221, 233, 235, 236, 241, 244–45, 247, 253, 259
Punisher, The, 240
purification, 17–18, 22, 25–29, 55, 117, 164, 187, 192, 250, 253, 258, 268n57. *See also* anti-Black racism; exclusion; hybridity
Pyle, Kai Minosh, 146, 154–55
Pynchon, Thomas, 133

QAnon, 109, 114, 160
Queen of Angels, 172
queerness, 15, 122, 134–35, 142, 155–56, 226, 230–31, 262

rabbit holes, 241–42
Rabid Puppies, 17
RaceFail '09, 17, 93
racism, 17, 28, 93, 104, 187, 242, 257; anti-Black, 4, 12–13, 23, 25, 29, 38–39, 92, 94–95, 142, 282n13; colonial, 89, 184
Raised by Unicorns, 226
rationality, 27, 29, 54–58, 107, 110–11, 117, 119, 147, 149, 159, 163, 170, 185–86, 193, 199, 204, 216, 227–31, 253, 294n41. *See also* irrationality; plausibility, rational

Ravetz, Jerome R., 246
Real, the (Lacan), 16, 21, 203, 267n43, 286n49
realism, 27, 61, 73, 121, 123, 132, 173, 244, 246, 258; aesthetic, 7, 12, 14, 19, 32, 108, 128, 151, 164, 234, 236; capitalist, 5–6, 244–45, 259–60; climate, 221; magical, 186; political, 7, 228–29; scientific, 7; statist, 5–6, 244–45, 260; subtractive, 6
Reclus, Elisée, 54
red pill, 241–42, 306n37, 306n43
reenchantment, 31, 36, 43, 50–51, 88, 253. *See also* disenchantment
Reich, Wilhelm, 53
replicants, 79
rhetoric, 90, 97, 101–19, 123, 141, 165–66, 168, 171, 177, 186, 195, 199, 202–3, 221–32, 239, 250–54, 259, 281n3, 282–83n28; of incredulity, 147–51, 155–56; of believing, 151–56
rhizomes, 70
Richter, Kathleen, 232
Ring Shout, or, Hunting Ku Kluxes In the End Times, 93
Roanhorse, Rebecca, 150–51
Robinson, Andrew, 92
Robinson, Kim Stanley, 64
Robinson, Lilian S., 140
Roček, Miriam Rosenberg, 243–44
Rohy, Valerie, 132
roleplaying games (RPGs), 63, 234–35, 239–43, 245, 304n8
Romanticism, 29, 48, 56, 58, 132, 231, 235, 250, 285n14
Rosencrantz & Guildenstern Are Dead, 179, 218
Rosny aîné, J.-H., 213
Rousseau, Jean-Jacques, 197
Rucker, Rudy, 83, 88
Rumsfeld, Donald, 41–42
Russ, Joanna, 111
Russian Empire, 188, 204
Ryner, Han, 83, 89–90

Sad Puppies, 17
Samman, Khaldoun, 180, 293n4
San Diego Comic-Con, 234, 238–39
Sanchez-Taylor, Joy, 12
Sapir-Whorf Hypothesis, 90, 279n24
Sartre, Jean-Paul, 88, 111
satire, 29, 94, 102, 111, 148–51, 258

Saxton, Josephine, 123–25
science, 4–7, 29, 40, 44–55, 66–68, 71–72, 80–81, 87, 90, 92, 94–96, 101, 105–6, 108, 110, 117–18, 128, 132–33, 147, 152, 159–65, 167–68, 175–77, 179–80, 182–83, 185–86, 191, 194–95, 198–99, 203, 216–17, 227, 229, 246, 250–53, 255, 260; fantasy, 27, 259; fictionality, 43, 234, 304n8; fictive, 175–76; junk, 163–64. *See also* pseudoscience; plausibility, psychological
Science Wars, 6, 162–63, 177
scientism, 6
Scott, Damion Kareem, 113
Searle, John R., 38–43, 90, 272n34
Sedgwick, Eve Kosofsky, 276–77n16, 280n46
Sense8, 252
Separation Wall, 244
Serge, Victor, 216, 218
SETI, 94–95
settler readers, 149, 152
sexuality, 17, 19, 27–29, 38, 92–93, 102, 104, 111, 122, 125, 132, 139–40, 142, 153, 155, 200, 203–4, 222, 226–27, 229–30, 262, 272–73n34, 297n118
Shawl, Nisi, ix, 4, 29, 97, 141–44, 249, 259
Sheldon, Rebekah, 28
Shelley, Mary, 50. *See also Frankenstein* (Shelley)
Shelley, Percy Bysshe, 30, 48, 235
Ship Who Sang, The, 129
Shipley, Maynard, 50
Siddiqi, Asif, 212–13
Silicon Valley, 164, 228, 241
Silver Chair, The, 210
Singer, Joseph William, 246
Snyder, Sharon L., 129–30
Society for Creative Anachronism (SCA), 242
sociomagic, 6, 177, 185, 189, 191, 194, 195–96, 253, 256–59
sociomorphism, 183, 185, 203
sociotechnics, 130, 189, 195–96, 256, 259, 269n84; disasters, 6, 22–23, 25
Sokal, Alan, 162–63, 172
Solomon, Rivers, 4, 29
Sontag, Susan, 236, 259
sovereignty, rhetorical, 145–56
Spade, Dean, 134
Spinoza, Baruch, 161, 170, 185
Stableford, Brian, 247, 251–52
Star Trek, 129–30, 212, 233, 257–58

Star Wars, 27, 236, 240
steampunk, 143, 244. *See also Difference Engine, The*; *Everfair*
STEM disciplines, 160
Sterling, Bruce, 15, 172, 235
Stetson, Charlotte Perkins, 57–58, 133, 285n14
Stillness, the, 13–16, 114, 133–37, 140–41, 266n20
Stirner, Max, 35, 204, 208, 291n24
Stoppard, Tom, 179
Story of Your Life, The, 90
Stranger in a Strange Land, 176
Strete, Craig, 149–51
Stryker, Susan, 105
sublime, 105, 201; biotechnical, 6, 35–37, 44, 59, 77; geological, 62; plasmogenic, 55
supercrip, 129–30
Superman, 108, 115, 121, 237
survivance, 5, 97, 187, 287n80, 293n4
Suvin, Darko, 11, 102, 126, 135, 145–49, 182, 266n14, 295n57, 306n40
Svyatogor, 92
Swann, Thomas, 75
Swift, Taylor, 226
Symbolic, the (Lacan), 86–87, 92, 109, 125, 172–73

Taciturn Indian stereotype, 149
Tárrida del Mármol, Fernando, 165, 167–69, 175, 292n24
Tate, Greg, 12
Taylor, Drew Hayden, 148–49
technics, 27, 33, 65–66, 69–70, 72, 77, 102, 105, 113, 116–17, 128–30, 133–34, 149, 152, 164, 168, 175, 177, 180, 182, 184–88, 192, 194–95, 198, 200, 202–3, 206, 209–13, 228–29, 244–45, 259, 269n84, 282n28, 297n108. *See also* Clarke's Third Law; sociotechnics
Thacker, Eugene, 12, 286n49
Theranos, 228
They Live, 109
Thomas, Ebony Elizabeth, 134
Thompson, E. P., 144
Three-Body Problem, The, 87
Time Machine, The, 6, 180
time travel, 63, 122, 127–29, 238, 243, 246
Tiptree, James, Jr., 129–30
Tiv people, 197
Tolkien, J. R. R., 14, 91, 132, 169, 236

Tolstoy, Leo, 188, 204–8, 213
Tomorrow, Tom, 101–2
Topash-Caldwell, Blaire, 147
Tosh, Daniel, 231
transgender people, 13, 15, 19, 153, 226, 262
tricksters, 148, 150
True Blood, 282n13
Trump, Donald, 19, 78, 81, 114–16, 228, 239–42, 307n59
Tsiolkovskii, Konstantin Eduardovich, 53–54, 96
Tuck, Eve, 259
Turing, Alan, 43, 171
Turing Test, 37–38, 117
Two-Spirit people, ix, 155
Tyler the Creator, 227

umet', 185–86, 256
unconscious, 35, 76, 129, 171; cognitive, 42, 173
Under the Pendulum Sun, 20, 83–86, 197
Unicorn Riot, 230
Unicorn Store, 226
Unite the Right rally, 242
University Discourse, 36, 52–53, 75–80, 215, 274n65
Updike, John, 235
Urey, Harold, 44, 50, 52
Üstündağ, Nazan, 261
utopia, 5–6, 19, 63, 97, 121–23, 130, 137, 139–42, 155, 164, 169, 173–74, 181–89, 194–96, 198, 200, 207, 214–16, 232, 258, 276n16, 285n14, 291n19, 298n139; critical, 110; flawed, 103, 109–10, 196

V for Vendetta, 10, 127, 237, 240
Vampire Diaries, The, 135
Van Camp, Richard, 151–53
Varley, John, 91, 129–30, 297n108
Verne, Jules, 26, 63, 202
Vess, Charles, 118
video games, 10, 17, 19, 240, 306n43
Vizenor, Gerald, ix, 5, 150, 188–89, 287n80

Wachowski, Lana, 127, 204, 241
Wachowski, Lilly, 127, 204, 241, 306n37
Walking Dead, The, 126–27
"Wall, The" (Saxton), 123–24
wampumpeag, 75, 83, 92, 97, 142
War of the Worlds (Wells), 95

Warburton, Theresa, 145, 259, 310n43
Ward, Colin, 70
Warhammer 3000, 239
Warner, Michael, 235
Warren, Calvin O., 11, 135–37
Weber, Max, 39, 191
Wells, H. G., 6, 26, 95, 108, 110, 126, 180, 202–3
Wetherbee, Ben, 282–83n28
Whitehead, Colson, 126, 135, 271n108
Who Fears Death, 259, 271n108
Why? Or, How a Peasant Reached the Land of Anarchy, 206–7
Wiindigoog, 146–47
Wilderson, Frank, III, 134, 137
Wilkins, John, 191–92
"Winter Market, The" (Gibson), 129–30
Wolf, Mark J. P., 113–14
Wolfenstein II: The New Colossus, 240, 243, 245
wonders, 152–53
Workers' Solidarity Alliance (WSA), 229–30
world (vs. Earth), 137–38
World Health Organization, 233
Wynter, Sylvia, ix, 29, 89

X-Men, 129

Yang, K. Wayne, 259. *See also* la paperson
"Yellow Wall-Paper, The" (Stetson) 57–58
Young, Kevin, ix, 74–75, 117, 159–60
Yu, Charles, 18

Zapatistas (EZLN), 75, 255, 261
Zeno of Elea, 123
Zionism, 188–89, 204, 217
Žižek, Slavoj, 16, 21, 41, 65–66, 91, 126, 176, 203, 293n52
Zola, Émile, 164
Zoline, Pamela, 131
zombies, 117, 127, 222–23, 233, 238, 241. See also *Girl with All the Gifts, The*; *Walking Dead, The*
Zucca, Antioco, 53

About the Author

Photo by Darlene Cohn

Jesse S. Cohn lives, works, and parents on Cession 180 in so-called Northwest Indiana. He is the author of *Underground Passages: Anarchist Resistance Culture, 1848-2011* and is translator of numerous works, including Daniel Colson's *Little Philosophical Lexicon of Anarchism From Proudhon to Deleuze*.

www.ingramcontent.com/pod-product-compliance
Lightning Source LLC
Chambersburg PA
CBHW031847220426
43663CB00006B/520